Nano-tribology and Materials in MEMS

Sujeet K. Sinha · N. Satyanarayana
Seh Chun Lim
Editors

Nano-tribology and Materials in MEMS

Editors
Sujeet K. Sinha
Department of Mechanical Engineering
Indian Institute of Technology
Kanpur
India

N. Satyanarayana
Seh Chun Lim
Department of Mechanical Engineering
National University of Singapore
Singapore

ISBN 978-3-642-36934-6 ISBN 978-3-642-36935-3 (eBook)
DOI 10.1007/978-3-642-36935-3
Springer Heidelberg New York Dordrecht London

Library of Congress Control Number: 2013934533

Printed on acid-free paper

Springer is part of Springer Science+Business Media (www.springer.com)

Preface

The field of nanotribology has advanced to a great extent thanks to the phenomenal growth of information storage industry. The magnetic hard disks used for recording and retrieving digital data require extremely thin nanolubricant for the protection of the disk from mechanical damages and wear by the slider which flies just above the surface of the disk with flying height only a few nanometer. Even though the slider is designed not to touch the disk, contacts between the slider and the disk are inevitable and hence we require the protection of the disk by the nanolubricant. A similar requirement, at least in length scale, is experienced in microsystems such as micro-electro-mechanical systems (MEMS) where micron-sized components, usually made of Si, are made to move about just like their macro-machine counterparts. Sliding, contact and impact between the components lead to the problems of adhesion, friction and wear. Because of the small length scales involved, the problems of tribology faced in microsystems differ drastically from those of the traditional macro-scale machines. Therefore, it is important to address these issues taking into considerations the materials, micro-fabrication process, lubricants and the lubrication methods.

A symposium titled "Nano-tribology and Related Materials Issues in MEMS" was organized by the Department of Mechanical Engineering, National University of Singapore from 13 to 14 May 2010. A number of invited talks were presented covering the fundamental nanotribology concepts, applications of new materials, surface modifications of Si and polymer substrates and simulations of the friction phenomenon under light load conditions. This book is a collection of the papers that were submitted by the presenters with some additional contributions by the experts in this field. Each chapter has been carefully selected and edited to bring out current practices in the MEMS tribology field with the explicit aim of finding appropriate solutions to the tribological problems faced in MEMS and nano-scale machines.

The editors would like to express their deepest appreciations to the invited and poster presenters of this symposium without whose help this event would not have been a reality. We thank the Dean, Faculty of Engineering and the Head of the Department of Mechanical Engineering, NUS, who provided all the supports needed for the organization of this event. We are also grateful to the Singapore

National Research Foundation (NRF) for the generous research grant (Award no.: NRF-CRP 2-2007-04) to our team which helped to support much of the research works that were presented in this symposium. Finally, we would like to thank the publisher and the authors of the chapters whose relentless effort through the manuscript preparation and editing has resulted in this compilation of very relevant works in the field of nanotribology and materials for MEMS. We earnestly hope that this edited book will positively add to the expanding literature in this field to help in current and future research.

April 2013

Sujeet K. Sinha
Nalam Satyanarayana
Seh Chun Lim

Contents

Chapter 1
Nanotribological Phenomena, Principles and Mechanisms for MEMS

Biswajit Saha, Erjia Liu and Shu Beng Tor

Abstract Nanotribology is referred to as a branch of tribology, which involves the interactions between two relatively moving materials in contact at a nanometer or an atomic scale. Nanotribology was stimulated by the fabrication of micro-electro-mechanical systems (MEMS). With the advent of scanning force microscopy (SPM), experimental approach to nanotribological regimes has been substantially advanced. Most common examples of nanotribological phenomena are in hard disk drives, MEMS, and nano-electro-mechanical systems (NEMS). Tribology is a surface phenomenon, which can be significantly affected by a very high surface-to-volume ratio in a micro or nanostructure. Small mass, light load, elastic deformation, and slight wear or absence of wear are typical of nanotribology. It has been widely perceived that various tribological test conditions, such as load, velocity, temperature, surface free energy, surface topography, environment, etc. play major roles in nanotribology. The experimental study of nanotribology is made possible by using surface force apparatus (SFA), atomic force microscope (AFM), friction force microscope (FFM), and ball-on-disk nanotribometer. Tribology research exceedingly needs broadened knowledge in various fields such as physics, chemistry, mechanics, materials science, etc. This chapter discusses the various aspects of nanotribology including nanofriction, nanowear and nanolubrication for MEMS. The nanotribological measurement methodologies and mathematical relationships between friction, bending and torsion forces based on AFM/LFM (lateral force microscopy) are comprehensively reviewed. The influences of various experimental conditions on nanotribology are described with various examples. Simulation techniques for nanotribology are also highlighted.

B. Saha · E. Liu (✉) · S. B. Tor
School of Mechanical and Aerospace Engineering,
Nanyang Technological University, 50 Nanyang Avenue,
Singapore 639798, Singapore
e-mail: mejliu@ntu.edu.sg

S. K. Sinha et al. (eds.), *Nano-tribology and Materials in MEMS*,
DOI: 10.1007/978-3-642-36935-3_1,

Contents

Introduction

The scientific term "tribology" is related to the interface of moving surfaces, which mainly deals with friction, wear and lubrication. Friction is resistance to relative motions of any state of materials in contact, wear is loss of materials from one solid body or another or both during their relative motions in contact, and lubrication is to reduce friction and wear between two moving materials by means of a lubricant present at the interface. Though tribological phenomena have existed since the very beginning of the world, the term "tribology" was only introduced by Jost in the modern age, i.e. 1966 [1]. This term has been more on technological use in industrial operation for economy, durability, lifetime and maintenance calculation.

Great progresses have been achieved in all the three fields of tribology during last few decades. Da Vinci, Amonton and Coulomb [1] have found that friction is proportional to applied load but independent of contact area and velocity, which is more applicable for hard solid materials. However, many researchers have found that friction of soft materials, such as polymeric materials, is significantly affected by real area of contact, surface adhesion and temperature. Hammerschmidt et al. [2] have reported that polymeric materials can show a special kind of friction behavior because of the structural relaxation of their molecular chains governed by energy dissipation and correlated frictional response with glass-to-rubber transition and secondary relaxation mechanisms. Enachescu et al. [3] have found that the tribological behavior of materials is a function of applied load, based on an observation that friction is directly proportional to the real area of contact. Exceptions of Amonton's law ($F = \mu.W$) have also been encountered by many other researchers [4–6].

Nanotribology is referred to as a branch of tribology, which involves the interactions between two relatively moving materials in contact at a nanometer or atomic scale, where atomic forces are one of the most dominating forces. Tribology is a surface phenomenon, which can be significantly affected by a very high surface-to-volume ratio in a micro or nanostructure.

The demand of society's particular needs in advanced technologies has been gradually increasing. The advent of micro and nanotechnologies has greatly motivated researchers' interest in designing novel and complex architectures. A proper understanding of nanotribological behavior is very important to improve the applicability and lifetime of tiny architectures and to protect micro and nanosystems from severe friction and wear.

Most common examples of nanotribological phenomena are in magnetic recording disk drives, micro-electro-mechanical systems (MEMS), and nano-electro-mechanical systems (NEMS). Small mass, light load, elastic deformation, and slight wear or absence of wear are typical of nanotribology. The experimental study of nanotribology is made possible by surface force apparatus (SFA), atomic force microscope (AFM), friction force microscope (FFM), and ball-on-disk nanotribometer.

Tribological properties of materials depend on the true area of contact between the contacting surfaces. Thus, roughness is one of the important affecting parameters and is also found to be a function of various parameters such as surface energy, velocity, humidity, environment, and temperature [2, 7–11].

This chapter deals with the various aspects of nanotribology including nano-friction, nanowear and nanolubrication for MEMS. The nanotribological measurement methodologies and mathematical relationships between friction, bending and torsion forces based on AFM/LFM (lateral force microscopy) are comprehensively presented. The influences of various parameters on nanotribology are described with various examples. Simulation techniques for nanotribology are also overviewed.

Measurement Techniques

Limitation of scanning tunneling microscopy (STM) leads to the invention of AFM by Binning et al. [12] in 1986. The basic working principle of AFM is schematically shown in Fig. 1.1, in which a laser light is reflected from the rear side of a cantilever to the position sensitive photodiode detector to record the movement of the cantilever. Any deflection or torsion of the cantilever causes a change in the laser path so does in the photodiode detector as shown in Fig. 1.2.

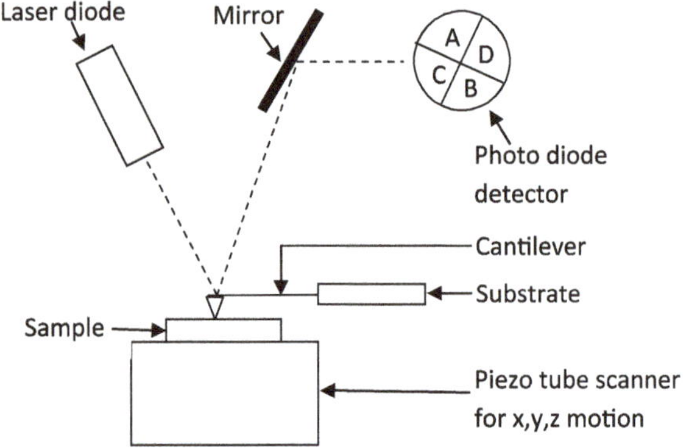

Fig. 1.1 Schematic operating principle of a typical AFM. The scanning position of a sample is controlled by a piezo controller. A laser light is reflected from the rear side of the cantilever to a photodiode detector to detect the bending and torsion of the cantilever

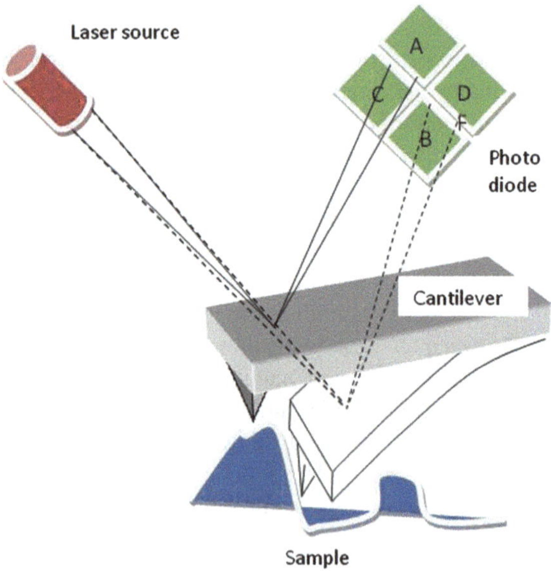

Fig. 1.2 A schematic AFM tip scanning over a sample. The acting force on the tip is determined by using a quadrant photodiode detector to measure the deflection and torsion of the cantilever

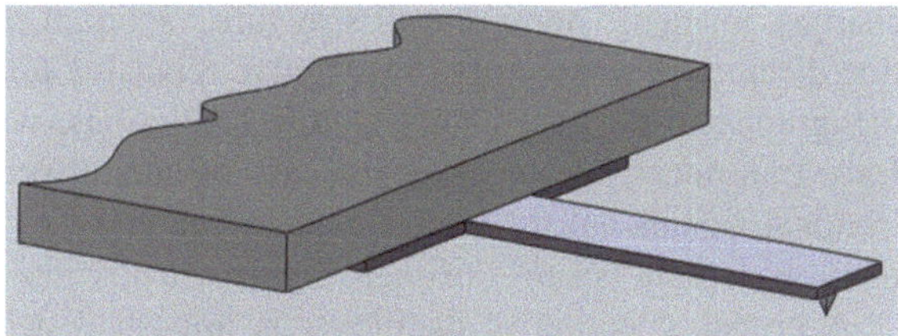

Fig. 1.3 Schematic diagram of rectangular cantilever

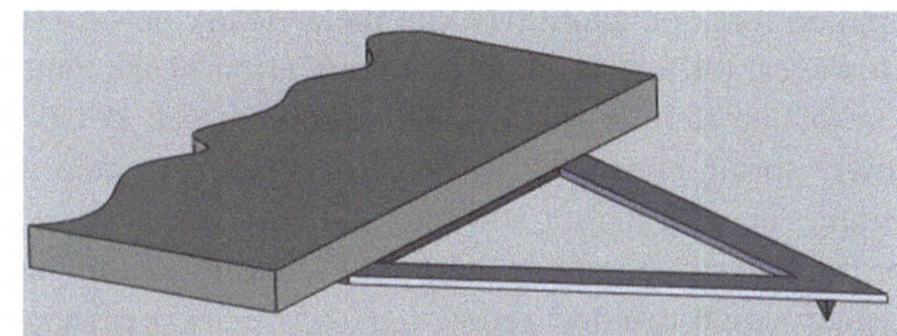

Fig. 1.4 Schematic diagram of V-shape cantilever

Importance of AFM Cantilever

Most AFM cantilevers are made of Si, Si_3N_4 or SiO_2, which work based on the knowledge of their spring constants (k_s). The k_s and resonance frequency (f_0) of a cantilever beam with a rectangular cross-section loaded at its flexible end can be determined by the following equations:

$$k_s = \frac{Ebt^3}{4l^3} \tag{1.1}$$

$$f_0 = 0.162\sqrt{\frac{E}{\rho}} \times \frac{t}{l^2} \tag{1.2}$$

where *E, b, t, l* and ρ are Young's modulus, width, thickness, length and density of the cantilever, respectively.

A simple method of static deflection of a cantilever under the force of a known mass is usually used to determine its spring constant and commercially available cantilevers are generally supplied with this information. Rectangle and triangle (V) shaped cantilever beams are two basic cantilever geometries as shown in Figs. 1.3 and 1.4, both of which are suitable for contact, intermittent contact and non-contact modes of operation. A higher mechanical stability with respect to lateral distortion makes V-shape cantilevers more preferable for contact mode AFM. In the contact mode AFM, repulsive and attractive forces acting on the cantilever depend upon the distance between AFM tip and sample surface. In the attractive regime, dominating forces are van der Waals, electrostatic, magnetic and capillary forces, which give the information about surface topography, distribution of charges, magnetic domain wall structure, or liquid film distribution. At a separation distance of a few angstroms, active repulsive forces allow the sample surface topography to be traced with atomic resolutions, where the cantilever acts as a force transducer. However, in the case of lateral force microscopy (LFM), where tortional signals are the main domain of interest, rectangular cantilevers are thus more preferable because of their higher sensitivities to lateral forces.

A proper selection of cantilever is important, which depends on a particular application for which the cantilever will be used. In the contact mode, a large applied force on cantilever can inelastically deform or modify the sample surface. However, for a hard sample, modification of the cantilever tip is more likely than that of the sample surface. This problem can be solved by using a cantilever of lower spring constant. However in noncontact and intermittent contact modes, a cantilever with higher stiffness and resonance frequency is preferable, which can prevent from contacting with and sticking to the sample surface and result in a higher signal-to-noise ratio.

The critical design criteria of contact mode AFM cantilevers include small mass (e.g. <0.1 μg), short lever length (e.g. <200 μm), small beam thickness (e.g. <0.5 μm), low bending stiffness for higher flexibility and sensitivity, high thermal stability, mechanical stability (e.g. resonant frequency >2 kHz) and torsion elastic constant, a sharp protruding tip, and a mirror or electrode for reflecting deflection and torsion signals.

The thermal distortion of a cantilever beam is

$$\frac{d\varepsilon}{dx} = \frac{\alpha dT}{dx} = \left(\frac{\alpha}{\lambda}\right) q \tag{1.3}$$

where ε, α, T, λ and q $\left(= \lambda \frac{dT}{dx}\right)$ are the thermal strain, thermal expansion coefficient, temperature, thermal conductivity, and heat input, respectively.

$\frac{d\varepsilon}{dx}$ of a beam is proportional to α and inversely proportional to λ of the beam. Therefore, $\frac{d\varepsilon}{dx}$ of a beam can be minimized by maximizing $\left(\frac{\lambda}{\alpha}\right)$.

The mass of a rectangular cantilever beam having a uniform cross-section can be represented by

$$m = Al\rho \ (\mathrm{Kg}) \tag{1.4}$$

where m, A, and ρ are the mass, cross-section area and density, respectively.

Relations among the deflection, bending force and spring constant of a cantilever beam are

$$\delta = F_{bend} l^3 / C_1 EI \ (\mathrm{m}) \tag{1.5}$$

$$k = F_{bend}/\delta \ (\mathrm{N/m}) \tag{1.6}$$

$$\text{with } I \propto A^2 \ (\mathrm{m}^4) \tag{1.7}$$

Here δ, F_{bend}, C_1, I, and k are the beam deflection, bending force, constant, second moment of area, and spring constant, respectively.

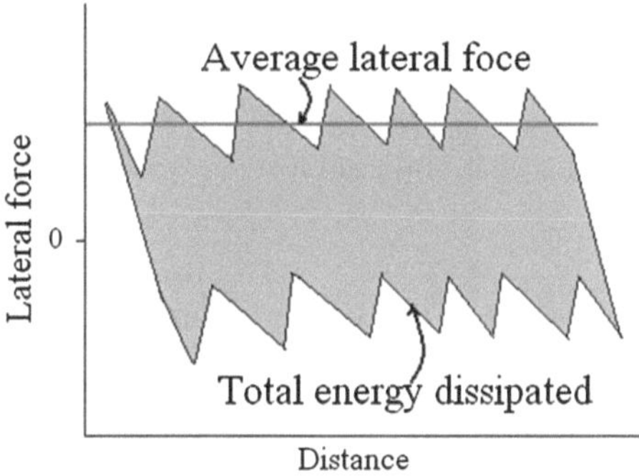

Fig. 1.5 Schematic diagram of lateral force versus scan distance by LFM

Thus, the mass of the beam can be expressed as

$$m = \left(\frac{12k}{C_1 l}\right)^{1/2} l^3 \left(\frac{\rho}{E^{1/2}}\right)\ (\mathrm{Kg}) \tag{1.8}$$

which can be minimized by maximizing

$$\left(E^{1/2}/\rho\right) \tag{1.9}$$

Lateral Force Microscope

One of the capabilities of contact mode AFM is to study the lateral forces or tribological properties between sample and AFM tip in contact and relative motion [13]. In this regard, AFM is known as lateral force microscope (LFM) or FFM. LFM has some advantages over conventional tribometer. LFM has an ability to quantify nanotribological properties at nano or atomic scale, such as interaction forces, i.e. attractive and repulsive forces. A big advantage of LFM is that most materials can be investigated under their natural conditions, including biological samples in an aqueous environment. In the schematic diagram shown in Fig. 1.5, a lateral force loop is used to determine the frictional behavior of a material surface in contact with a LFM tip, and the area inside the loop indicates a total dissipated energy. The measured lateral forces depend on the applied normal forces and the local and average friction coefficients can be calculated by dividing the respective local and average lateral forces by the applied normal forces. The specialty of LFM is that as the distance scale can be very small it is able to identify the friction properties of materials with atomic resolutions, which allows LFM to map the composition distribution of a material surface.

The four-sector photodiode detector of LFM as shown in Fig. 1.1 can simultaneously detect both deflection and torsion signals of the cantilever. A schematic measuring principle of vertical and lateral movements of the tip is shown in Fig. 1.6, where the vertical motion of the tip is detected by the photodiode pair A and B while the lateral motion of the tip is monitored by the pair C and D [14].

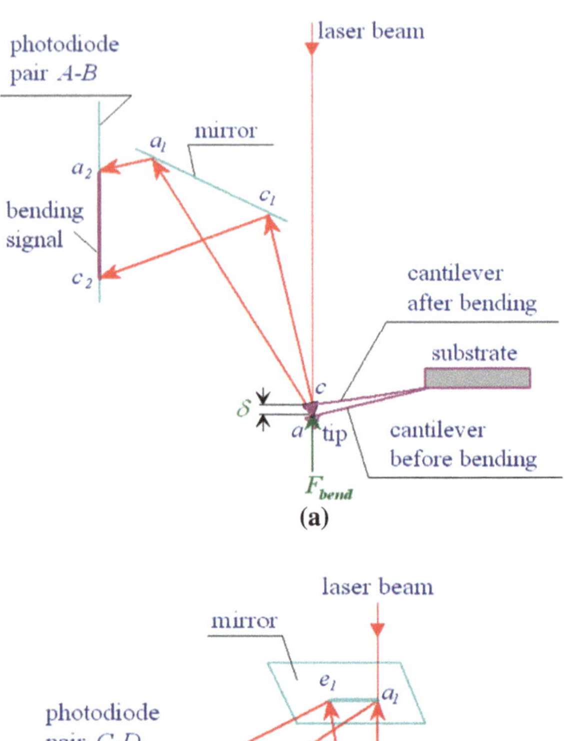

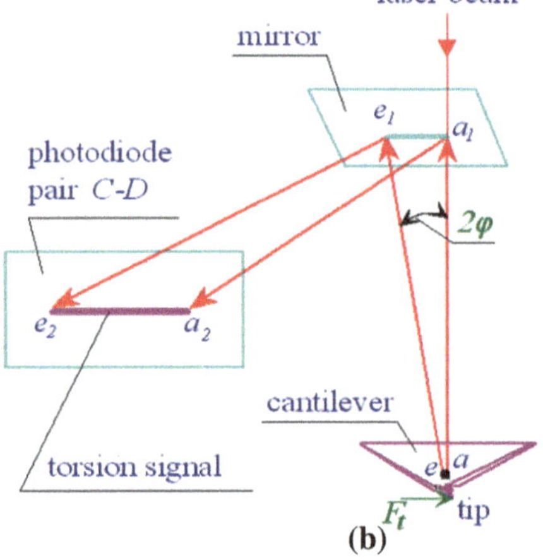

Fig. 1.6 Schematic diagrams of **a** bending and **b** torsion signals of cantilever. (Reprinted with permissions from Liu et al. [14] and Liu [15])

A voltage signal measured by the photodiodes C and D, which is a result of torsion, is converted to a lateral tangential displacement of the cantilever tip as schematically shown in Fig. 1.6b. The resulted lateral tangential force is then calculated from the product of the torsion elastic constant of the cantilever and the lateral tangential displacement of the tip.

Bending of Rectangular Cantilever

Figure 1.7a and b show the front and side views of a rectangular cantilever. The deflection of the cantilever beam caused by the bending force F_{bend} is shown in Fig. 1.8a, where the curvature of the beam along the x-direction can be derived as

$$\frac{d^2z}{dx^2} = \frac{M_y}{EI_y} \tag{1.10}$$

with

$$M_y = F_{bend}x \text{ (Nm)}, \ (0 \le x \le l) \tag{1.11}$$

and

$$I_y = \frac{bt^3}{12} \text{ (m}^4\text{)} \tag{1.12}$$

where M_y is bending moment of the cantilever beam about the y-axis, I_y is the second moment of area of the beam, E is Young's modulus of the beam material, and l, b and t are the length, width and thickness of the beam.

Integration of Eq. 1.10 is represented by Eq. 1.13,

$$\begin{aligned} z' &= \int \frac{M_y}{EI_y} dx = \int \frac{F_{bend}x}{EI_y} dx \\ &= \frac{F_{bend}}{EI_y}\left(\frac{x^2}{2} + C_1\right) \end{aligned} \tag{1.13}$$

with the slop of the beam at $x = l$ being zero.

Therefore, $z'|_{x=l} = \frac{F_{bend}}{EI_y}\left(\frac{l^2}{2} + C_1\right) = 0$, from which

$$C_1 = -\frac{l^2}{2}. \tag{1.14}$$

Thus,

$$z' = \frac{F_{bend}}{2EI_y}(x^2 - l^2) \tag{1.15}$$

Integration of Eq. 1.15 is presented as Eq. 1.16,

$$\begin{aligned} z &= \int z' dx = \int \frac{F_{bend}}{2EI_y}(x^2 - l^2) dx \\ &= \frac{F_{bend}}{2EI_y}\left(\frac{x^3}{3} - l^2 x + C_2\right) \text{ (m)} \end{aligned} \tag{1.16}$$

Because of the zero deflection of the beam at $x = l$, $z|_{x=l} = \frac{F_{bend}}{2EI_y}\left(\frac{l^3}{3} - l^3 + C_2\right) = \frac{F_{bend}}{2EI_y}\left(-\frac{2l^3}{3} + C_2\right) = 0$,

$$C_2 = \frac{2l^3}{3} \tag{1.17}$$

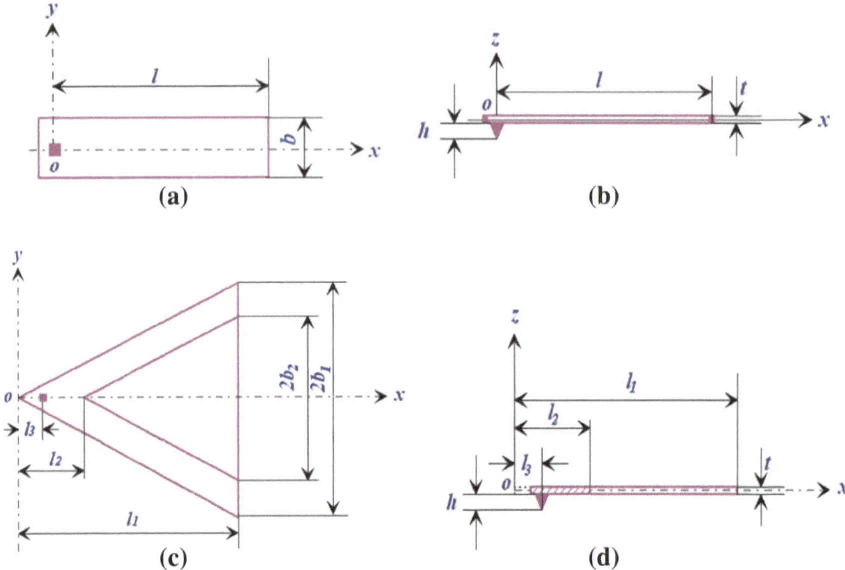

Fig. 1.7 Schematic diagrams of **a** front and **b** side views of rectangular cantilever and **c** front and **d** side views of V-shaped cantilever. (Reprinted with permissions from Liu et al. [14] and Liu [15])

Therefore, Eq. 1.16 is rewritten as,

$$z = \frac{F_{bend}}{2EI_y}\left(\frac{x^3}{3} - l^2x + \frac{2l^3}{3}\right)\ (\mathrm{m}) \tag{1.18}$$

The deflection of the cantilever beam (δ) at the tip position (x = 0) is

$$\delta = z|_{x=0} = \frac{F_{bend}}{2EI_y}\left(\frac{2l^3}{3}\right) = \frac{F_{bend}l^3}{3EI_y}\ (\mathrm{m}) \tag{1.19}$$

Thus, the bending elastic constant of the rectangular cantilever is

$$k = \frac{F_{bend}}{\delta} = \frac{3EI_y}{l^3} = \frac{3Ebt^3}{12l^3} = \frac{Ebt^3}{4l^3}\left(\frac{\mathrm{N}}{\mathrm{m}}\right) \tag{1.20}$$

Bending of V-Shaped Cantilever

The schematic diagrams of the front and side views of a V-shaped cantilever are shown in Fig. 1.7c and d and the corresponding deflected and distorted ones are shown in Fig. 1.8c and d.

The curvature of the cantilever at any point along the x-axis is

$$\frac{d^2z}{dx^2} = \frac{M_y}{EI_y} \tag{1.21}$$

with

$$M_y = F_{bend}(x - l_3) \text{ (Nm)}, \ (0 \le x \le l_1) \tag{1.22}$$

$$I_{y1} = \frac{(b_1 - b_2)t^3}{6} \text{ (m}^4\text{)}, \ (l_2 \le x \le l_1) \tag{1.23}$$

and

$$I_{y2} = \frac{b_1 t^3 x}{6 l_1} \text{ (m}^4\text{)}, \ (l_3 \le x \le l_2) \tag{1.24}$$

where b_1 and b_2 are the beam widths of different segments and l_1, l_2 and l_3 are the beam lengths of different segments.

After integrating Eq. 1.21,

$$\begin{aligned} z &= z_1 + z_2 \\ &= \iint\limits_{(l_2 \le x \le l_1)} \frac{M_y}{EI_{y1}} dx^2 + \iint\limits_{(l_3 \le x \le l_2)} \frac{M_y}{EI_{y2}} dx^2 \\ &= \frac{6F_{bend}}{E(b_1 - b_2)t^3} \iint\limits_{(l_2 \le x \le l_1)} (x - l_3) dx^2 + \frac{6F_{bend} l_1}{E b_1 t^3} \iint\limits_{(l_3 \le x \le l_2)} \frac{(x - l_3)}{x} dx^2 \end{aligned} \tag{1.25}$$

The boundary conditions for Eq. 1.25 are: $z_1 = z_1' = 0$ at $x = l_1$, $z_1 = z_2$ at $x = l_2$, and $z_1' = z_2'$ at $x = l_2$. Therefore the deflection of the V-shaped cantilever at the tip position ($x = l_3$) is

$$\begin{aligned} \delta &= z|_{x=l_3} \\ &= \frac{F_{bend}}{Et^3} \left\{ \frac{3l_1 \left[l_2^2 - 4l_2 l_3 + 2l_3^2 \ln(l_2/l_3) + 3l_3^2 \right]}{b_1} + \frac{2\left[l_1^3 - l_2^3 - 3l_3 (l_1^2 - l_2^2) + 3l_3^2 (l_1 - l_2) \right]}{b_1 - b_2} \right\} \end{aligned} \tag{1.26}$$

Thus, the normal bending elastic constant of V-shaped cantilever can be expressed as

$$\begin{aligned} k &= \frac{F_{bend}}{\delta} \\ &= Et^3 \left(\left\{ \frac{3l_1 \left[l_2^2 - 4l_3 l_2 + 2l_3^2 ln(l_2/l_3) + 3l_3^2 \right]}{b_1} + \frac{2\left[l_1^3 - l_2^3 - 3l_3 (l_1^2 - l_2^2) + 3l_3^2 (l_1 - l_2) \right]}{b_1 - b_2} \right\} \right)^{-1} \end{aligned} \tag{1.27}$$

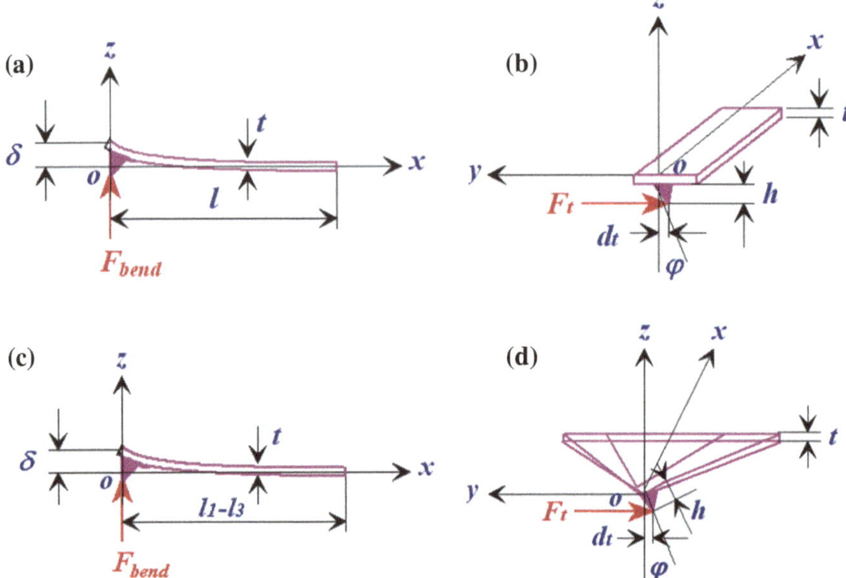

Fig. 1.8 Schematic diagrams of **a** rectangle cantilever deflected by normal bending force F_{bend} and **b** distorted by lateral force F_t, and **c** V-shaped cantilever deflected by F_{bend} and **d** distorted by F_t. (Reprinted with permissions from Liu et al. [14] and Liu [15])

Torsion of Rectangular Cantilever

When the rectangular beam is distorted by a lateral force F_t (Fig. 1.8b), the first derivative of torsion angle is expressed as

$$\frac{d\varphi}{dx} = \frac{T}{GJ}, (0 \leq x \leq l) \tag{1.28}$$

with

$$T = F_t(h + t/2) \text{ (Nm)} \tag{1.29}$$

and

$$J = b^3t^3/3.6\left(b^2 + t^2\right) \text{ (m}^4\text{)} \tag{1.30}$$

where: φ, T, h, G, and J are torsion angle, torsion moment, tip height, shear modulus, and polar moment of inertia, respectively.

By integrating Eq. 1.28, the torsion angle of the cantilever beam at the tip position ($x = 0$) is shown as

$$\begin{aligned}\varphi &= \int_l^0 \frac{T}{GJ} d(l-x) = \int_0^l \frac{F_t(h+t/2)\times 3.6(b^2+t^2)}{Gb^3t^3} dx \\ &= \frac{3.6F_t l(h+t/2)(b^2+t^2)}{Gb^3t^3} \text{ (rad)}\end{aligned} \tag{1.31}$$

The lateral tangential displacement of the tip is thus derived as

$$\begin{aligned}d_t &= \left(h+\frac{t}{2}\right)\sin\varphi \approx \left(h+\frac{t}{2}\right)\varphi \\ &= \frac{3.6F_t l(h+t/2)^2(b^2+t^2)}{Gb^3t^3} \text{ (m)}\end{aligned} \tag{1.32}$$

and, the torsion elastic constant of the rectangular cantilever is

$$k_t = \frac{F_t}{d_t} = \frac{Gb^3t^3}{3.6l(h+t/2)^2(b^2+t^2)} \left(\frac{\text{N}}{\text{m}}\right) \tag{1.33}$$

Torsion of V-Shaped Cantilever

The distortion of the V-shaped cantilever caused by the lateral force F_t as shown is Fig. 1.8d can be represented, in a similar way to that of the rectangular cantilever, by Eq. 1.34.

$$\frac{d\varphi}{dx} = \frac{T}{GJ} \tag{1.34}$$

where the expression of torsion (T) is similar to Eq. 1.29. The polar moment of inertia of the beam can be represented as

$$J_1 = \frac{2t^3(b_1-b_2)}{3} \text{ (m}^4\text{)}, \ (l_2 \le x \le l_1) \tag{1.35}$$

and

$$J_2 = \frac{2t^3 b_1 x}{3l_1} \text{ (m}^4\text{)}, \ (l_3 \le x \le l_2) \tag{1.36}$$

where φ, T, G and J have their usual meanings.

The torsion angle of the cantilever beam at the tip position ($x = l_3$) is determined by integrating Eq. 1.34, which is presented as

$$\begin{aligned}
\varphi &= \int_{l_2}^{l_1} \frac{T}{GJ_1} dx + \int_{l_3}^{l_2} \frac{T}{GJ_2} dx \\
&= \int_{l_2}^{l_1} \frac{3T}{2Gt^3(b_1 - b_2)} dx + \int_{l_3}^{l_2} \frac{3Tl_1}{2Gt^3 b_1 x} dx \\
&= \frac{3T}{2Gt^3(b_1 - b_2)} \int_{l_2}^{l_1} dx + \frac{3Tl_1}{2Gt^3 b_1} \int_{l_3}^{l_2} \frac{1}{x} dx \\
&= \frac{3F_t(h + t/2)}{2Gt^3} \left[\frac{1}{(b_1 - b_2)} \int_{l_2}^{l_1} dx + \frac{l_1}{b_1} \int_{l_3}^{l_2} \frac{1}{x} dx \right] \\
&= \frac{3F_t(h + t/2)}{2Gt^3} \left[\frac{l_1 - l_2}{b_1 - b_2} + \frac{l_1}{b_1} \ln\left(\frac{l_2}{l_3}\right) \right] \text{ (rad)}
\end{aligned} \tag{1.37}$$

The tip lateral tangential displacement of the V-shaped cantilever for a very small φ is expressed as

$$\begin{aligned}
d_t &= \left(h + \frac{\mathrm{t}}{2}\right) \sin\varphi \approx \left(h + \frac{\mathrm{t}}{2}\right) \varphi \\
&= \frac{3F_t(h + t/2)^2}{2Gt^3} \left[\frac{l_1 - l_2}{b_1 - b_2} + \frac{l_1}{b_1} \ln\left(\frac{l_2}{l_3}\right) \right] \text{ (m)}
\end{aligned} \tag{1.38}$$

Therefore, the torsion elastic constant of the V-shaped cantilever is

$$k_t = \frac{F_t}{d_t} = \frac{2Gt^3}{3(h + t/2)^2 \left[\frac{l_1 - l_2}{b_1 - b_2} + \frac{l_1}{b_1} \ln\left(\frac{l_2}{l_3}\right) \right]} \left(\frac{\mathrm{N}}{\mathrm{m}}\right) \tag{1.39}$$

Chemical Mapping and Topography Using Nanofriction

LFM is also a powerful tool to map the chemical composition and inhomogeneity of a sample surface. For a smooth surface, the variations in lateral distortion of the cantilever arise from the changes in friction sources. As schematically shown in

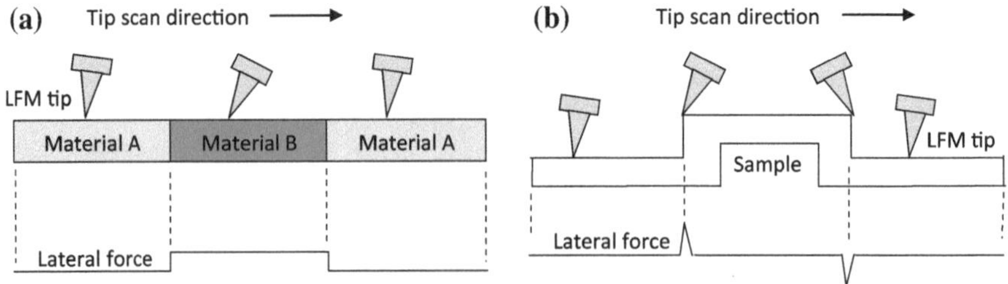

Fig. 1.9 Schematic diagrams of variations of lateral force due to variations in **a** composition and **b** topography of sample surfaces

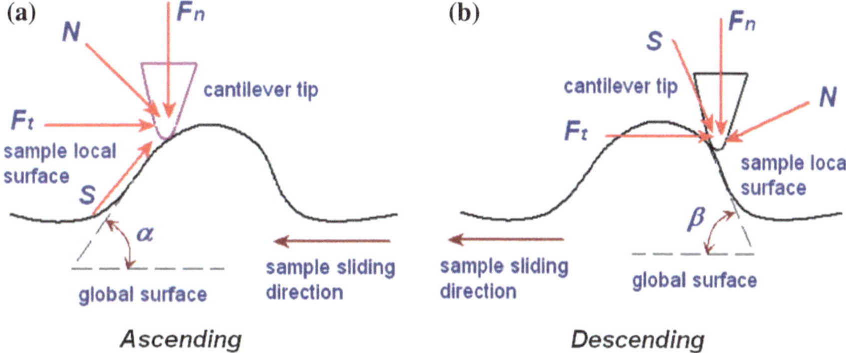

Fig. 1.10 Schematic diagrams of scanning LFM tip at a local asperity during **a** upward and **b** downward movements. (Reprinted with permissions from Liu [15] and Bhushan [16])

Fig. 1.9a a compound comprises different materials A and B. In the compound, if material B has a higher friction than material A against the tip, the friction by material B will cause a higher deflection of the cantilever when the tip scans through material B. The variations of the cantilever deflection are recoded as a LFM image corresponding to the different compositions in the compound and thus create a chemical map.

For a homogeneous material with a rough surface, lateral forces sensed by the cantilever vary according to the locations on the material surface as shown in Fig. 1.9b. Sudden changes in lateral force occur when the tip scans a local surface feature depending on whether it is an upward or downward traveling. Therefore for chemical mapping, a comparison with a topographical plot is very important.

The variations of local friction due to variations in surface topography can be calculated by incorporating the plowing effect in the calculation. An assumption is made for the tip radius smaller than the size of asperities. An upward movement of the tip causes an additional torsion in the cantilever beam, which induces a higher friction. However, a downward motion causes almost no additional torsion in the beam, which lowers the friction.

The local coefficient of friction during the upward movement of the tip under a normal load F_n at an asperity on the sample surface having a slope of α with respect to the global surface can be calculated using Eq. 1.40 based on the schematic shown in Fig. 1.10a.

$$\begin{aligned}\mu_1 &= \frac{F_t}{F_n} = \frac{S\cos\alpha + N\sin\alpha}{N\cos\alpha - S\sin\alpha} \\ &= \frac{\mu_0 + \tan\alpha}{1 - \mu_0\tan\alpha} \approx \mu_0 + \tan\alpha\end{aligned} \tag{1.40}$$

for a very small angle α, $\mu_0 \tan\alpha \ll 1$.

Here, S, N, F_t, and F_n are the shear force perpendicular to the local surface, normal force perpendicular to the local surface, friction force along the global surface, and normal force perpendicular to the global surface, respectively. It is noted that the uphill plowing component (tan α) has a positive contribution to the local friction.

Similarly, the local coefficient of friction during the downward sliding at the same asperity having a slope of β with respect to the global surface can be determined based on the schematic shown in Fig. 1.10b:

$$\begin{aligned}\mu_2 &= \frac{F_t}{F_n} = \frac{S\cos\beta - N\sin\beta}{N\cos\beta + S\sin\beta} \\ &= \frac{\mu_0 - \tan\beta}{1 + \mu_0\tan\beta} \approx \mu_0 - \tan\beta\end{aligned} \tag{1.41}$$

for a very small angle β, $\mu_0 \tan\beta \ll 1$.

It is noted that the downhill plowing component (*tan* β) has a negative sign in Eq. 1.41.

For a symmetric asperity with $\beta = \alpha$, the average local coefficient of friction across the asperity is

$$\begin{aligned}\mu_{ave} &= \frac{\mu_1 + \mu_2}{2} \\ &= \frac{\mu_0(1 + \tan^2\alpha)}{1 - \mu_0^2\tan^2\alpha} \\ &\approx \mu_0(1 + \tan^2\alpha)\end{aligned} \tag{1.42}$$

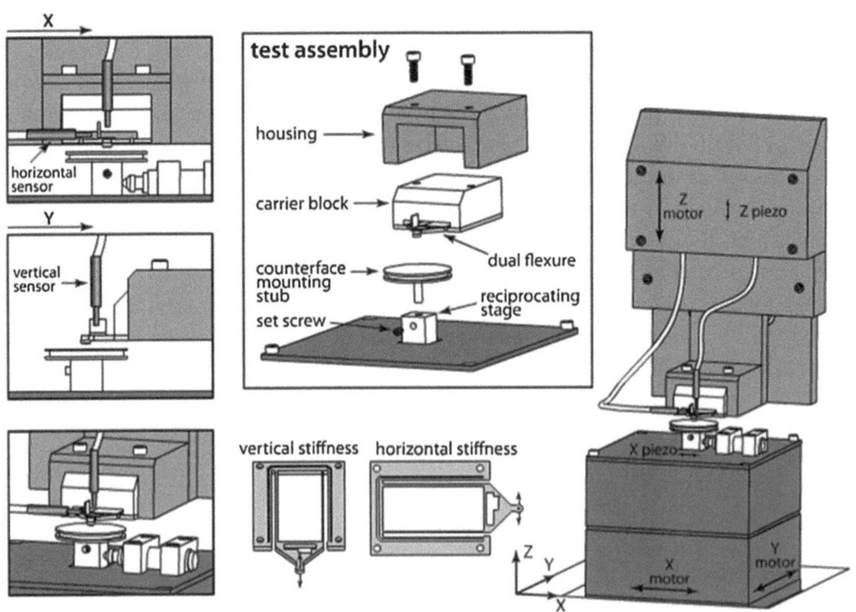

Fig. 1.11 A schematic diagram showing a nanotribometer along with its various components

for a very small angle α, $\mu_0^2 \tan^2 \alpha \ll 1$.

The plowing component of the local coefficient of friction is $\pm \tan \alpha$ in either moving direction depending on the scanning direction of the tip relative to the sample surface.

Nanotribometer

Nanotribometer is also used to measure friction coefficient of materials at nanoscales. A schematic diagram of a typical nanotribometer (CSM) together with its various components is shown in Fig. 1.11. Such a tribometer has certain advantages over LFM, such as high speed, wide load range and large traveling distance with both rotating and reciprocating operation modes. In such a system a specimen is mounted at the bottom and a flexible cantilever is mounted at the top. A z-piezo is used to control the movement of the cantilever up and down. Generally, two

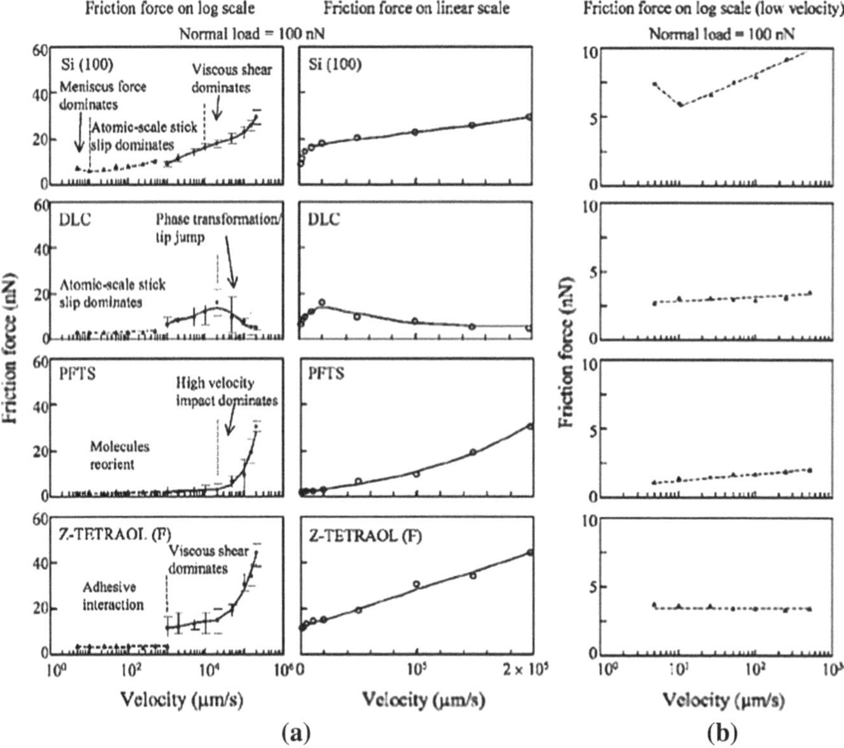

Fig. 1.12 Frictional force versus velocity at **a** both log and linear scales and **b** log scale in a range of lower velocities. (Reprinted with permission from Tao and Bhushan [17], Copyright 2007, American Vacuum Society) [17]

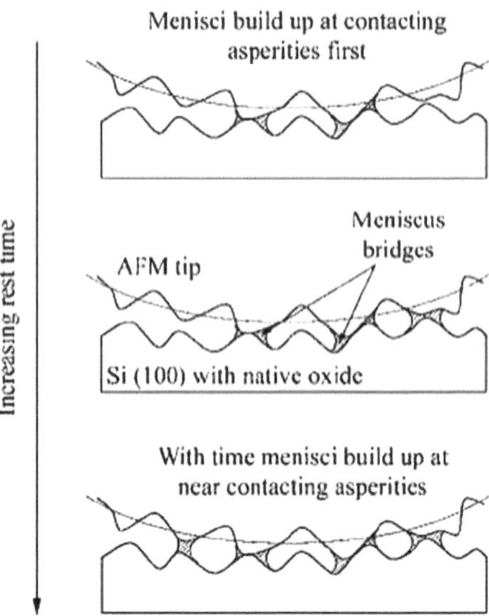

Fig. 1.13 Schematic diagram showing the effect of resting time on adhesive force due to formation of meniscus bridges in contact zone. (Reprinted with permission from Tambe and Bhushan [18])

force sensors are connected to the cantilever holder to measure the deflections of the cantilever in the x and z-directions. Two mirrors are attached to the flexible end of the cantilever, from which the force sensors detect the optical signals reflected from the mirrors. The measured reflected light intensity depends on the distances between the sensors and mirrors, which are proportional to the deflections of the cantilever caused by friction or adhesion. The lateral and vertical deflections of the cantilever are induced by the frictional and adhesion forces, respectively. Therefore, the corresponding spring constants of the cantilever are used to calculate friction or adhesion forces.

Controlling Parameters for Friction Coefficient

Effect of Velocity

A great influence of relative moving velocity of two bodies in contact on friction force or friction coefficient is shown in Fig. 1.12. Many researchers have studied friction force as a function of velocity for Si, diamond-like carbon (DLC) and many other materials [17, 18]. In the ambient condition a thin film of moisture can be formed on material surfaces, which affects friction coefficient in a range of low velocities. For example, the friction force between Si surface and AFM tip decreases with increasing velocity in a range of very low velocities, which is because the meniscus bridges developed around the tip are a function of equilibrium time as shown in Fig. 1.13. During the movement of the tip, it breaks the meniscus bridges, which generates a friction force. The generated friction force depends on the strengths of the meniscus bridges. However, the meniscus bridges cannot be fully reformed when velocity is increased, which results in a lower meniscus force and thus a lower friction force as shown in Fig. 1.13.

At very high velocities, the meniscus bridges do not have enough time to form and thus their effect becomes negligible and the stick–slip becomes more dominating for controlling the friction coefficient. The stick–slip causes a linear increase in friction force with velocity, which can be explained by viscous shear.

The motion of the AFM tip is expressed by a spring mass model (Eq. 1.43) to explain the atomic scale stick–slip mechanism of friction [19]:

$$m\ddot{x}_t = -\eta\dot{x}_t - k(x_M - x_t) - F \tag{1.43}$$

with

$$x_m = v_m t \tag{1.44}$$

where m, η, k, x_t, F, x_m, v_m, and t are the effective mass of the system, viscous damping coefficient, spring constant of the cantilever, position of the tip, external force, equilibrium position of the cantilever, scan velocity, and rest time, respectively.

The lateral force is expressed as

$$F_l = k(x_M - x_t) \tag{1.45}$$

which is the friction force.

Tomlinson has assumed a periodic surface with a potential $V(x)$ as presented in Eq. 1.46 [20].

$$V(x) = V_0(1 - cos\frac{2\pi x}{a}) \tag{1.46}$$

where V_0 and a are the surface barrier potential height and lattice constant of the surface, respectively.

Therefore, the force F is expressed as

$$F = V'(x) = \frac{2\pi V_0}{a}\sin(\frac{2\pi x}{a}) \tag{1.47}$$

The Tomlinson model has taken into account the effect of thermal activation or elastic energy stored in the cantilever to derive the relationship between friction force and velocity, which is presented by Eq. 1.48:

$$F_{stick-slip} = F_0 + c\, ln\, v \tag{1.48}$$

where F_o and c are the constants and υ is the sliding velocity.

If a thin film of any liquid like water or moisture is present, the friction force is related to the velocity of the tip and the viscosity of the film as presented below:

$$F_{friction} = \mu N + \eta \dot{\gamma} A \approx \mu N + \frac{\eta v A}{d} \quad (1.49)$$

where μ, N, η, $\dot{\gamma}$, v, A, and d are the coefficient of friction at dry condition, applied load, viscosity of the film, velocity gradient, sliding velocity, real contact area, and thickness of the film, respectively.

Here, a noticeable thing is that a thin film of moisture on a material surface has been considered. This moisture layer depends on the surface nature of the material concerned such as hydrophilicity or hydrophobicity, which will be discussed later. Because of the formation and breaking of the menisci at lower velocities and the stick–slip mechanism at higher velocities, the friction force first decreases and then increases with the increase in velocity as shown in Fig. 1.12.

A moisture layer can be easily formed and the meniscus contribution is significant on a hydrophilic surface such as Si. In the contrast, the effect of meniscus bridges is not significant on some hydrophobic surfaces where the stick–slip is the dominating factor, which causes a continuous increase in coefficient of friction with the increase in velocity for these surfaces.

Effect of Surface Force

The nature of material surfaces plays a key role in controlling friction. An AFM tip can experience various kinds of forces, such as van der Waals, capillary, chemical bonding, magnetic, hydration and electrostatic forces, when it comes close to the counterbody. A basic equation of friction coefficient is the ratio of lateral force to normal force. Acting forces that increase or decrease the normal force according to the direction of action affect the friction force and friction coefficient. When the tip comes near the sample surface it feels an attractive force and when it almost touches the sample surface it feels a repulsive force, both of which contribute to the friction force accordingly.

Van der Waals forces are short-range attractive forces acting between all atoms, molecules, nanoparticles, etc. An AFM tip usually experiences van der Waals forces (F_{vdw}) when it is brought into close proximity to the sample surface, which is a function of distance (r_d) and tip radius (R_{tip}) as represented by Eq. 1.50:

$$F_{vdw} = A_{ham} R_{tip} / (6 r_d^2) \quad (1.50)$$

where A_{ham} is the Hamaker constant that depends on the polarizability of the material.

The acting electrostatic force (F_{el}) can be attractive or repulsive depending upon the charges, which is expressed as

$$F_{el} = (q_1 q_2) / (4 \pi \varepsilon D_d^2) \quad (1.51)$$

where q_1 and q_2, ε, and D_d are the charges of two materials, dielectric constant and distance between charges, respectively.

The nature of a surface, such as hydrophobicity, hodrophilicity, or surface energy, forms other friction controlling parameters. For example, surface energy can be used to monitor attractive force through Eq. 1.52:

$$W_A = 4\left[\frac{\gamma_1^p \gamma_2^p}{\gamma_1^p + \gamma_2^p} + \frac{\gamma_1^d \gamma_2^d}{\gamma_1^d + \gamma_2^d}\right] \tag{1.52}$$

where W_A, γ_i^p and γ_i^d are the work of adhesion, and the polar and dispersive components of surface energy of material i, respectively.

The surface energy and its polar and dispersive components of a surface can be determined by measuring the contact angles using two different liquids on the material surface.

A relation between surface energy and contact angle is presented by the well-known Young Eq. [21]:

$$\gamma_{sv} = \gamma_{sl} + \gamma_{lv}\cos\theta + \pi_e \tag{1.53}$$

where γ_{sv} and γ_{lv} are the free energies of the solid and liquid surfaces against the saturated vapor, respectively, γ_{sl} is the free energy of the interface between liquid and solid, π_e is an equilibrium pressure of the adsorbed vapor on the solid that can be assumed to be zero, and θ is the contact angle between solid and liquid.

The dispersive (d) and polar (p) components of surface and interface energies are as follows

$$\gamma_{sv} = \gamma_{sv}^d + \gamma_{sv}^p \tag{1.54}$$

$$\gamma_{lv} = \gamma_{lv}^d + \gamma_{lv}^p \tag{1.55}$$

$$\gamma_{sl} = \gamma_{sl}^d + \gamma_{sl}^p \tag{1.56}$$

The dispersive and polar components of the interfacial tension between solid and liquid are related to those of the solid and liquid surface energies by the following Eq. [21]:

$$\gamma_{sl}^d = \gamma_{sv}^d + \gamma_{lv}^d - 2(\gamma_{sv}^d \times \gamma_{lv}^d)^{1/2} \tag{1.57}$$

$$\gamma_{sl}^p = \gamma_{sv}^p + \gamma_{lv}^p - 2(\gamma_{sv}^p \times \gamma_{lv}^p)^{1/2} \tag{1.58}$$

After combining Eqs. 1.54–1.58 with Eq. 1.53,

$$(1 + \cos\theta)(\gamma_{lv}^d + \gamma_{lv}^p) = 2(\gamma_{sv}^d + \gamma_{lv}^d)^{1/2} + 2(\gamma_{sv}^p + \gamma_{lv}^p)^{1/2} \tag{1.59}$$

where γ_{sv}^d and γ_{sv}^p can be determined by measuring θ values of two different liquids with known γ_{lv}^d and γ_{lv}^p.

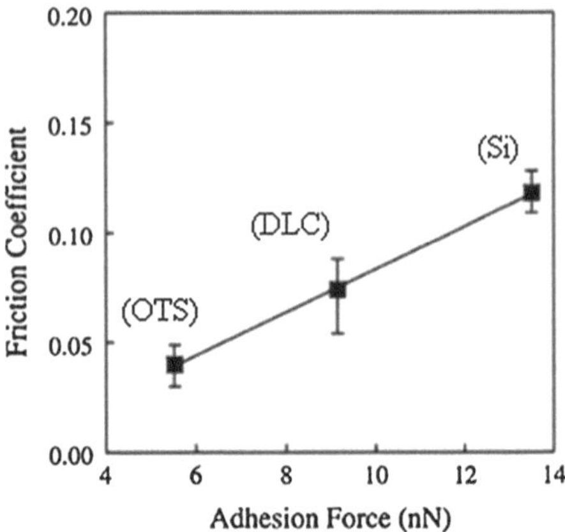

Fig. 1.14 Friction coefficient versus adhesive force of uncoated and OTS and DLC coated Si-wafer surfaces measured by LFM using a Si_3N_4 tip. (Reprinted with permission from Yoon et al. [22])

Yoon et al. have shown that the adhesive force and coefficient of friction of Si-wafer, DLC coated Si and OTS coated Si measured using a Si_3N_4 tip are related to the water contact angles on these surfaces, and the friction coefficient increases with increased adhesion force as shown in Fig. 1.14 [22].

A layer of liquid deposited on the surface of a material plays an important role according to Eq. 1.47, which depends on the polar and dispersive components of the surface energies. Bhushan and Burton have shown that adhesion and coefficient of friction are functions of contact angle and nature of materials as shown in Fig. 1.15 [23]. In ambient condition adhesive force and friction coefficient increase with the decrease in contact angle because of the formation of higher meniscus bridges and the increase in surface energy. However in the presence of a liquid film with a lower viscosity, adhesion and friction decrease for those materials that have lower contact angles. This is because a more uniform layer is created on the top of a material surface with a lower contact angle, which acts as a lubricant. A viscous liquid film increasing the adhesion and friction of same material surface is attributed to its higher viscosity.

Effect of Temperature

Friction of a hard material may increase or decrease depending on the testing environment. Under an atmospheric condition, desorption of moisture on the material surface increases with the decrease in temperature, which acts as a lubricant and causes a decrease in friction. In most cases, the viscosity of contact liquid (lubricant) increases with increased temperature, in which the friction of the material surface increases accordingly as per Eq. 1.49.

Friction of soft polymeric materials depends on their viscoelastic behaviors that vary with temperature. Friction force is caused by interfacial or adhesive force together with deformation force or mechanical loss. Therefore, inclusion of both

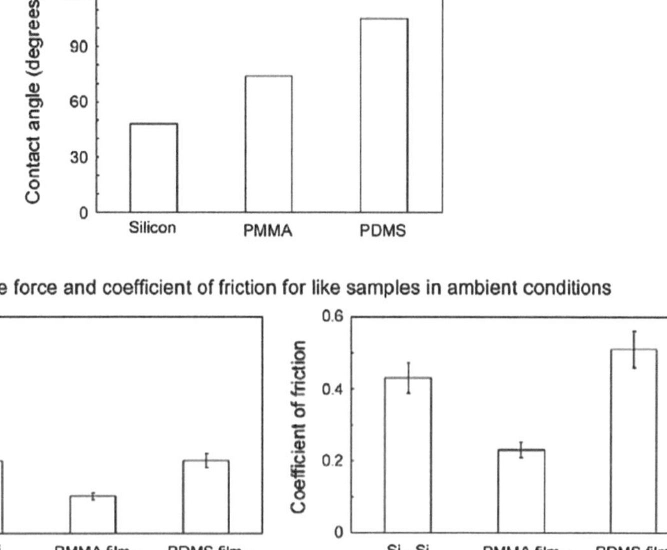

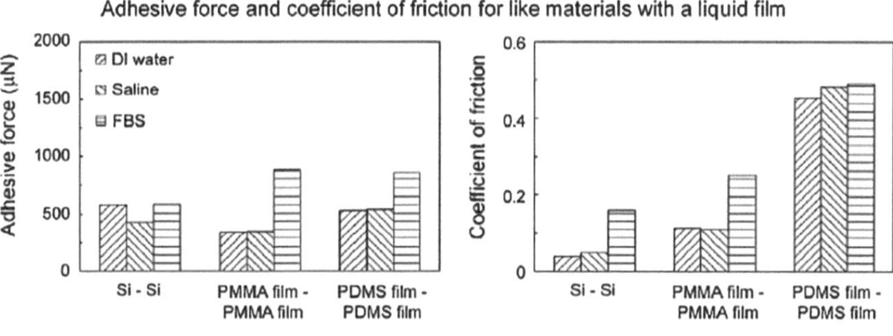

Fig. 1.15 Contact angles, adhesive forces and coefficients of friction of different materials under different conditions. (Reprinted with permission from Bhushan and Burton [23])

forces is equally important for the analysis of friction coefficient. Though many reports have stated that the pull-off or adhesive force of a polymer remains constant at a temperature below the glass transition temperature (T_g) of the polymer [2, 24] a variation in friction coefficient is observed at a temperature much lower than T_g [2], indicating that a deformation loss is another controlling factor of friction. Hammerschmidt and Gladfelter have studied the friction behavior of PMMA, PS and PET and shown the variations of friction with the variations of viscoelastic properties of those materials, which are closely related to testing temperature [2] as shown in Fig. 1.16. A good agreement of the overall shape of viscoelastic loss tangent (tan δ) data with respect to friction coefficient for PMMA, PET and PS is observed as the friction of the polymers is proportional to their loss

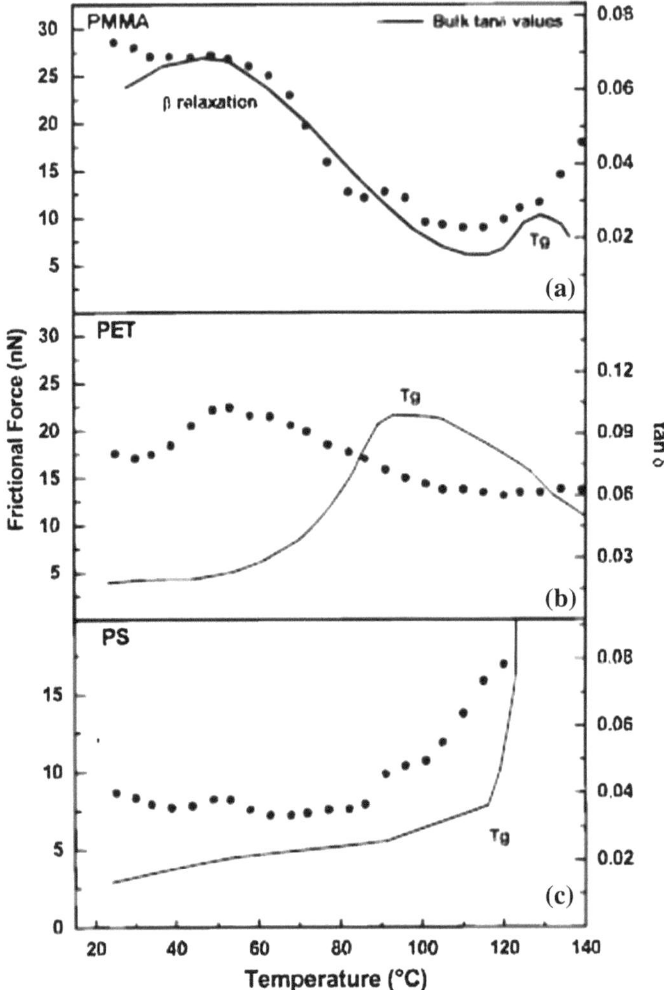

Fig. 1.16 Friction force and *tan δ* of different polymers with respect to testing temperature. (Reprinted with permission from Hammerschmidt et al. [2])

moduli. Around T_g, the friction rapidly increases due to the phase change from solid to rubber state.

Effect of Load

The increase in applied load acts in different ways to increase coefficient of friction and as a result the friction force no longer remains proportional to the applied load. In an actual process, friction force $(F) = F_0 + \mu P$, where F_0 is the contribution of adhesive force that increases with the contact force, μ is the coefficient of friction according to Amonton's law and P is the applied normal load.

Adhesion based friction force (F_0) can be expressed as [25]:

$$F_0 = A_r \tau_a \tag{1.60}$$

where τ_a and A_r are the average shear strength during sliding and the real contact area, respectively.

Therefore, adhesion based friction coefficient is

$$\mu_a = \frac{A_r \tau_a}{P} \tag{1.61}$$

The radius of contact area (a) can be calculated based on the JKR theory as presented in Eq. 1.62.

$$a^3 = \frac{R}{K}\left\{P + 3W_{12}\pi R + \sqrt{\left[6W_{12}\pi RP + (3W_{12}\pi R)^2\right]}\right\} \tag{1.62}$$

with

$$R = \frac{r_1 r_2}{(r_1 + r_2)} \tag{1.63}$$

$$K = \frac{4}{3\pi(k_1 k_2)} \tag{1.64}$$

$$k_1 = \frac{1 - \upsilon_1^2}{\pi E_1} \tag{1.65}$$

$$k_2 = \frac{1 - \upsilon_2^2}{\pi E_2} \tag{1.66}$$

where R, K, W_{12}, r_1, r_2, υ_1, υ_2, k_1, k_2 and E_1 and E_2 are the equivalent radius, composite spring constant, interfacial energy, radii of contacting surfaces, and Poisson ratios, elastic constants and Young moduli of contacting materials, respectively.

For an elastic spherical indenter, homogeneous half-space contact area is presented as

$$A_r = \pi\left(\frac{3PR}{4E^*}\right)^{\frac{2}{3}} \tag{1.67}$$

where $E^* = \left[\left(1-\upsilon_1^2\right)/E_1 + \left(1-\upsilon_2^2\right)/E_2\right]^{-1}$ is the composite Young modulus.

By combining Eqs. 1.61 and 1.67:

$$\mu_a = \pi\tau_a\left(\frac{3R}{4E^*}\right)^{2/3} P^{-1/3} \tag{1.68}$$

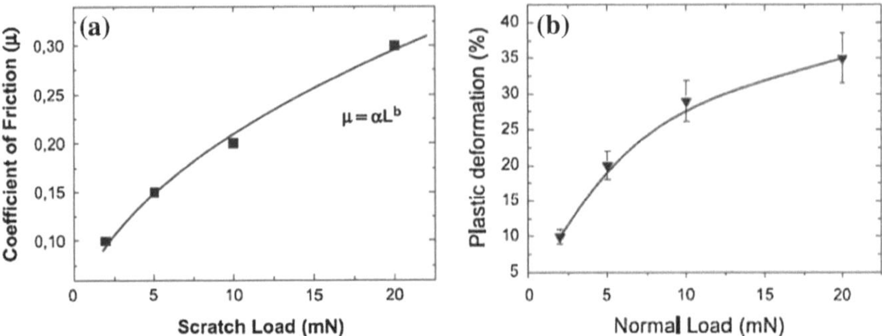

Fig. 1.17 Effect of scratch load on **a** coefficient of friction and **b** plastic deformation of CN_x thin film. (Reprinted with permission from Charitidis and Logothetidis [25])

Equation 1.68 indicates that the relation between normal load and friction coefficient attributed to adhesion is not linear. At the same time the increase in load increases the plowing depth of the indenter. The plowing effect promotes the friction coefficient. Therefore, the increase in load increases the coefficient of friction due to the plowing effect. Adhesion is the dominating mechanism for low loads, whereas plowing effect and deformation are the main controlling parameters for high loads.

Charitidis and Logothetidis have studied the tribological behavior of amorphous carbon nitride (CN_x) thin films deposited on Si substrates with respect to applied load [26] and noticed that the effect of load is not significant at low loads. At high loads plastic deformation and friction coefficient increase with the increase in load as shown in Fig. 1.17.

Effect of Contact Area

It has been discussed previously that different surface forces affect coefficient of friction and modern theories of friction have proposed that the friction force due to adhesion is proportional to the real area (A_r) of contact according to Eq. 1.60. Therefore, the friction coefficient increases with the increase in contact area. Yoon et al. have used different sizes of balls attached to an AFM tip to study the effect of contact area on the coefficients of friction of bare and DLC coated Si surfaces and observed that the friction coefficients of the both surfaces increase with the increase in ball radius as shown in Fig. 1.18 [27]. This phenomenon is equally applicable to microtribological test in which the effect of contact area is more prominent for these materials, which can induce higher adhesive energies.

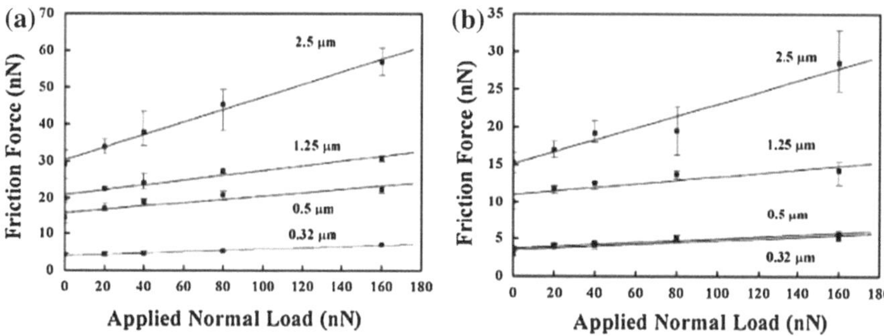

Fig. 1.18 Effect of contact area on friction force measured by AFM with attached balls of radii of 0.32, 0.5, 1.25 and 2.5 μm for **a** uncoated and **b** DLC coated Si surfaces. (Reprinted with permission from Yoon et al [27])

Effect of Environment

It has been reported by Bhushan and Burton that some environmental conditions such as ambient atmosphere, dry air, dry N_2 and vacuum have no significant influence on the coefficient of friction of different wear couples like Si on Si, PMMA and PDMS as shown in Fig. 1.19 [23], except a higher coefficient of friction of the Si–Si couple measured in ambient atmosphere, which is due to the presence of moisture that forms meniscus bridges on the two Si surfaces in contact.

Applications of Nanotribology

Micro-Electro-Mechanical System

Small and light weight micro-electro-mechanical systems (MEMS) and nano-electro-mechanical system (NEMS) are gaining increasing importance in aerospace applications, integrated sensors and intelligent control systems. In these systems of highly integrated electronics, mechanical components are co-fabricated on planar wafers and subsequently etched free for mechanical movements in three dimensions. Si is a well established material for fabrication of MEMS and NEMS but the reliability of Si moving components has been in doubt as Si is a brittle material in nature. MEMS and NEMS usually have a large surface area-to-volume ratio that makes them particularly vulnerable to surface damage due to high friction and adhesion of the components to the adjacent structures during operation or use. A major design limitation for these systems is their inability to withstand a prolonged sliding surface contact due to high friction. A short functional lifetime of these devices is attributed to high coefficient of friction and excessive wear rate of the devices. Due to the small dimensions of MEMS and NEMS, wear debris

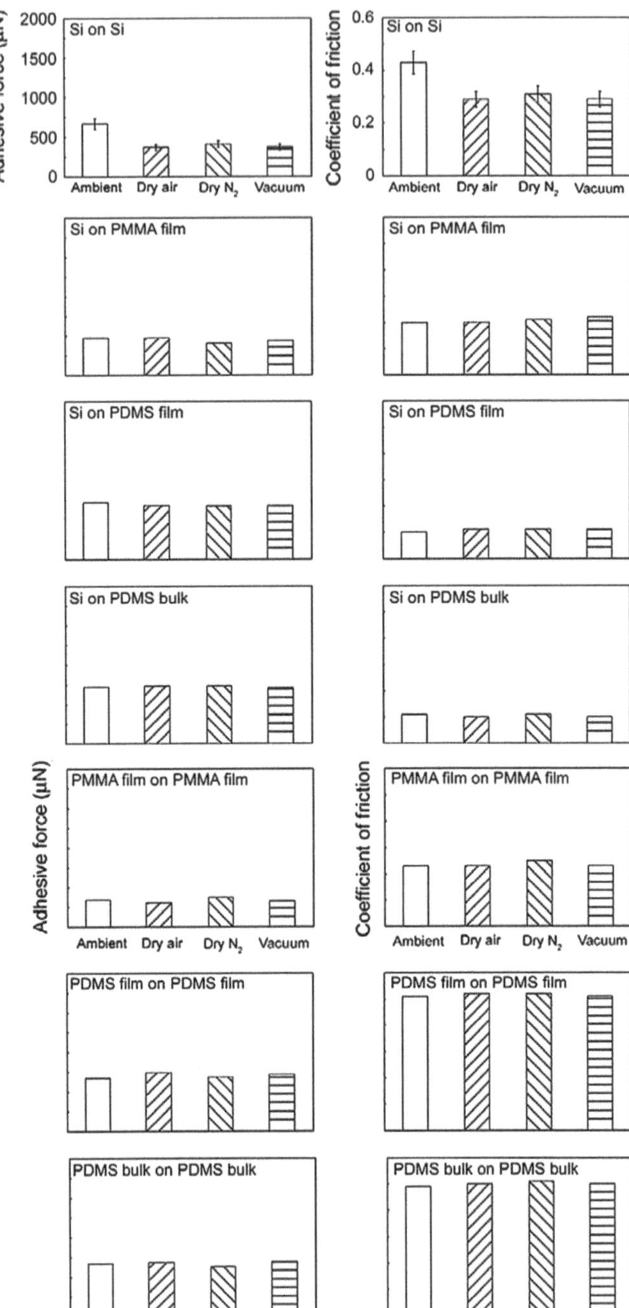

Fig. 1.19 Effect of environmental condition on adhesive force and coefficient of friction of different wear couples. (Reprinted with permission from Bhushan and Burton [23])

accumulated at the interface can cause jamming. Wear can also lower the corrosion resistance of the components in a harsh environment. It is also noticed that the friction coefficient in a micro or nanodevice is greater than that in a macro device due to size effect under similar operation conditions. Therefore, friction is a more critical issue for NEMS and MEMS. Another important issue associated with MEMS and NEMS is adhesion that arises during the physical contact of the components when a MEMS or NEMS device is in operation. To overcome those problems many researchers have used surface modification techniques or dry/liquid lubrication methods.

The fabrication of MEMS and NEMS also suffers from the adhesion problem. During the last step of a micro or nano fabrication process, the Si device that is fabricated is rinsed with water and exposed to the atmosphere, during which a native oxide layer is formed at the top of the Si device, which can induce a severe adhesion between its parts in contact and cause difficulties in separation of the parts. Usually, freezing-drying or supercritical CO_2 drying methods can be used to overcome the adhesion problem.

It is observed from the tests of MEMS and NEMS actuators that continuous replenishment of an adsorbed water layer and operation at a moderate relative humidity can improve the lifetime of the devices [28]. However, excessive moisture leads to capillary induced adhesion that prevents the actuators from running.

Deposition of an organic layer on material surfaces is one of effective methods to reduce adhesion as surface energy and friction can be reduced by this process. But the durability of this type of organic lubricant coating can be poor if the coating is not good under normal storing condition and during operation [29]. Nevertheless, a chemically bonded organic film can significantly reduce the friction and adhesion of material surfaces as the mobile part of this type of lubricant can spread over the newly exposed area due to wear and freshly cover the surface. In general, a simple dip coating or vapor phase deposition process is employed for this type of coatings. Zhuang et al. have studied the performance of different types of anti-sticking self-assembled monolayers (SAMs) such as Tridecafluoro-1,1,2,2-tetra-hydrooctyltrichlorosilane (FOTS), 1H, 1H, 2H, 2H-perfluorodecyltrichlorosilane (FDTS), Octadecyltrichlorosilane (OTS), etc. for MEMS applications and observed that these layers can significantly reduce the adhesion and friction of Si as shown in Fig. 1.20 [30]. Among the SAMs, FDTS shows the best performance.

Adhesive force is very high for bare Si–Si contact, which can be reduced when one or both Si surfaces are coated with an OTS coating as shown in Fig. 1.21 [31], among which the OTS–OTS contact shows the lowest adhesive force.

Popular protective hard coatings for high wear resistance and low friction and adhesion are DLC, TiN, TiAlN, WS_2, TiC, SiC, MoS_2, etc., which can be produced by different deposition techniques such as magnetron sputtering, cathodic arc, electron beam and chemical vapor deposition, etc. As the above materials and processes have their own advantages and limitations, proper processes and materials can be selected according to design requirements.

Ashurst et al. employed different types of coatings to protect MEMS devices from high friction and wear and have observed that the lifetime and coefficient of

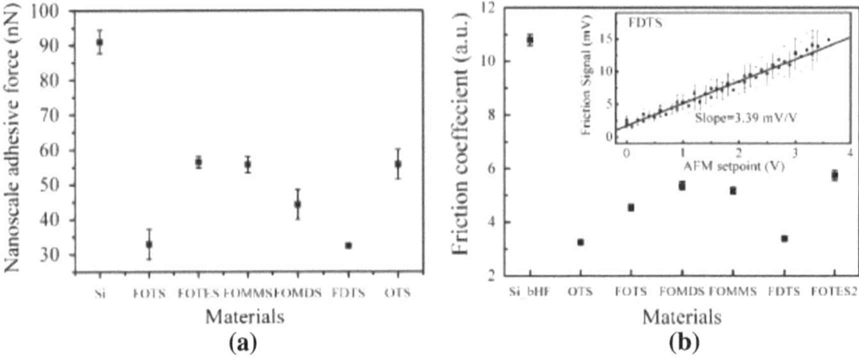

Fig. 1.20 Adhesion forces and friction coefficients of Si with or without SAM coatings of different lubricants. (Reprinted with permission from Zhuang et al. [30])

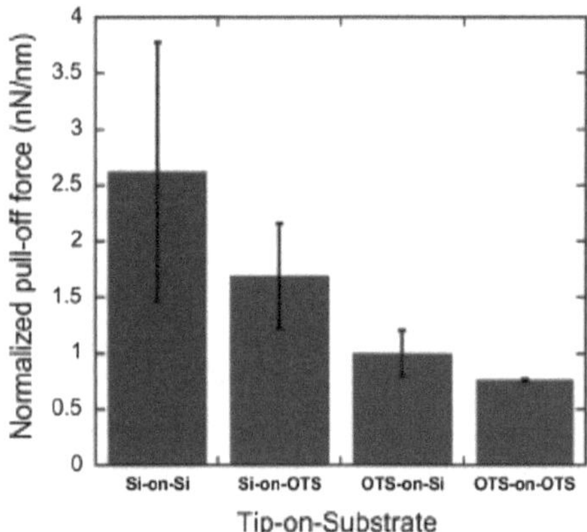

Fig. 1.21 Effect of octadecyltrichlorosilane (OTS) coating on adhesive force between AFM Si tip and Si sample surfaces without and with one or both surfaces coated with OTS. (Reprinted with permission from Zhuang et al. [31])

friction of the devices are different with respect to different coatings as shown in Fig. 1.22 [32]. Figure 1.22 shows that the oxide layer coated surface has a higher coefficient of friction along with a shorter lifetime, whereas the vapor based dichlorodimethylsilane (DMS) and SiC coated surfaces have lower coefficients of friction. It was observed that the SiC coated device was still intact even after one million cycles. The wear of the side walls of the device with respect to wear cycles are shown in Fig. 1.23.

Microfluidic Device and Micromold

In the early days almost all microfluidic devices were made of Si. Due to the limitations of Si such as non-transparency, use of harmful chemicals during fabrication, higher cost and difficulty in sealing of channels, etc., polymers have been

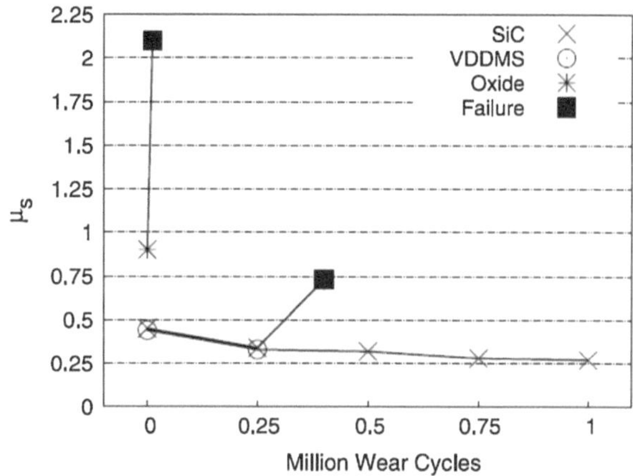

Fig. 1.22 Wear cycle versus coefficient of friction of MEMS devices under different surface conditions. (Reprinted with permission from Ashurst et al. [32])

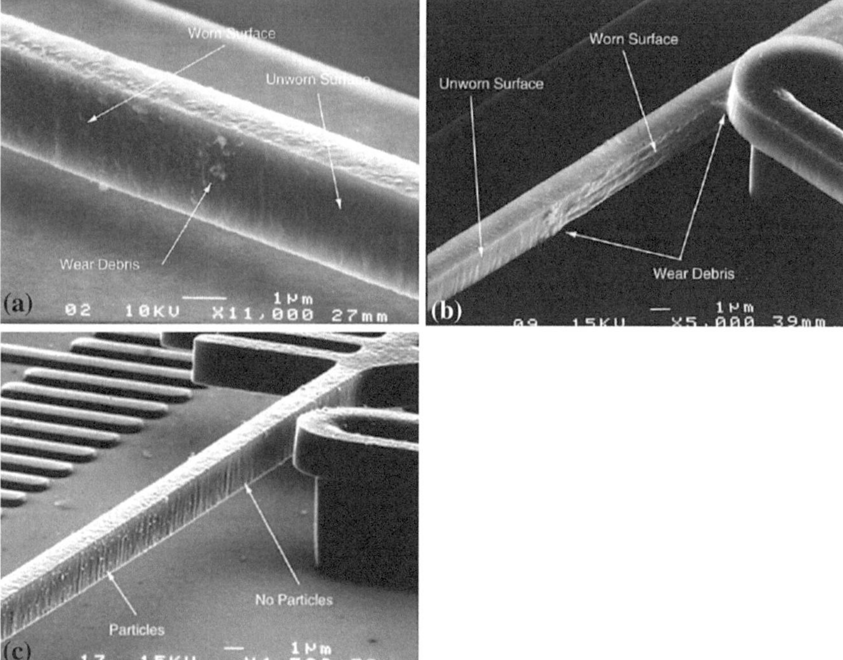

Fig. 1.23 SEM micrographs of side walls coated with **a** oxide layer after 1.1×10^4 cycles, **b** DMS layer after 2.5×10^5 cycles and **c** SiC layer after 10^6 cycles. (Reprinted with permission from Ashurst et al. [32])

the most popular materials for microfluidic devices nowadays. Nevertheless, Si has been widely used to fabricate micromolds for production of microfluidic devices in hot embossing, injection molding, micro powder injection molding, and nanoimprint lithography processes [33, 34]. Si has poor tribological properties that can

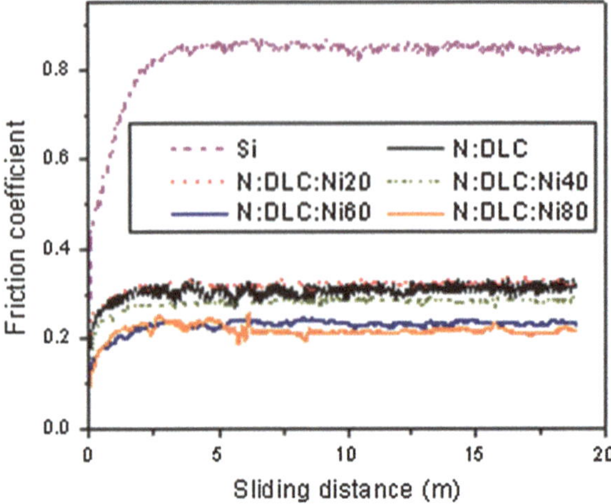

Fig. 1.24 Coefficients of friction of uncoated and various DLC coated Si micromolds with respect to sliding distance. (Reprinted with permission from Saha et al. [21])

cause defects in a part when it is separated from the mold (demolding) [35]. Some other challenges are that micromolds should have good thermal resistance and chemical inertness. Several researchers have attempted to improve the tribological performance of Si micromolds by applying surface coatings, including fluoropolymers and hard coatings.

The surface of an ideal micromold should have low adhesion and friction along with high hardness and wear resistance. In this regard, Saha et al. have used different types of DLC coatings to improve the performance and lifetime of Si micromolds [21]. Friction coefficient is one of the dominating characteristics of the demolding process in micro-embossing as the surface-to-volume ratio of a microdevice is very high. Therefore, the friction coefficient of micromolds can be minimized by using DLC coatings. The friction coefficients between coated or uncoated Si micromolds and PMMA counterballs with respect to sliding distance are presented in Fig. 1.24, from which it is observed that the steady state friction coefficient first slightly increases with the increase in Ni sputtering power from 0 to 20 W, and then continuously decreases with the further increase in Ni target power. The increased surface roughness of the coatings in addition to Ni content could be another reason for the increase in friction coefficient. A higher Ni content in DLC coatings promotes the formation of more sp^2 bonds that act as a solid lubricant and thus cause a continuous decrease in friction coefficient. The wear scars on the PMMA balls against the micromolds are in the order of versus Si (740 μm) > N:DLC:Ni20 (545 μm) > N:DLC (456 μm) > N:DLC:Ni40 (441 μm) > N:DLC:Ni60 (399 μm) > N:DLC:Ni80 (322 μm). Figure 1.25 shows the wear scars of the bare Si and N:DLC:Ni80 coated sample surfaces.

The uncoated and N:DLC:Ni60 coated Si micromolds are used to evaluate the performance of the DLC coatings and the effect of surface parameters. A higher

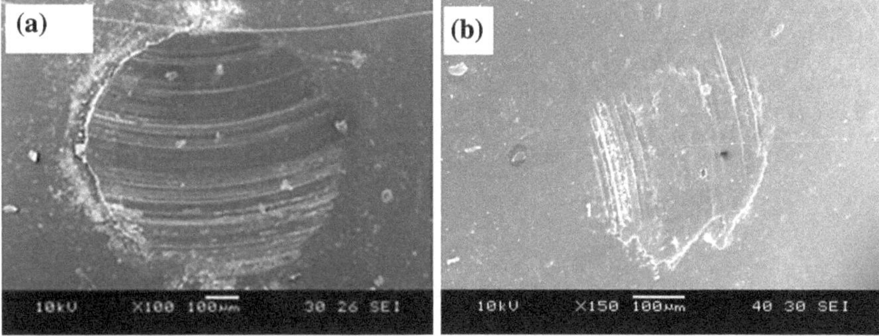

Fig. 1.25 Wear scars of PMMA balls against **a** uncoated and **b** N:DLC:Ni80 coated Si samples. (Reprinted with permission from Saha et al. [21])

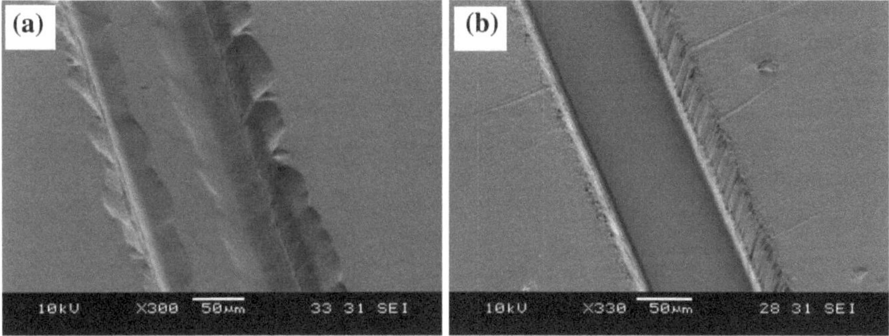

Fig. 1.26 Channel profiles of microfluidic devices fabricated using **a** uncoated and **b** N:DLC:Ni60 coated Si micromolds. (Reprinted with permission from Saha et al [21])

distortion of the sidewalls of the channels in a micorfluidic device embossed using the bare Si micromold is observed as compared to that embossed by using the DLC coated Si micromold as shown in Fig. 1.26, which indicates that the replication quality of the DLC coated Si micromold is much better than that of the uncoated Si.

Magnetic Recording System

In a magnetic recording system a slider flies over a rigid disk as shown in Fig. 1.27 [36], in which an extremely small gap is required to increase its data recording density. Ideally, a zero spacing between read/write transducer and magnetic disk is desired, however, which could cause the slider to crash due to vibration. With an extremely small clearance, several problems can occur due to adhesion, friction and wear. Therefore, to increase the recording density and lifetime of hard disk drives, a special care of tribological issues must be properly exercised. Researchers

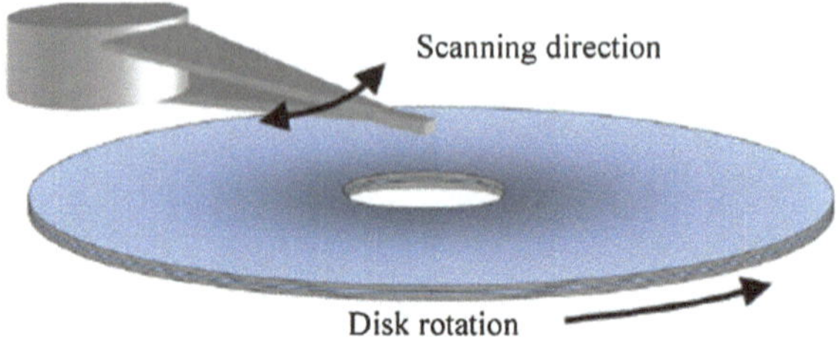

Fig. 1.27 Schematic diagram of head-disk interface, (Reprinted with permission from Vakis and Polycarpou [36])

have tried different kinds of lubricants to suppress the tribological effect. Mate has shown that friction and adhesive force decrease for lubricated disks and the friction also decreases with the increase in lubricant thickness as shown in Fig. 1.28 [37].

Khurshudov and Waltman have studied the wear of magnetic disk and slider surfaces [38], in which they have evaluated the effect of lubricant on the corrosion behavior of the disk surface. Figure 1.29 shows a wear depth in carbon overcoat in a 3 mm section in width analyzed by optical surface analyzer (OSA) using S and P polarized light. Significant wear of the disk after a 72 h run has generated lots of wear particles that are especially harmful during the movement of the slider.

The thickness of the carbon overcoat layer can be decreased by wear during sliding, because of which, the corrosion of the disk surface becomes a significant problem [38].

Nanowear

Removal of materials during relative motion of two mating surfaces is called wear, which depends on the process parameters and the properties of the materials in contact. Most of the time, wear generated during the nanotribological operation of a micro/nanosystem hampers the process and reduces the lifetime and efficiency of the system. Therefore, understanding of wear process, rate and mechanism as well as its remedies is very important. The wear depths and coefficients of various materials are summarized in Table 1.1 [39], where the wear coefficients vary between 0.10 and 10^{-5} that can cause significant changes in dimensions and severe problems in micro/nanosystems.

The measurement of nanowear behavior and volume is not straightforward as the amount of wear is usually very small. The surfaces in contact are often burnished by each other, in which the surface roughnesses change and thus the wear volumes cannot be quantified. Therefore, more data need to be gathered and analyzed for a better understanding of the nanoscale wear behavior. Researchers have tried different methods to measure wear depth. For example, groove depths below 1 nm of thin films can be examined using AFM with a diamond-tipped cantilever [40]. If a wear depth on a surface is very large and significant with respect to its roughness, a direct AFM scan can determine the wear rate. However, if the wear depth is only a few nm or below, the sample surface has to be scanned

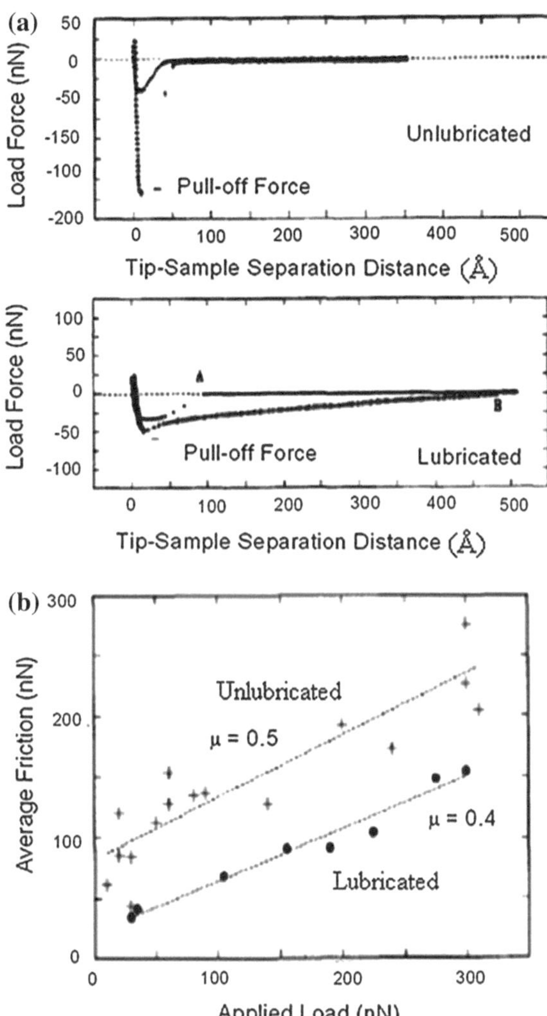

Fig. 1.28 Effect of perfluoropolyether lubricant on **a** adhesive force and **b** friction at head-disk interface, (Reprinted with permission from Mate [37])

using AFM before and after the wear test so that the wear depth can be determined by the subtraction of the two scans. It means that the surface profile of a worn surface subtracted by that of the same surface before the wear test can indicate the wear volume [41].

Kelvin probe microscopy has been used to detect wear precursors at ultralow loads by Bhushan and Goldade [42], which is to detect the change in surface potential due to a chemical or structural change because of wear. Very often field emission scanning electron microscopy (FESEM) and high-resolution transmission electron microscopy (HRTEM) are used to image the microstructure of a sample before and after a wear test in order to identify the changes in the microstructure due to nanowear. HRTEM can also provide information about nanofracture and

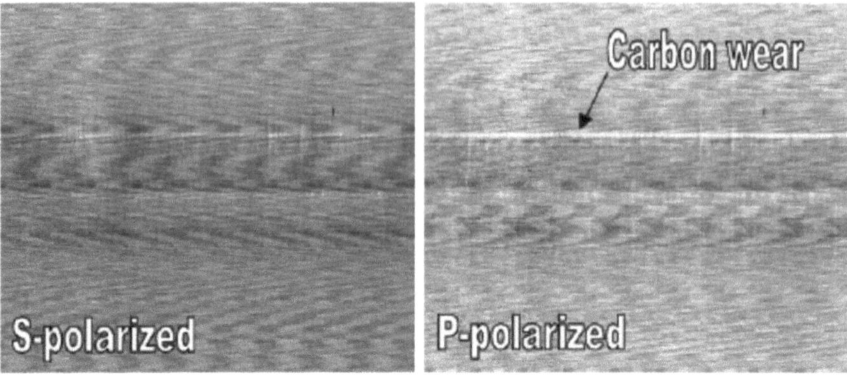

Fig. 1.29 A worn disk surface after a 72 h test with a low-flying positive pressure slider, measured using S and P polarized light. (Reprinted with permission from Khurshudov and Waltman [38])

change in crystalinity due to wear. Sometimes, nanoindentation or focused ion beam (FIB) can be used to mark on a sample surface before it is subjected to a wear test. Then the change in dimension due to wear is used to quantify the wear rate [43]. Raman spectroscopy can also be used for this purpose when a sample surface is coated with a diamond or DLC coating with which the calibrated intensities of the Raman G and D peaks with respect to the change in thickness of the carbon coating are used to determine the wear rate of the sample [44].

It is reported that wear is strongly related to coefficient of friction as it is significantly influenced by the energy dissipation mechanism that occurs at the mating interface during wear test [45]. The wear behavior of most of the materials is similar, which increases with increased normal load and number of wear cycles. Sundararajan and Bhushan have reported that wear rates of a number of materials are a function of load and number of cycles [46]. Among the tested materials, SiC has the highest wear resistance as well as scratch resistance as shown in Fig. 1.30. The high hardness of SiC is one of the factors for its high wear and scratch resistances. The presence of plastic deformation can cause an additional fracture, which induces abrasive wear. Thus, the wear rate increases more quickly with increased load and number of cycles for those materials that have higher abrasive wear. It is also observed that the wear volume of a micro-nano contact system increases with the increase in relative humidity due to increased adhesive force [39]. Though the effect of sliding velocity on wear rate is not significant [47], heat generated during wear test can cause significant changes in microstructure or mechanical properties of material surfaces, which leads to a significant increase in wear rate of the samples [48]. Song and Zhang have found significant changes in wear lifetime of polyfluorowax (PFW)/polyurethane (PU) coatings embedded with 5 wt % nano-SiC or nano-ZrO_2 with variations in speed as shown in Fig. 1.31.

Table 1.1 Wear depths/rates and wear coefficients of different materials under varying loading conditions [39]

Test method	Material	Load (max contact pressure)	Wear depth/rate	Calculated wear coefficient
Specimen on disk	DLC	10–30 μN (50–420 Pa)	80 nm	~ 10^{-5}
	Si (100)		100 nm	~ 10^{-5}
	SiO_2		260 nm	~ 10^{-4}
	Si_3N_4		130 nm	~ 10^{-4}
	Polysilicon		120 nm	~ 10^{-4}
AFM	Si (100)	100 μN (~ 10 GPa)	~190 nm	~ 10^{-1}
	p-Si (100)		~220 nm	~ 10^{-1}
	Polysilicon		~220 nm	~ 10^{-1}
	n-Polysilicon		~160 nm	~ 10^{-1}
	SiC		~20 nm	~ 10^{-2}
Pin-on-disk	Si (111)	100 mN (160 MPa)	2.3×10^{-5} mm^3/Nm	~ 10^{-4}
	Carbon nitride coating		2.9×10^{-6} mm^3/Nm	~ 10^{-5}
	Carbon coating		5.5×10^{-6} mm^3/Nm	~ 10^{-5}
AFM	a-C	3–13 μN	2 nm/cycle	–
	Si (100)		0.01–0.36 nm/cycle	–
Pin-on-disk	Si	100 mN	7.4–8.8 $\times 10^{-4}$ mm^3/Nm	~ 10^{-3}
	SiC		0.2–1.8 $\times 10^{-5}$ mm^3/Nm	~ 10^{-5}
	Polycrystalline diamond		6.1–8.6 $\times 10^{-7}$ mm^3/Nm	~ 10^{-5}
AFM	DLC	100–800 nN (4–7 GPa)	8.5–8.7 $\times 10^{-9}$ mm^3/Ncycle	
	Si (100)			
	Cu		8.1×10^{-8} mm^3/Ncycle	
	Au		6.3×10^{-8} mm^3/Ncycle	
	Si tip	10–325 nN (5–10 GPa)	–	10^{-1}–10^{-3}
	Si_3N_4 tip		–	10^{-4}–10^{-5}

Lubricants for Nanotechnology

A lubricant is a substance that is used to reduce the friction and wear of mating surfaces moving relative to each other. A lubricant can be in solid, liquid or gas state. In most cases, the thickness of a liquid lubricant layer at a nanometer level is difficult to maintain and thus it becomes comparable with the dimensions of a micro/nanosystem concerned. In this small system it is difficult to continuously replace the liquid lubricant. Therefore, liquid lubricants are not preferable for micro/nanodevices. A lubricant layer for a micro or nanosystem must be tenacious, long-lasting and extremely thin. Thus, most suitable lubricants for micro and

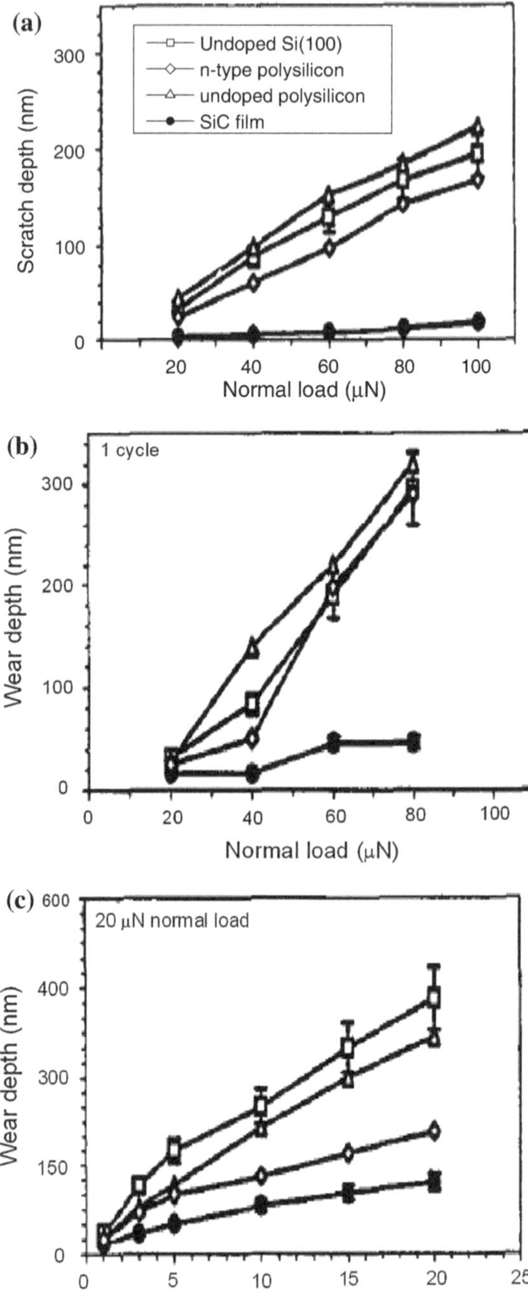

Fig. 1.30 **a** Scratch depths and **b** wear depths of various materials versus load or number of cycles. (Reprinted with permission from Sundararajan and Bhushan [46])

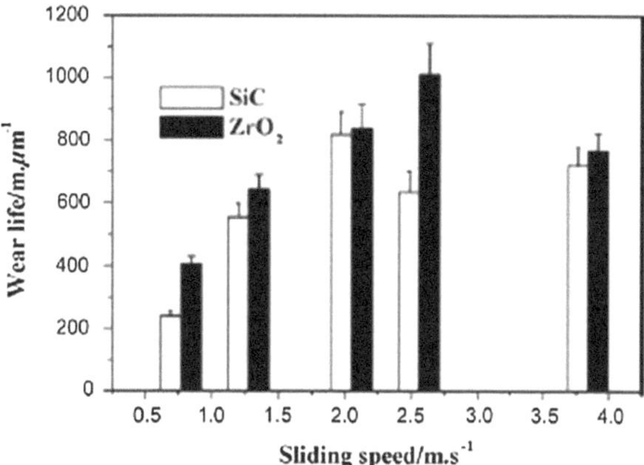

Fig. 1.31 Effect of sliding speed on wear life of PFW/PU coatings filled with 5 wt % nano-SiC or nano-ZrO_2 under a load of 320 N. (Reprinted with permission from Song and Zhang [48])

nanosystems are solid types. Other advantages of solid lubricants are that they have very low vapor pressures and less chance of contamination and can endure under high loads, function at low and high temperatures, and be stored for a long time. Some solid lubricants are also electrically conductive, which are preferable for special cases. However, solid lubricants have some disadvantages such as poor thermal conductivity and difficulty for replacement after a finite wear life, etc.

Various types of solid lubricants are commercially available such as metal and ceramic coatings, fluoropolymers, and self-assembled monolayers (SAMs), etc. with their properties and performances being described below.

Metallic and Ceramic Lubricants

The performance of single element metals as lubricants are generally poor as shown in Table 1.2 [49]. For some metals, their coefficients of friction remain unchanged in water vapor, ambient atmosphere or O_2 gas environment but getting

Table 1.2 Coefficients of friction of like metal wear couples under various environmental conditions [49]

Metal	Coefficient of friction after admitting		
	H_2 or N_2	Air or O_2	Water vapor
Al on Al	–	1.9	1.1
Cu on Cu	4	1.6	1.6
Au on Au	4	2.8	2.5
Fe on Fe	–	1.2	1.2
Mo on Mo	–	0.8	0.8
Ni on Ni	5	3	1.6
Pt on Pt	–	3	3
Ag on Ag	–	1.5	1.5

Table 1.3 Coefficients of friction of various solid lubricants under various loads

Lubricant	Counter material	Load	Environment	Test method	Sliding velocity	Coefficient of friction	References
B_4C	Si_3N_4	100–366 nN	RH 5–40 %	AFM	2–20 mm/s	0.3–0.5	[50]
BC_xN_y	Si_3N_4	100–366 nN	RH 5–40 %	AFM	2–20 mm/s	0.4–0.5	[50]
WSC	Diamond tip	0.5–4 mN		Scratch tester	1 μm/s	0.06–0.16	[51]
CN	Diamond stylus	100 mN	Ar + N_2	Scratch tester		~0.074	[52]
BCN	Diamond stylus	100 mN	Ar + N_2	Scratch tester		~0.054	[52]
CN	Diamond stylus	100 mN	N_2 + He	Scratch tester		~0.33	[52]
BCN	Diamond stylus	100 mN	N_2 + He	Scratch tester		~0.027	[52]
DLC	Borosilicate glass	0–160 nN	RH 45 ± 5 %	AFM	5 μm/s	0.01–0.025	[27]
MoS_2	Si_3N_4	100–200 nN	RH 3–95 %	AFM		0.005–0.04	[53]
SiC	WC	20 mN	RH 45 ± 5 %	Accelerated micro-tribometer	1 mm/s	0.17–0.2	[54]
CoCrMo	WC	20 mN	RH 45 ± 5 %	Micro-tribometer	1 mm/s	0.17–0.23	[54]
Ti-6Al-4 V	WC	20 mN	RH 45 ± 5 %	Micro-tribometer	1 mm/s	0.23–0.26	[54]
HOPG	Si_3N_4	0–20 nN	280 K	AFM	600 nm/s	~ 0.05	[55]

very high in H_2 or N_2 gas environment. This means that the adhesion of those materials is very strong in inert environment. Graphite, DLC, MoS_2, TiAlN, WS_2, hexagonal boron nitride (h-BN), H_3BO_3 and SiC are very common lubricants, among which graphite and h-BN are very good for high temperature applications and MoS_2 and TiAlN are preferable at high load carrying conditions with the performance of MoS_2 affected by relative humidity. Table 1.3 summaries the coefficients of friction of various metallic and ceramic lubricants at very low loads [27, 50–55].

The basic working principle of these lubricants is that their atoms are closely packed and strongly bonded in each atomic layer, which gives a high in-plane strength for a longer wear lifetime. But the atoms of different atomic layers are relatively far away from each other, which gives a weak interlayer bonding strength resulting in a relatively easy motion to provide effective lubrication. Metallic and ceramic lubricant coatings are usually produced by various vapor deposition processes such as magnetron sputtering, cathodic arc, electron beam, chemical vapor deposition, etc.

Fluoropolymer Lubricants

This type of polymer lubricants is very common because they are light in weight, relatively inexpensive, easy to fabricate, and good in performance, which can be produced by spraying, dipping, sputtering or chemical vapor deposition. Most popular fluoropolymer lubricants are polytetrafluoroethylene (PTFE), widely known as Teflon, octafluorocyclobutane (C_4F_8), perfluoroalkoxy (PFA), perfluoropolyether (PFPE) and fluorinated ethylene propylene (FEP), etc. Another advantage of this kind of lubricants is that their adhesive strengths are also very low. Therefore, these lubricants are popular for micro-nano injection molding, hot embossing, MEMS and NEMS applications. Nowadays, researchers have blended fluoropolymers with various ceramic materials to improve their wear resistance and thermal properties.

PFPE nanolubricant coatings have been widely used to increase the durability of hard disk drives [56], which are deposited on the disk surfaces simply by dipping the disks into a solution of Fomblin-Zdol or Fomblin-Z03 in hydrofluoroether. The friction coefficients of both Zdol and Z03 coatings are low and further decrease with the increase in surface coverage of the disks.

Si micromolds fabricated by deep reactive ion etching (DRIE) usually have very high coefficient of friction, surface energy and side wall roughness that increase the difficulties of demolding process. C_4F_8 plasma treatment of Si micromolds after etching can improve the replication performance of the molds [57], in which case C_4F_8 acts as a lubricant and lowers the friction and adhesive forces simultaneously.

Self Assembled Monolayer

Previously described lubricants are mainly produced via vapor deposition processes. SAM is a kind of boundary lubricant, which can be deposited on substrate surfaces spontaneously by simply immersing the substrates into a solution comprising an organic solvent and an active surfactant. The SAM coated surfaces can have very low friction coefficient and surface energy and good thermal stability and rupture strength due to the SAM's strong chemical bonding with the substrate surfaces. SAM coatings have attracted a great attention as a molecular lubricant for MEMS, nanoimprint lithography (NIL) and bio-technology to tackle the friction and adhesion related problems.

A SAM comprises highly organized amphiphilic molecules with their heads strongly bonded to a substrate surface and their tails having particular functional groups. It means that the hydrophilic groups of the molecules assemble together on the substrate surface, whereas the hydrophobic ends of the molecules assemble far from the substrate surface and control the surface properties of the coated sample. The backbones of the molecules are weakly bonded by van der Waals forces, so

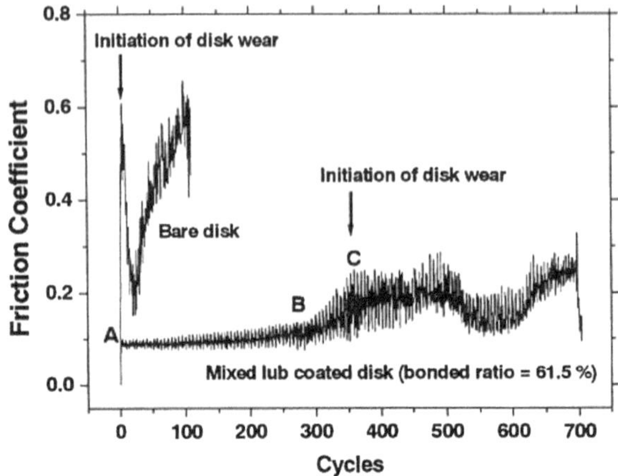

Fig. 1.32 Coefficients of friction of bare and lubricated disks with respect to number of cycles, where initiation of disk wear is indicated. (Reprinted with permission from Choi et al. [58])

Table 1.4 Properties of various SAM lubricant coatings

SAM	Thickness (nm)	Water contact angle (°)	Friction coefficient	Surface roughness (nm)	Thermal stability (°C)	References
PFDA	2.0	105	~0.17	0.447	–	[62]
FDTS	1.3–1.5	106–115.4	0.1–0.13	0.14	400–425	[60, 61, 63]
OTS	–	109–110	0.07–0.073	0.09–25	180–225	[59, 64, 65]
FOTS	–	107.6	–	0.07	~425	[64]
FOMDS	–	106.0	–	0.13	~425	[64]
FOTES	–	103.6	–	0.12	~425	[64]
FOMMS	–	104.3	–	0.12	~425	[64]
1- octadecene	–	104	~0.05	0.4	~225	[59]
DTS	1.5	104	~0.08	0.12	–	[61]
MAC-DTS	5.0	105	~0.075	–	–	[61]
MAC-FDTS	5.0	106	~0.12	–	–	[61]

the molecules have mobile characteristics sideways in addition to the strong chemical bonding between the head groups and substrate surface. The unbonded molecules can replenish the lubricant coating at the locations where the coating is depleted during the test. Figure 1.32 shows the wear profiles of the bare and the SAM lubricated disks. The coefficient of friction of the bare disk jumps to a high value of about 0.6 along with the initiation of wear just at the beginning of the wear test, whereas the coefficient of friction of the SAM coated disk is very low from the start of the test till wear appears at about 350 wear cycles, which has indicated the effectiveness of the SAM lubricant coating [58]. Table 1.4 summarizes the properties of various SAM lubricant coatings [59–65].

Simulation of Tribological Phenomena

Nowadays, simulation is a demanding part of research for various reasons. A main advantage of simulation is that it provides the designer a practical feedback when a real world system is under design, which allows the designer to establish the correctness and efficiency of the design with an actual process. Simulation also allows to explore the merits of various ideas without a physical system, in which the effects of various parameters can be predicted before dealing with the actual phenomena. Another benefit of simulation is that it permits the researchers to study a problem at several different levels of abstraction.

Owing to the above advantages, simulation is critical for nanotribological phenomena. Researchers have been able to measure friction forces at atomic levels since the advent of AFM in 1980's, which has promoted the knowledge in nanotribology about the origin of friction, wear and related phenomena at atomic scales. The acquired nanotribological knowledge has been used to develop theoretical models. In this particular field a popular simulation tool is molecular dynamic (MD) simulation, which simulates spatial and temporal motions of atoms and molecules by analysis of their relative positions, velocities and interaction forces. In a MD simulation, initial conditions are described and interatomic forces are calculated using classical potential energy functions from electronic structure calculations. Morse potential and Lennard-Jones (LJ) potential are the two common forms of the potential. More commonly used two-body potential for interactions between atoms or molecules with closed electron shells is the LJ potential (V) that is presented in Eq. 1.69:

$$V(x_{ij}) = 4\varepsilon\left[\left(\frac{\sigma}{x_{ij}}\right)^{12} - \left(\frac{\sigma}{x_{ij}}\right)^{6}\right] \tag{1.69}$$

and the LJ time unit is defined as

$$\tau_{LJ} = \left(\frac{m\sigma^2}{\varepsilon}\right)^{\frac{1}{2}} \tag{1.70}$$

where x_{ij}, ε, σ and m are the distance between particles i and j, *LJ* interaction energy, diameter, and mass, respectively.

MD simulations are used to study adhesion, friction, wear, scratch, etc. Song and Srolovitz have simulated amount of transferred materials from one body to another, in which the plastic deformation caused depends on the work of adhesion as shown in Fig. 1.33 [66].

The stick–slip phenomenon between a diamond slider and a Cu specimen is shown in Fig. 1.34 [67], In which a three dimensional structure of Cu is considered and it is assumed that a Morse potential exists between a pair of Cu atoms. An atomic array at 0 K is established by considering the principal minimum potential energy, while the atomic array is established by considering the thermal expansion

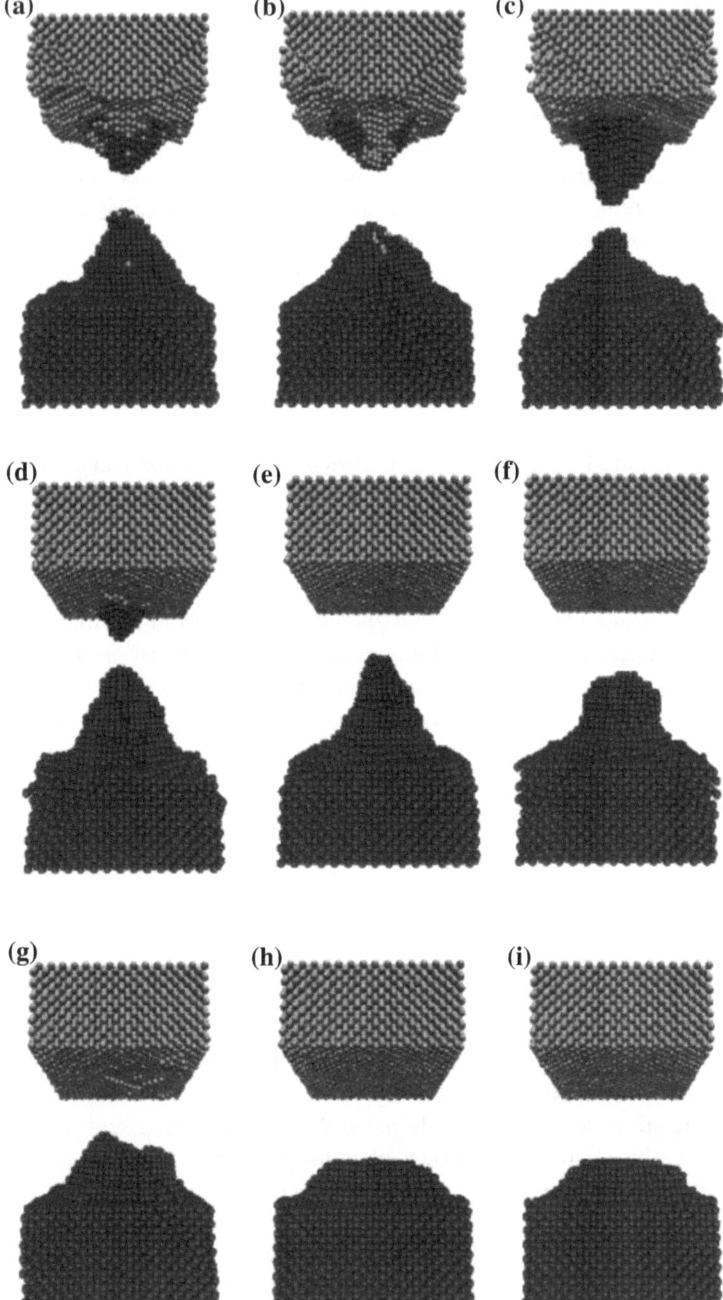

Fig. 1.33 Configurations of a contact system after compression and separation with different values of the work of adhesion: **a** 1366, **b** 1157.9, **c** 998.8, **d** 881.8, **e** 735, **f** 694.2, **g** 610.9, **h** 532.6 and **i** 388.4 erg/cm^2. (Reprinted with permission from Song and Srolovitz [66])

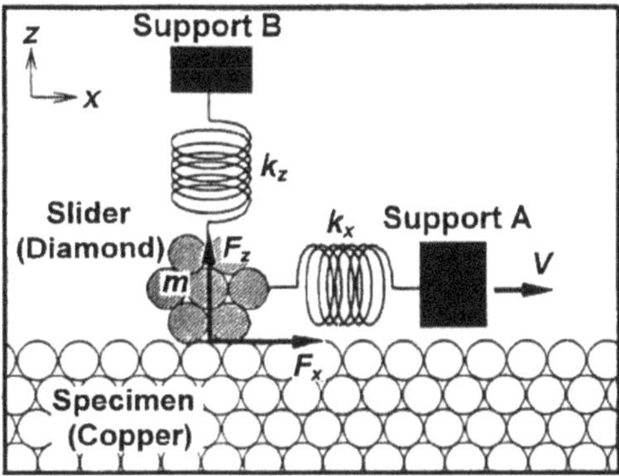

Fig. 1.34 A molecular dynamic simulation model for stick–slip phenomenon. (Reprinted with permission from Shimizu et al. [67])

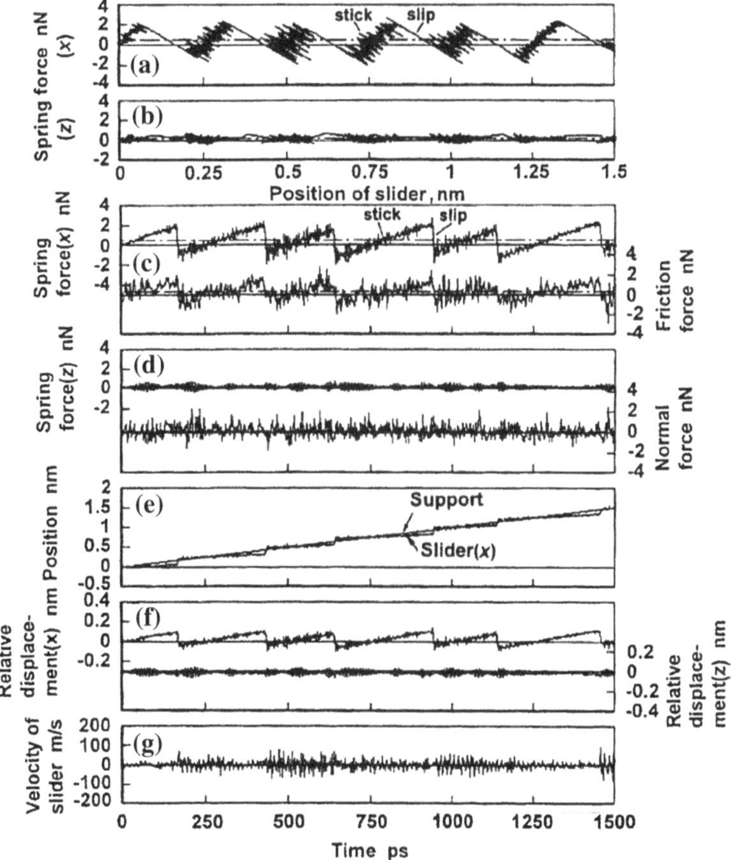

Fig. 1.35 Variations of spring, friction and normal forces exerted on a diamond slider as well as dynamics of the slider, where sliding speed $v = 1$ m/s, and spring constants $k_x = 20$ N/m and $k_z = 10$ N/m. (Reprinted with permission from Shimizu et al. [67])

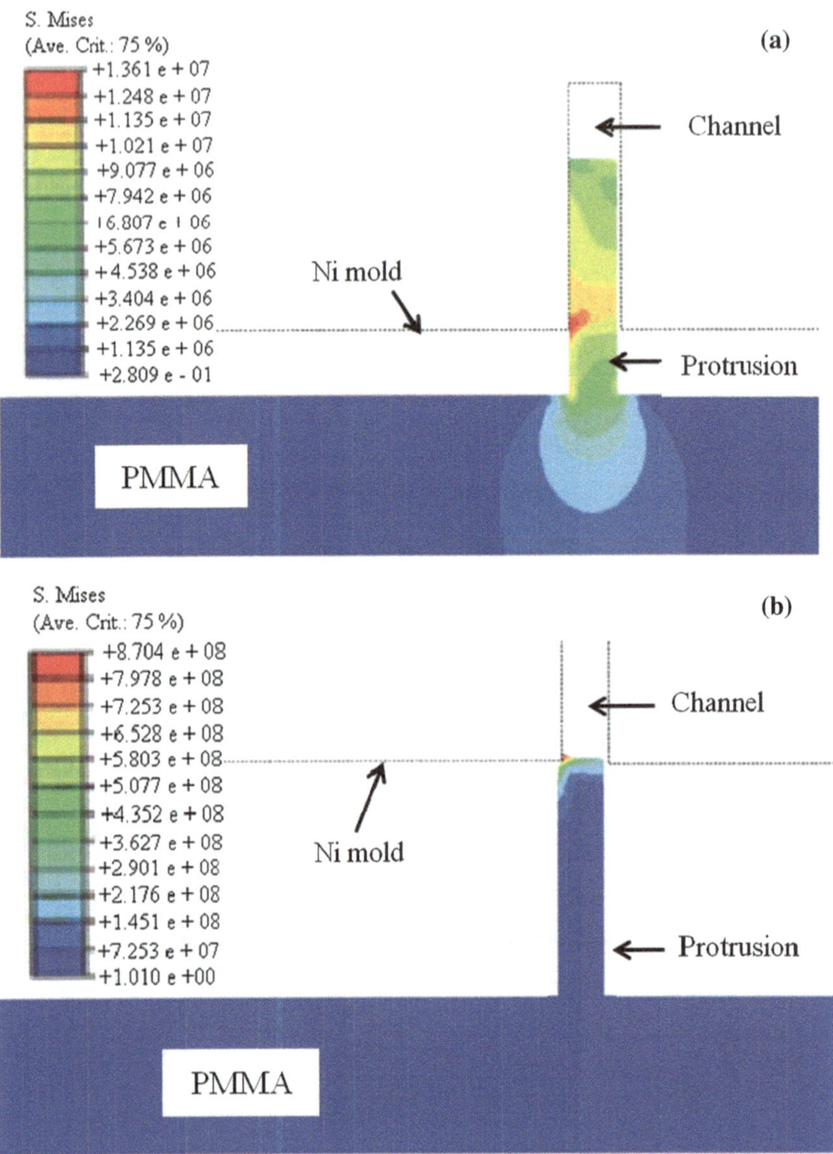

Fig. 36 Simulation of effects of adhesion and friction using a Ni mold insert during demolding at **a** 3.5 s and **b** 12 s, and using a PTFE film as a release agent during demolding at **c** 3.5 s and **d** 12 s (pressures in Pa). (Reprinted with permission from Guo et al. [68])

of the array and the mean velocity of the atoms in the array at room temperature. The spring constants k_x and k_z of an AFM diamond slider are introduced (Fig. 1.34) to consider the effect of the spring constants. The supports A and B

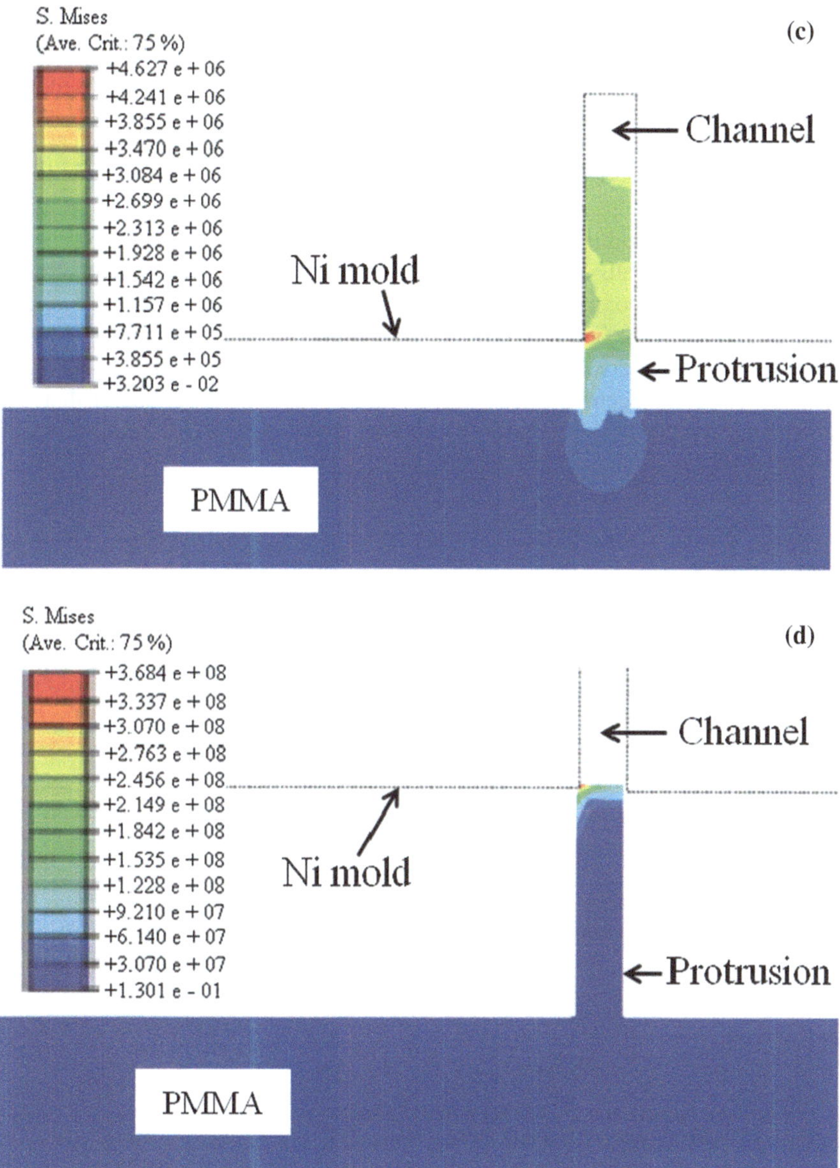

Fig. 36 (continued)

move at a constant velocity (v) and the friction force (F_x) is experienced by the slider. The simulated results of the spring, friction and normal forces exerted on the slider with respect to various parameters are illustrated in Fig. 1.35.

Effects of friction and adhesion have been studied for various micro/nanoprocesses. One of the potential areas is to reduce the stress and distortion of micro or nanostructures induced during the demolding of hot embossing process, which has been simulated by researchers as shown in Fig. 1.36 [68]. The two surfaces in contact undertake a shearing force when they are moved in the opposite directions, and this shearing force is considered as the friction force. As polymer is softened in the demolding process, adhesion at the interface between the mold and polymer material is another dominating force.

The simulated results as illustrated in Fig. 1.36 show that a maximum stress is encountered at the contact point and the stress increases with the progress in demolding as the force becomes concentrated within the reduced contact area. The simulation also depicts that the stress experienced by the nickel mold insert without a releasing agent is about 13–28.7 MPa, which is much higher than that encountered by the Ni insert with a releasing agent of a PTFE film, i.e. about 4.6–10.5 MPa.

Summary

Tribology research exceedingly needs broadened knowledge in various fields such as physics, chemistry, mechanics, materials science, etc. This chapter attempted to give a broad overview on the fundamentals, measuring techniques, effects of various conditions, applications and simulations of friction, wear and lubrication at nanoscales. It was widely perceived that tribological test conditions such as load, velocity, temperature, surface energy, surface topography, environment, etc. played major roles in nanotribology.

Acknowledgments Professor Yee Cheong Lam, School of Mechanical and Aerospace Engineering, Nanyang Technological University, offered his valuable advice for the chapter.

References

1. B. Bhushan: Introduction to Tribology, Wiley (2002)
2. Hammerschmidt, J.A., Gladfelter, W.L., Haugstad, G.: Probing polymer viscoelastic relaxations with temperature-controlled friction force microscopy. Macromolecules **32**, 3360–3367 (1999)
3. Enachescu, M., van den Oetelaar, R.J.A., Carpick, R.W., Ogletree, D.F., Flipse, C.F.J., Salmeron, M.: Observation of proportionality between friction and contact area at the nanometer scale. Tribol. Lett. **7**, 73–78 (1999)
4. Ishikawa, T., Kobayashi, M., Takahara, A.: Macroscopic Frictional Properties of Poly(1-(2-methacryloyloxy)ethyl-3-butyl Imidazolium Bis(trifluoromethanesulfonyl)-imide) Brush Surfaces in an Ionic Liquid. Acs Appl. Mate. Interfaces **2**, 1120–1128 (2010)
5. Wang, X.P., Tsui, O.K.C., Xiao, X.D.: Dynamic study of polymer films by friction force microscopy with continuously varying load. Langmuir **18**, 7066–7072 (2002)

6. Zhang, Q., Archer, L.A.: Interfacial friction of surfaces grafted with one- and two-component self-assembled monolayers. Langmuir **21**, 5405–5413 (2005)
7. Tian, F., Xiao, X.D., Loy, M.M.T., Wang, C., Bai, C.L.: Humidity and temperature effect on frictional properties of mica and alkylsilane monolayer self-assembled on mica. Langmuir **15**, 244–249 (1999)
8. Liu, E., Ding, Y.F., Li, L., Blanpain, B., Celis, J.P.: Influence of humidity on the friction of diamond and diamond-like carbon materials. Tribol. Int. **40**, 216–219 (2007)
9. Zhang, W., Tanaka, A., Wazumi, K., Koga, Y.: Effect of environment on friction and wear properties of diamond-like carbon film. Thin Solid Films **413**, 104–109 (2002)
10. Kalin, M., Novak, S., Vizintin, J.: Wear and friction behavior of alumina ceramics in aqueous solutions with different pH. Wear **254**, 1141–1146 (2003)
11. Sang, Y., Dube, M., Grant, M.: Dependence of friction on roughness, velocity, and temperature. Phys. Rev. E **77**, 036123 (2008)
12. Binnig, G., Quate, C.F., Gerber, C.: Atomic force microscope. Phys. Rev. Lett. **56**, 930–933 (1986)
13. Mate, C.M., McClelland, G.M., Erlandsson, R., Chiang, S.: Atomic-scale friction of a tungsten tip on a graphite surface. Phys. Rev. Lett. **59**, 1942–1945 (1987)
14. Liu, E., Blanpain, B., Celis, J.P.: Calibration procedures for frictional measurements with a lateral force microscope. Wear **192**, 141–150 (1996)
15. E. Liu, Chapter 5: Friction from reciprocating sliding of different scales, in Tribology Research Trends, Taisho Hasegawa (ed.), 2009, Nova, USA
16. Bhushan, B.: Tribology and Mechanics of Magnetic Storage Devices, 2nd edn. New York, Springer-Verlag (1996)
17. Tao, Z., Bhushan, B.: Velocity dependence and rest time effect on nanoscale friction of ultrathin films at high sliding velocities. J. Vac. Sci. Technol. A **25**, 1267–1274 (2007)
18. Tambe, N.S., Bhushan, B.: A new atomic force microscopy based technique for studying nanoscale friction at high sliding velocities. J. Phys. D **38**, 764–773 (2005)
19. Zworner, O., Holscher, H., Schwarz, U.D., Wiesendanger, R.: The velocity dependence of frictional forces in point-contact friction. Appl. Phys. A **66**, S263–S267 (1998)
20. Tomlinson, G.A.: A molecular theory of friction. Philos. Mag. **7**, 905–939 (1929)
21. Saha, B., Liu, E., Tor, S.B., Hardt, D.E., Chun, J.H., Khun, N.W.: Improvement in lifetime and replication quality of Si micromold using N:DLC:Ni coatings for microfluidic devices. Sens. Actuators B **150**, 174–182 (2010)
22. Yoon, E.S., Yang, S.H., Han, H.G., Kong, H.: An experimental study on the adhesion at a nano-contact. Wear **254**, 974–980 (2003)
23. Bhushan, B., Burton, Z.: Adhesion and friction properties of polymers in microfluidic devices. Nanotechnology **16**, 467–478 (2005)
24. Kim, K.S., Ando, Y., Kim, K.W.: The effect of temperature on the nanoscale adhesion and friction behaviors of thermoplastic polymer films. Nanotechnology **19**, 105701 (2008)
25. Charitidis, C.A., Logothetidis, S.: Effects of normal load on nanotribological properties of sputtered carbon nitride films. Diam. Relat. Mater. **14**, 98–108 (2005)
26. Charitidis, C.A., Logothetidis, S.: Effects of normal load on nanotribological properties of sputtered carbon nitride films. Diam. Relat. Mater. **14**, 98–108 (2005)
27. Yoon, E.S., Singh, R.A., Oh, H.J., Kong, H.: The effect of contact area on nano/micro-scale friction. Wear **259**, 1424–1431 (2005)
28. Patton, S.T., Eapen, K.C., Zabinski, J.S.: Effects of adsorbed water and sample aging in air on the mu N level adhesion force between Si(100) and silicon nitride. Tribol. Int. **34**, 481–491 (2001)
29. Baker, M.A., Li, J.: The influence of an OTS self-assembled monolayer on the wear-resistant properties of polysilicon based MEMS. Surf. Interface Anal. **38**, 863–867 (2006)
30. Zhuang, Y.X., Hansen, O., Knieling, T., Wang, C., Rombach, P., Lang, W., Benecke, W., Kehlenbeck, M., Koblitz, J.: Vapor-phase self-assembled monolayers for anti-stiction applications in MEMS. J. Microelectromech. Syst. **16**, 1451–1460 (2007)

31. Flater, E.E., Ashurst, W.R., Carpick, R.W.: Nanotribology of octadecyltrichlorosilane monolayers and silicon: Self-mated versus unmated interfaces and local packing density effects. Langmuir **23**, 9242–9252 (2007)
32. Ashurst, W.R., Wijesundara, M.B.J., Carraro, C., Maboudian, R.: Tribological impact of SiC encapsulation of released polycrystalline silicon microstructures. Tribol. Lett. **17**, 195–198 (2004)
33. Becker, H., Heim, U., Ieee, I.: Silicon as tool material for polymer hot embossing, in Mems '99: Twelfth IEEE International Conference on Micro Electro Mechanical Systems, Technical Digest. pp. 228-231 (1999)
34. Fu, G., Tor, S., Loh, N., Tay, B., Hardt, D.E.: A micro powder injection molding apparatus for high aspect ratio metal micro-structure production. J. Micromech. Microeng. **17**, 1803–1809 (2007)
35. Hirai, Y., Yoshida, S., Takagi, N.: Defect analysis in thermal nanoimprint lithography. J. Vac. Sci. Technol. B **21**, 2765–2770 (2003)
36. A.I. Vakis, A.A. Polycarpou, Head-disk interface nanotribology for Tbit/inch(2) recording densities: near-contact and contact recording. J. Phys. D: Appl. Phys. **43** (2010) 225301_1-13
37. Mate, C.M.: Nanotribology of lubricated and unlubricated carbon overcoats on magnetic disks studied by friction force microscopy. Surf. Coat. Technol. **62**, 373–379 (1993)
38. Khurshudov, A., Waltman, R.J.: Tribology challenges of modern magnetic hard disk drives. Wear **250–251**, 1124–1132 (2001)
39. Chung, K.H., Kim, D.E.: Fundamental investigation of micro wear rate using an atomic force microscope. Tribol. Lett. **15**, 135–144 (2003)
40. Wienss, A., Persch-Schuy, G., Vogelgesang, M., Hartmann, U.: Scratching resistance of diamond-like carbon coatings in the subnanometer regime. Appl. Phys. Lett. **75**, 1077–1079 (1999)
41. Gahlin, R., Jacobson, S.: A novel method to map and quantify wear on a micro-scale. Wear **222**, 93–102 (1998)
42. Bhushan, B., Goldade, A.V.: Kelvin probe microscopy measurements of surface potential change under wear at low loads. Wear **244**, 104–117 (2000)
43. Prabhakaran, V., Kim, S.K., Talke, F.E.: Tribology of the helical scan head tape interface. Wear **215**, 91–97 (1998)
44. Varanasi, S.S., Lauer, J.L., Talke, F.E., Wang, G., Judy, J.H.: Friction and wear studies of carbon overcoated thin films magnetic sliders: application of Raman microspectroscopy. J. Tribol. **119**, 471–475 (1997)
45. Prioli, R., Reigada, D.C., Freire, F.L.: Correlation between nano-scale friction and wear of boron carbide films deposited by dc-magnetron sputtering. Appl. Phys. Lett. **75**, 1317–1319 (1999)
46. Sundararajan, S., Bhushan, B.: Micro/nanotribological studies of polysilicon and SiC films for MEMS applications. Wear **217**, 251–261 (1998)
47. Bhushan, B., Sundararajan, S.: Micro/nanoscale friction and wear mechanisms of thin films using atomic force and friction force microscopy. Acta Mater. **46**, 3793–3804 (1998)
48. Song, H.J., Zhang, Z.Z.: Investigation of the tribological properties of polyfluo wax/ polyurethane composite coating filled with nano-SiC or nano-ZrO_2. Mater. Sci. Eng. A **426**, 59–65 (2006)
49. Bowden, F.P., Tabor, D.: The Friction and lubrication of solids. Oxford University Press (1996)
50. Freire, F.L., Reigada, D.C., Prioli, R.: Boron carbide and boron-carbon nitride films deposited by DC-magnetron sputtering: Structural characterization and nanotribological properties. Phys. Stat. Sol. A **187**, 1–12 (2001)
51. Roy, M., Koch, T., Pauschitz, A.: The influence of sputtering procedure on nanoindentation and nanoscratch behaviour of W-S-C film. Appl. Surf. Sci. **256**, 6850–6858 (2010)
52. Miyake, S., Wang, M.: Mechanical properties of extremely thin B-C-N protective layer deposited with helium addition. Jpn. J. Appl. Phys. **43**, 3566–3571 (2004)

53. Zhao, X.Y., Perry, S.S.: The Role of Water in Modifying Friction within MoS_2 Sliding Interfaces. ACS Appl. Mate. Interfaces **2**, 1444–1448 (2010)
54. Li, X.D., Wang, X.N., Bondokov, R., Morris, J., An, Y.H.H., Sudarshan, T.S.: Micro/nanoscale mechanical and tribological characterization of SiC for orthopedic applications. J. Biomed. Mater. Res. B Appl. Biomater. **72B**, 353–361 (2005)
55. Zhao, X.Y., Hamilton, M., Sawyer, W.G., Perry, S.S.: Thermally activated friction. Tribol. Lett. **27**, 113–117 (2007)
56. Choi, J., Kawaguchi, M., Kato, T.: The surface coverage effect on the frictional properties of patterned PFPE nanolubricant films in HDI. IEEE Trans. Magn. **39**, 2492–2494 (2003)
57. Gao, J.X., Yeo, L.P., Chan-Park, M.B., Miao, J.M., Yan, Y.H., Sun, J.B., Lam, Y.C., Yue, C.Y.: Antistick postpassivation of high-aspect ratio silicon molds fabricated by deep-reactive ion etching. J. Microelectromech. Syst. **15**, 84–93 (2006)
58. Choi, J.H., Kawaguchi, M., Kato, T.: Nanoscale lubricant with strongly bonded phase and mobile phase. Tribol. Lett. **15**, 353–358 (2003)
59. Ashurst, W.R., Yau, C., Carraro, C., Lee, C., Kluth, G.J., Howe, R.T., Maboudian, R.: Alkene based monolayer films as anti-stiction coatings for polysilicon MEMS. Sens. Actuators B **91**, 239–248 (2001)
60. Srinivasan, U., Houston, M.R., Howe, R.T., Maboudian, R.: Alkyltrichlorosilane-based self-assembled monolayer films for stiction reduction in silicon micromachines. J. Microelectromech. Syst. **7**, 252–260 (1998)
61. Ma, J.Q., Liu, J.X., Mo, Y.F., Bai, M.W.: Effect of multiply-alkylated cyclopentane (MAC) on durability and load-carrying capacity of self-assembled monolayers on silicon wafer. Colloids Surf. A **301**, 481–489 (2007)
62. Mo, Y.F., Zhu, M., Bai, M.W.: Preparation and nano/microtribological properties of perfluorododecanoic acid (PFDA)-3-aminopropyltriethoxysilane (APS) self-assembled dual-layer film deposited on silicon. Colloids Surf. A **322**, 170–176 (2008)
63. Kushmerick, J.G., Hankins, M.G., de Boer, M.P., Clews, P.J., Carpick, R.W., Bunker, B.C.: The influence of coating structure on micromachine stiction. Tribol. Lett. **10**, 103–108 (2001)
64. Zhuang, Y.X., Hansen, O., Knieling, T., Wang, C., Rombach, P., Lang, W., Benecke, W., Kehlenbeck, M., Koblitz, J.: Thermal stability of vapor phase deposited self-assembled monolayers for MEMS anti-stiction. J. Micromech. Microeng. **16**, 2259–2264 (2006)
65. Ashurst, W.R., Yau, C., Carraro, C., Maboudian, R., Dugger, M.T.: Dichlorodimethylsilane as an anti-stiction monolayer for MEMS: A comparison to the octadecyltrichlosilane self-assembled monolayer. J. Microelectromech. Syst. **10**, 41–49 (2001)
66. Song, J., Srolovitz, D.J.: Adhesion effects in material transfer in mechanical contacts. Acta Mater. **54**, 5305–5312 (2006)
67. Shimizu, J., Eda, H., Yoritsune, M., Ohmura, E.: Molecular dynamics simulation of friction on the atomic scale. Nanotechnology **9**, 118–123 (1998)
68. Guo, Y.H., Liu, G., Xiong, Y., Tian, Y.C.: Study of the demolding process—implications for thermal stress, adhesion and friction control. J. Micromech. Microeng. **17**, 9–19 (2007)

Chapter 2
Biomimetic Inspiration Regarding Nano-Tribology and Materials Issues in MEMS

Ille C. Gebeshuber

Abstract Tribology is omnipresent in living nature. Blinking eyes, synovial joints, white blood cells rolling along the endothelium and the foetus moving in a mother's womb—tribological problems with evolutionary optimized solutions! This chapter introduces biology for tribologists, highlights the benefits of biomimetics (i.e., knowledge transfer from living nature to engineering), first for tribology in general and subsequently specifically for nano-tribology and materials issues in MEMS. The outlook deals with perspectives of green and sustainable nanotribology for a liveable future for all.

Contents

I. C. Gebeshuber (✉)
Institute of Microengineering and Nanoelectronics (IMEN),
Universiti Kebangsaan Malaysia, Bangi, Kuala Lumpur, Malaysia
e-mail: gebeshuber@iap.tuwien.ac.at

I. C. Gebeshuber
Institute of Applied Physics, Vienna University of Technology, Vienna, Austria

I. C. Gebeshuber
Austrian Center of Competence for Tribology, Wiener Neustadt, Austria

S. K. Sinha et al. (eds.), *Nano-tribology and Materials in MEMS*,
DOI: 10.1007/978-3-642-36935-3_2, © Springer-Verlag Berlin Heidelberg 2013

Introduction

> Producing each of its creations… nature intermingled the harmony of beauty and the harmony of expediency and shaped it into the unique form which is perfect from the point of view of an engineer. (M. Tupolev)

Biomimetics is especially inspiring when it comes to MEMS. This has the reason that in organisms and in MEMS, just very few base materials are used, and variations in the structure are used to achieve certain functionalities. The relationship between structure and function is highly distinct in most biological entities. One prominent example for structure-based approaches at achieving certain functionalities is structural colours. These colours are generated by the structures alone, no chemical dyes are needed (examples: rainbow, thin oil film on water, soap bubble, CD, DVD, certain butterfly wings, iridescent slime moulds, blue tropical understory ferns, …) (Kinoshita [1] Gebeshuber and Lee [2]). The usage of structures rather than material is one of the basic principles of biological systems. Organisms are also excelling at just slightly changing the chemistry and thereby achieving altered functionalities—in this regard out rechnology is just at the beginning with our current material science.

Tribological and material issues prevent successful implementation of 3D MEMS in current technology. Some of these issues can be addressed by improved structures of the MEMS. Conveniently, there is a best practice system in nature where rigid parts on the hundreds of nanometers scale occur in relative motion: diatoms. Diatoms are single celled algae that biomineralize an outer shell of silica [3]. This silica shell is nanostructured, and—for some of the tens of thousands of different diatom species that exist—exhibit hinges and interlocking devices on the micro— and nanoscale [4–6], (Fig. 2.1). Normally, biotribological systems are rather optimized regarding friction than regarding wear. However, and this makes the diatoms so interesting concerning MEMS development (where friction is the major issue), diatoms are optimized regarding wear—one reason for this being that normally, living tissue can be repaired, therefore the focus is rather on friction than on wear, whereas in the case of the diatoms, the shells are built just once and generations of these single celled organisms have to live with them (since at cell division, each daughter cell receives one shell from its parent cell, and biomineralizes the other).

No sign of wear has ever been detected in diatoms (Richard Crawford, personal communication), even when exposed to rough environments or when they were lying in the ground for millions of years, as for example the Eocene fossil diatom *Solium exscuptum* in Fig. 2.1 that was alive 45 millions of years ago.

Nanotribology and MEMS are highly interesting, interdisciplinary research areas. This calls for well educated people who can think deeply and who not only possess book-learnt knowledge, but who can understand and connect knowledge, and construct realities from few, scattered inputs with lots of unknowns. Current education in most cases does not promote such an approach to knowledge. Various

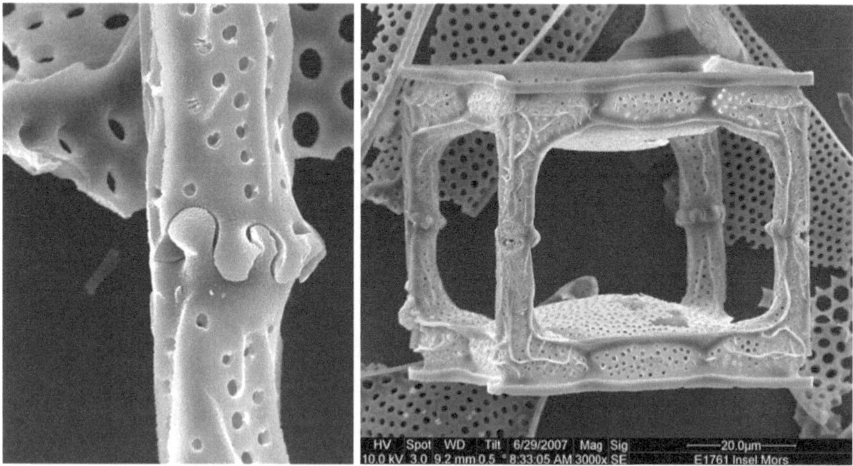

Fig. 2.1 The diatom *Solium exsculptum* lived 45 millions of years ago. This fossil diatom from the island of Mors in Denmark is from the Hustedt Collection in Bremerhaven, Germany, # E1761. It beautifully shows a nanostructured shell, reinforcement ribs, connections and primary mechanical structures. The image on the *left* is a zoom into the most *left* junction in the *right* hand image. Image © F. Hinz, AWI Bremerhaven. Image reproduced with kind permission

undergraduate programs in nanoscience and nanotechnology are underway in many countries, and it remains to be seen if the people produced by these programs are the independent thinkers that are currently needed to propel our society forward towards sustainability [7].

The large degree of interdisciplinarity in nanoscience, in nanotribology and in bioinspired MEMS approaches calls for interdisciplinary scientists who have access to the frameworks of thoughts in more than just one discipline. Concerning nanomaterials engineering, Gebeshuber and Majlis introduced in 2011 a novel concept for innovation [8]. This concept can be translated to nanotribology and MEMS: the basic idea is to apply the 3D method (3D stands for discover, develop and design, Fig. 2.2) to the engineering problem one is currently working on. Ecosystems with high biodiversity such as virgin tropical rainforests are used as treasure box full of ideas and best practices, and the engineers team up with local biologists, designers and materials scientists and spend intense time discussing the problem and watching nature from a functional point of view on rainforest walks. Besides the virgin tropical rainforests, coral reefs can serve as inspirational environment for the nanotribologist, due to their exquisite biodiversity and the high complexity of the interactions (c.f., tribosystem).

This chapter gives an overview of work in our group and of others over the last 10 years in the field of micro— and nanotribology, biomimetics, learning from biology for the tribologists and bioinspired MEMS development.

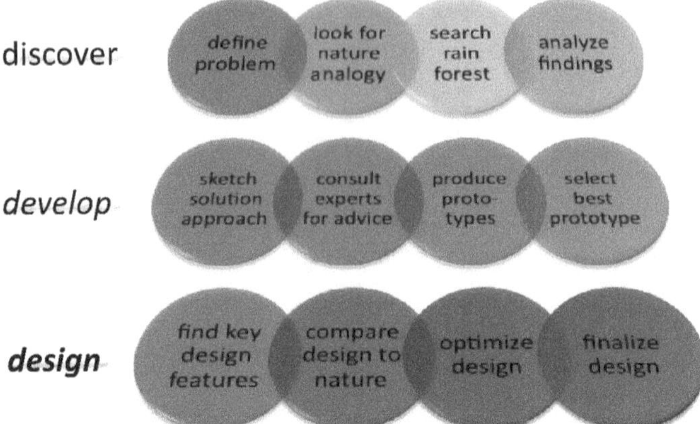

Fig. 2.2 The three pillars in the 3D concept for innovation in nanomaterials engineering [8]. Copyright © 2011 Inderscience Enterprises Ltd. Own image reused with permission; Inderscience retains copyright of the original figure and the article [8]

Biology for Tribologists

It is not easy for tribologists to appreciate the world of biologists. Complicated Latin names for the organisms and highly descriptive style of published work for centuries separated the two fields. Only just recently, collaboration for specialists from the respective fields became increasingly easier. The reason for this change is that tribology as well as biology went through major changes, and additionally, biology and nanotechnology converge on the nanoscale [9]: in both fields, the amount of descriptive knowledge decreases, while the amount of causal knowledge increases, proving a promising area of overlap in terms of ideas, goals, visions, approaches, concepts and language [10], (Fig. 2.3).

This overlap results in an increased number of joint research projects and publications related to biotribology (Fig. 2.4). Starting from 2001, the amount of related papers in the ISI database has increased manifold. Note that 2001 was the publication year of the book "Biological Micro— and Nanotribology: Nature's Solutions" [11] by Scherge and Gorb.

Normally, engineers and biologists do not see very much overlap in their professions. There are nevertheless various synergies that result from collaborations of these two fields. This has for example been shown in intriguing ways by George de Mestral, the inventor of Velcro (which in fact was inspired by a plant with hooks), by a bioinspired bale-straw screw and by the self-cleaning effects of the lotus leaf found by Barthlott and collaborators (summarized by Gebeshuber and Drack [12]). Biomimetics can happen in two directions, both of which are important and yield new results (Fig. 2.5). In "Biomimetics by Analogy" the tribologist formulates the problem, looks in nature for inspiration, identifies best

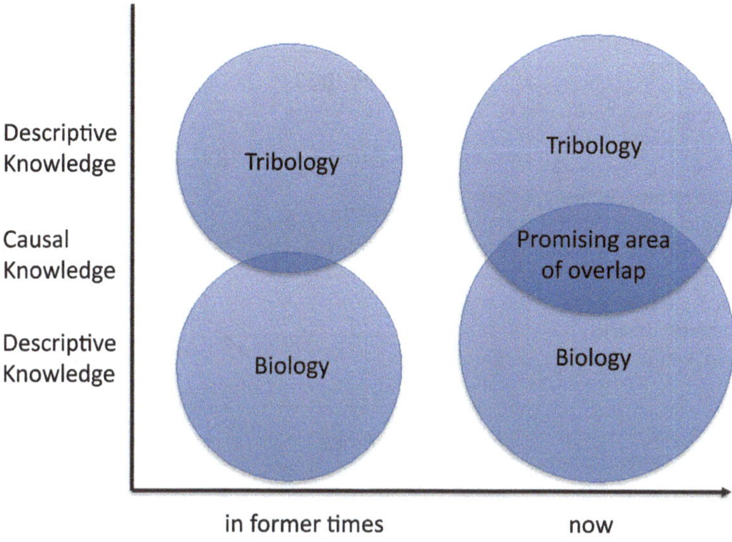

Fig. 2.3 The increasing amount of causal laws in biology generates promising areas of overlap with tribology [10]. © Springer 2011. Own image reused with permission

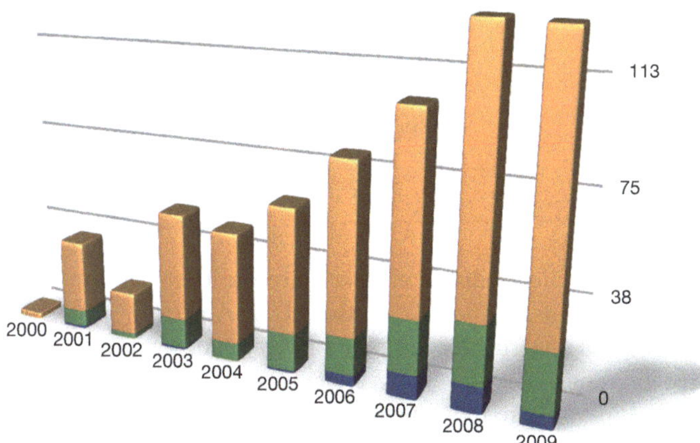

Fig. 2.4 Number of ISI publications on biotribology in the period from 2000 to 2009 (from Gebeshuber and Majlis [34]). *Yellow* bio* and tribolog* in topic, *green* biology* and tribolog' in topic, *blue* biomim* and tribology* in topic. © 2010 W. S. Maney and Son Ltd. Own image reused with permission

practices and their deep principles and transfers them to engineering. In this way, the bale-straw screw was invented [13].

In our increasingly converging society and with the major attempts being undertaken regarding nanotechnology (where all the natural sciences meet), engineers are expected to have basic knowledge in biology. Some engineers went

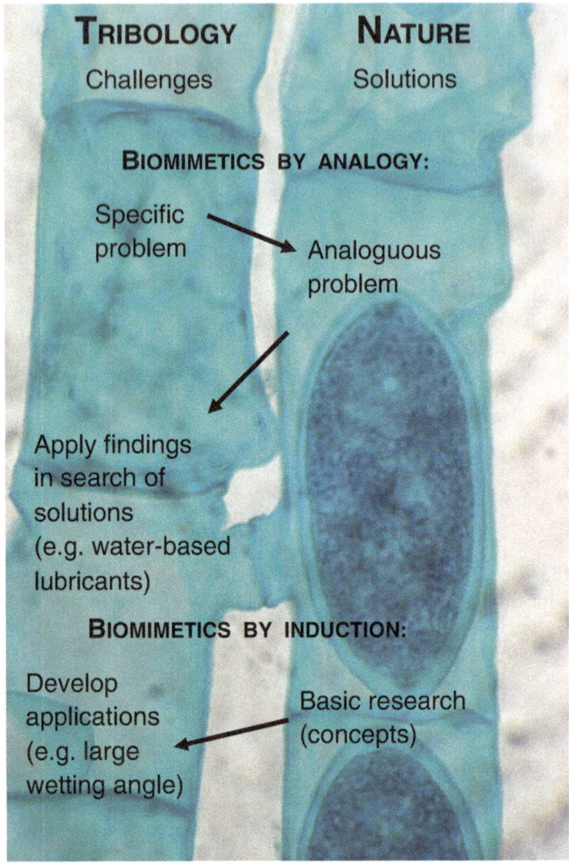

Fig. 2.5 Ways of inspiration by nature for tribologists. *Background image* conjugation in the alga Spirogyra where two originally independent cells form a connection, resulting in exchange of intracellular material, symbolizing fruitful exchange between nature and tribology. © David Polcyn, California State University, San Bernardino. Image reproduced with permission

through a purely technical education, and therefore even lack most basic ideas about biological systems. Arthur T. Johnson, U.S. American professor in a Biological Resources Engineering Department in Maryland, went trough major efforts when writing his book 'Biology for Engineers' [14]. On nearly 1,000 pages he deals with principles from the sciences (physics, chemistry, mathematics and engineering sciences as well as biology), responses of living systems, scaling factors and how to utilize living systems. This book is highly recommended for the beginner-in-biology engineer.

In 2001 Scherge and Gorb's meanwhile classical book on nature's solutions regarding biological micro— and nanotribology appeared—they treat treefrogs, insects, the gecko, and many other organisms with increased or decreased adhesion or friction. Before this date, most of the biotribological literature was concentrated just on a handful of examples from nature, and some decades ago, biotribology solely indicating work related to synovial joints (e.g., hip and knee joints), such as the highly influential work published in Proc. IMechE Part C ([15–17] revisited by Gebeshuber in [18]).

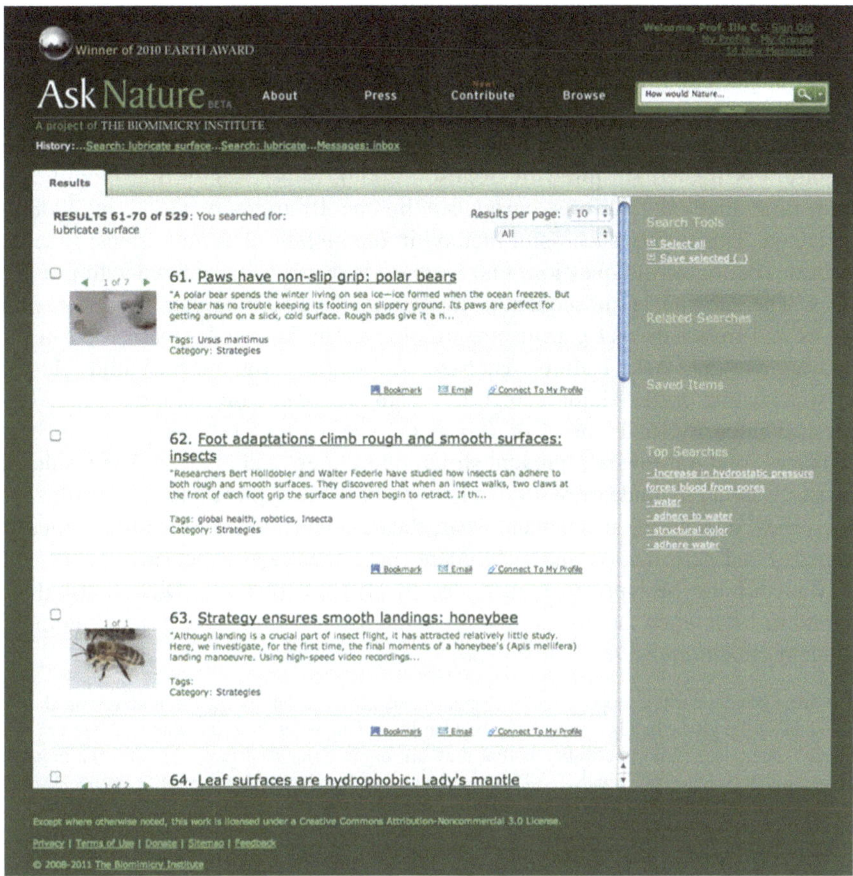

Fig. 2.6 The Ask Nature database from the US American Biomimicry Guild is a helpful tool to start reading into strategies of nature regarding inspiration for biomimetics applications. Tribologists find a plenitude of natural materials, structures and processes that provide input for subsequent sustainable development of biomimetic devices. http://www.asknature.org/

The U.S. American Biomimicry Guild established the Ask Nature Platform, a free web-based application that allows browsing and searching "strategies of nature" (Fig. 2.6). This is an indispensable database for tribologists who want to start to read on biotribological systems or to deepen their knowledge on them (http://www.asknature.org).

One whole section in the book chapter "Biomimetics in Tribology" [9] in a recent a Springer publication on biomimetics [19] is denoted to Johnson's biology for engineers, where the authors predominantly concentrate on biological responses and their possible extrapolation to technical systems. Biology is evolving as new Leitwissenschaft, with more and more causation and natural laws being uncovered. Biomimetics is a field that has the potential to drive major technical advances and that might change the research landscape and the engineering culture dramatically, by the blending of disciplines. Biomimetics might substantially support successful mastering of current tribological challenges concerning friction, adhesion, lubrication and wear in devices and systems from the meter to the nanometer scale. In the introduction of the chapter, the authors highlight the historical background and current developments, then biology for engineers is treated (inspired by Johnson 2009). Subsequently, the biomimicry innovation method that was introduced by the Biomimicry Guild in 2008 is explained. This highly successful method in biomimetics has three steps: Identify function, biologize the question, find nature's best practices and generate product ideas. The method is subsequently applied to identify biological systems, processes and materials that can inspire tribology. The knowledge base used in the analysis is the Ask Nature database from the Biomimcry Guild (http://www.asknature.org/): Major categories from this database comprise 'maintain physical integrity' and 'move or stay put', with various sub-categories such as 'manage structural forces', 'prevent structural failures' and 'attach'. The results of the study are a plenitude of best practices and possible applications (incl. extensive references) concerning mechanical wear, shear, tension, buckling, fatigue, fracture (rupture) and deformation and attachment (permanent and temporary).

Many of the best practice examples in [10] are diatoms. Diatom tribology started in 1999 when the first atomic force microscopy images of living diatoms in ambient conditions were obtained [20].

> Girdle bands can telescope as cells elongate and grow.... The bead-like features on the edges of the girdle bands ... are yet to be identified. This is the first time that such features have been seen. One possibility is that they are organic material that lubricates the connection between girdle bands. ... This suggests to us that the beads are a lubricant because they only occur on the new bands. [21]

Biotribologists gather information about the tribology of biological systems and subsequently apply this knowledge to technological innovation as well as to the development of environmentally sound products [22]. Three key examples for biological model systems of possible interest to the tribologists are synovial joints (low friction coefficient), rolling switchable adhesion of white blood cells in the blood vessels and diatom (Figs. 2.1 and 2.7) tribology.

Regarding joint lubrication, friction coefficients as low as 0.001 have been reported [23–25]. However, especially in biological systems, friction and wear are not simply related phenomena [26, 27]: low friction systems do not necessarily result in low levels of wear. Low friction is in many cases more preferable than low wear since worn material can generally be replaced during the life-time of the single organism (but not in the diatoms).

The adhesion of leukocytes is mediated by switchable adhesion molecules and also by the force environment present in the blood vessel (leukocyte rolling, [28]). Initially, the leukocytes move freely along with the blood stream. Leukocyte adaptive adhesion involves a cascade of adhesive events commonly referred to as initial tethering, rolling adhesion (an adhesive modality that enables surveillance for signs of inflammation), firm adhesion, and escape from blood vessels into

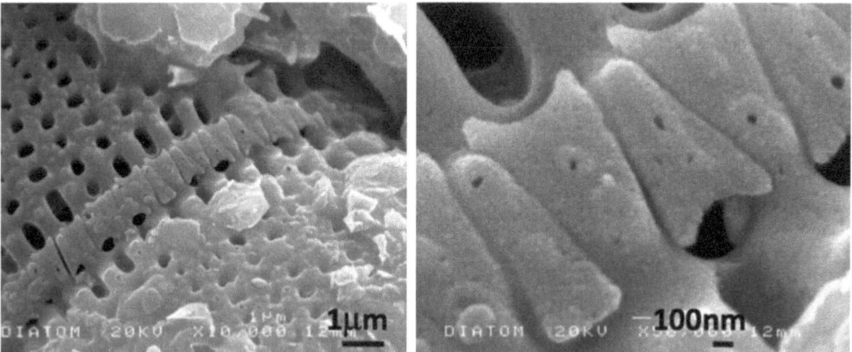

Fig. 2.7 *Left* Zipper-like biogenic silica structure in a diatom. *Right* Zoom into the same image. © Duncan Waddell, XTAL Enterprises, Australia. Image reproduced with permission

tissue [29]. Their rolling velocity is typically 10 to 100 times lower than a non-adherent leukocyte moving next to the vessel wall. The rolling is mediated by integrins, the most sophisticated adhesion molecules known.

In January 2005 the AIP Journal of Nanoscience and Nanotechnology published the special issue "Diatom Nanotechnology" [30]. There, *Ellerbeckia arenaria* who forms stringlike colonies that—when stretched and then released—swing back like springs, *Melosira* sp. who is an exquisite example for interlocking devices in single cells and between single cells, and *Bacillaria paxillifer*, a diatom that as colony actively moves through the water, were introduced as promising species regarding tribology on the micro- and nanoscale [31, 32].

Work on the interface between micro— and nanotribology and biology can be highly inspiring in completely unforeseen directions. The GEMS concept (GEMS stands for Group Electro-Mechanical Systems) for a MEMS-based virtual innervated system in materials was proposed in 2011 (Fig. 2.8), [33]. The fact that a huge body of tribologically relevant information is published in biology papers that are hard to access (in terms of language and concepts) for tribologists lead to the draft of a concept for new ways of scientific publishing and accessing human knowledge inspired by transdisciplinary approaches [34]. This new approach shall help to make biology more accessible for tribologists (and vice versa).

Biomimetics

The inventor of the Schmitt trigger, the American engineer, and biophysicist Otto H. Schmitt, coined the word 'biomimetics' in the 1980s [35]. Biomimetics denotes a method in science, engineering and the arts that gains inspiration from nature [36]. Nature is not blindly copied, but the basic principles are identified and transferred to the respective field. One very successful example for biomimetics and in fact the work that forms the basis of the current highly positive connotation

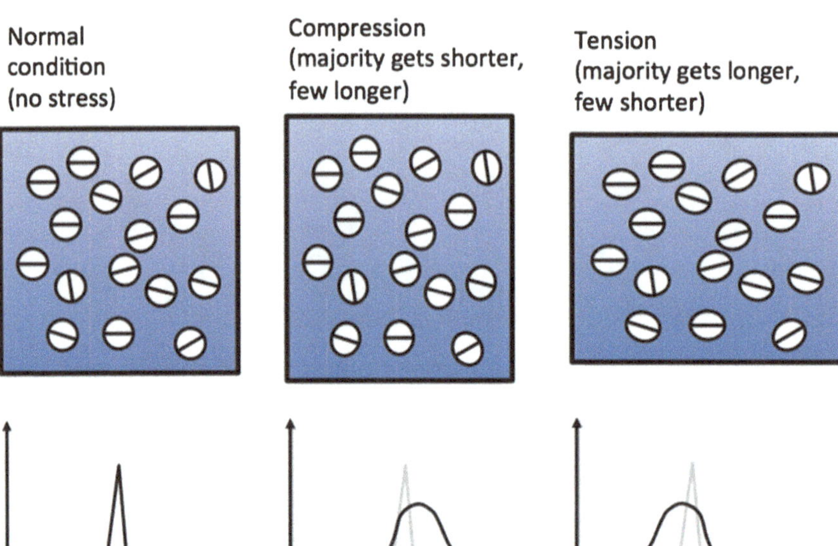

Fig. 2.8 Characteristic responses of a material innervated with Group Electro-Mechanical Systems (GEMS). Normal condition (*left*), compression condition (*middle*) and tension condition (*right*) can easily be discerned by measuring the external induction signal. The *top* images give a sketch of the GEMS embedded in the matrix, and the *bottom* images the characteristic resulting induction versus time curves [33]. © 2011 Trans Tech Publications Ltd. Switzerland. Own image reused with permission

of biomimetics in the general public is technological applications inspired by the self-cleaning property of the lotus leaf. Minuscule wax structures on the leaf surface yield a high contact angle for water droplets, which subsequently, if the surface is just slightly tilted, roll off, taking impurities with them [37]. Buddhists admire the purity of the lotus, and it is one of their sacred plants. Inspired by the surface structure of the lotus leaf, currently, various self-cleaning surface coatings and paints are on the market, and millions of liters of the facade paint Lotusan® by the STO company have been used in the recent years.

Biological materials are sophisticated, have a high degree of miniaturization and hierarchical organization, are resistant and adaptive. An example for a remarkably resistant and adaptive material is human skin—it protects the inside of our body, serves as "packaging", it is a barrier for hazardous molecules in the air but nevertheless allows water vapour to leave the body. It can self-repair and grow or shrink, be elongated by a certain amount and compressed. It controls the temperature of the body by adjusting the amount of vapour and/or sweat leaving the body, providing a cooling effect when needed. Generally, the hydrodynamic, aerodynamic, wetting and adhesive properties of biological materials are remarkable. Biomimetic principle-transfer to engineering paves the way for more reliable, efficient and environment-respecting materials [38]. The increased interest of engineers in the field is exemplified by various publications, e.g., the special issue on biomimetics in engineering in the prestigious British Proc. IMechE Part C Journal in 2007 [39].

Words that are increasingly important in engineering are "green" and "sustainability". Also regarding tribology first attempts have been made to define the respective fields and to establish a framework of tasks that need to be accomplished so that the resulting tribology is "green" or "sustainable" [40–43]. Si-wei Zhang, past chairman of the Chinese Tribology Institution introduced the term "Green Tribology" in 2009 and Peter H. Jost stressed this field as of high importance for tribology at the 2009 Tribology World Congress in Kyoto, Japan.

Biomimetics is not inherently sustainable [44]. It is simply a method, and as such it is not connected with any value. Not surprisingly, some biomimetic materials are way less sustainable than their commonly produced counterparts—this comes from the fact that in living nature, sustainability emerges for the whole system, and not necessarily its single subsystems are sustainable. If we extract a principle from such a subsystem, and apply it to our technology, the result might even be dangerous. Therefore, especially in nanoscience, in nanotribology, in learning from nature on the nanolevel—which is the very level of nature's language—accompanying technology assessment is of such high importance [45].

Biomimetic work has advantages for both biology and engineering, and there are various motivations for a dialogue [12]. It is necessary to establish a common language of biologists and engineers, in which descriptions at different level of detail are more compatible. The engineering approach with established knowledge about structure–function relationships and its application for technological optimization (lighter, faster, and cheaper) meets the biologists approach who gather basic knowledge to enhance the understanding of the principles of living systems: both have to do with constructions, processes and developments. Examples for successful biomimetic engineering come both from biomimetics by induction, i.e., solution based biomimetics, and biomimetics by analogy, i.e., problem based biomimetics, (Fig. 2.5) and technical biology. An intriguing example for biomimetics by analogy is winglets on airplanes. The engineering problem is the large turbulences induced by the wingtips of airplanes. The biological best practice example are big gliders such as storks. The solution in nature is that the feathers at the wingtips of these birds are arranged in a way that the lift-induced drag caused by wingtip vortices is minimized by dividing the large vortex into several smaller vortices. The principle transfer in engineering yields the winglets that are currently applied to many commercial airplanes. Further abstraction of the principle behind the winglets results in spiroids (split wing loops) [46] that considerably cut fuel consumption. A second example for biomimetics by analogy is the bioinspired straw bale screw [13]. Trees feature load-adaptive growth and maintain a uniform stress distribution at their surface. The surface stresses that arise from gravity and external sources are kept constant during their growth. When branches are cut or the environment changes (e.g., the tree tilts because the ground tilts) the tree continues to grow according to the principle of minimising surface stress peaks and this results in regaining uniform surface stresses. The biomimetically optimised screw for mounting façades and other items to straw-bale buildings was developed by applying a tree-inspired method to optimize work pieces [47]. Material use in this screw is minimised and toughness is increased in certain relevant cases of load by more than a third in comparison to a non-optimized geometry. An intriguing example for biomimetics by induction is Velcro. The Swiss engineer George de Mestral walked his dog in 1949, and was amazed by the fact how closely the fruits of a burdock stuck to the fur of his dog. Under the microscope he saw hundreds of tiny flexible but strong hooks that where thus able to reversibly attach themselves to textile and hairy structures. His bioinspired hook-and-loop fastening system was filed for patent in 1951 (Swiss patent # 295638) and de Mestral became a millionaire. A second example for biomimetics by induction is the self-cleaning surfaces and paints inspired by the lotus leaf (Barthlott and Neinhuis [37], US patent # 6660363). The example that [12] give concerning technical biology is the investigations on spiders that have been performed by Barth and co-workers for many decades (for review, see Barth [48, 49]. Biological sensors are amazing, concerning what they can sense and also concerning their bandwidth [50]. Single photon detectors [51] and mechanoreceptors with sub-nanometer sensitivity [52] are two examples. Biologists such as Barth use engineering methodology and approaches to elucidate the function of exquisite spider sensors such as the slit sensilla organs on the legs that act as biological strain gauges, the Trichobothria, hairs that measure medium flow with a threshold for the work driving a single receptor hair over one oscillation cycle of 2.5×10^{-20} J [48, 49] and spider tactile hairs, touch receptors that act as nonlinear sensors. This kind of approach is termed technical biology.

Karman et al. [53] propose a novel approach to MEMS that is based on biological sensory systems, to assist, enhance and expand human sensory perceptions. Current MEMS cover the range of the human sensory system, and additionally provide data about signals that are too weak for the human sensory system (in terms of signal strength) and signal types that are not covered by the human sensory system (Fig. 2.9).

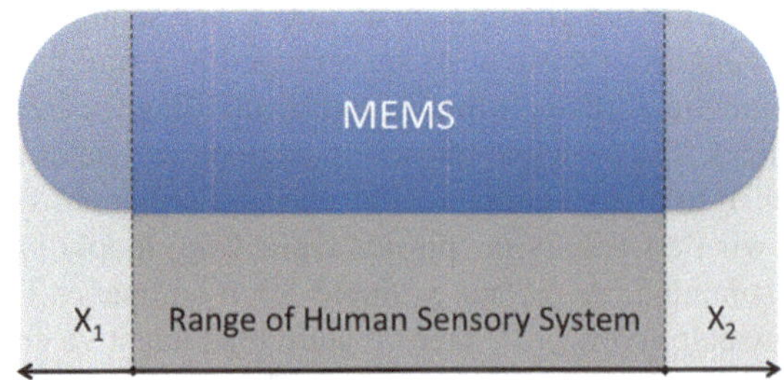

Fig. 2.9 Functional regions of smart MEMS sensors compared to the human sensory system [53]. © 2011 Trans Tech Publications Ltd. Switzerland. Own image reused with permission

Fig. 2.10 "The Soul of the Rose" by John William Waterhouse, public domain image. The sense of smell is one of the oldest senses in humans. Assistance, enhancement and expansion of this sense via MEMS open a whole new universe for engineering applications

Senses in organisms comprise vision (the human visible range is from 390 to 750 nm; various animals can see in the UV range, from 300 nm), temperature sensing (forest fire seeking beetles; snakes who "see" a temperature image with 250 × 250 pixels), hearing and echolocation (man: 20–20 kHz, mice up to 85.5 kHz, bats up to 120 kHz), smelling (Fig. 2.10; people have single molecule sensitivity; sharks can smell one drop of blood in the water 0.4 km away!), feeling touch and vibration, the magnetic sense (in more than 50 species such as migratory birds, honeybees, butterflies, snails, fish, sea turtles, cows, deer, salamanders, geckos, earthworms) and electroreception.

MEMS sensors on the market concerning these senses comprise MEMS that detect electromagnetic waves, temperature, sounds, smells, touch and vibration as well as flavours.

The technological and societal potential of biomimetics is treated in Stachelberger, Gruber and Gebeshuber [54]. The micro— and nanoscale is of utmost importance in relation to biological materials, structures and processes, and relating biomimetics. There is no guarantee that a technical solution based on biomimetics will be eco-friendly. However, biomimetics as an interdisciplinary scientific subject is thought to contribute to some extent to sustainable innovation [55]. Emergence as one of the key properties in biology can be put into relation with sustainable innovation, via a bridge of necessity to go well beyond the frontiers of classical disciplines, thought patterns and organizational structures. In order to introduce innovation principles into societal practice there is a need for ingenious and well-educated people as well as a proactive environment.

MEMS

Wear, friction and adhesion are major reliability issues in MEMS. Reasons are the high surface to volume ratio and the small mass of these devices. Human technology has just started to develop and utilize such small devices, and there are still various tribological issues that need to be addressed [56–60]. High performance microactuators, surface characterization, biosensors and microfluidic systems, MEMS materials and structures, the influence of water adsorption on the microtribology of micromachines, rheological modelling of thin film lubrication, nanotribology of vaporphase lubricants, the tribology of hydrocarbon surfaces investigated using molecular dynamics, surface force induced failures in microelectromechanical systems, reliability and fatigue testing of MEMS and general tribology issues and opportunities in MEMS are current hot topics of research. Especially concerning three-dimensional MEMS, we have not yet progressed very much since 1987, when Texas Instruments Digital Micromirror Devices were introduced to the market.

Various species of diatoms are excellent biological systems from which we can learn concerning tribology. Diatoms have rigid surfaces in relative motion, they are small and experience various forces from the environment [44, 61, 62]. For example, a most exquisite click-stop mechanism on the microscale in diatoms can be envisaged to have high inspirational potential regarding 3D MEMS, especially in constructing 3D MEMS from 2D structures, by folding and fixing them [4]. A zipper-like nanoclasp with high potential regarding 3D MEMS had recently been described by Tiffany and co-workers [63].

Learning from Nature for Tribology: A General Perspective

Although the worlds of tribologists and biologists increasingly get closer, some tribologists choose not to deal directly with animated nature. Even for such researchers, without having to deal with biology at all, some very general principles can be extracted and systematic technology transfer from biology to engineering thereby becomes generally accessible. The biologist Werner Nachtigall, the initiator of biomimetics in Germany, introduced ten general principles. These principles in biology comprise integration instead of additive construction, optimization of the whole instead of maximization of a single component feature, multi-functionality instead of mono-functionality, fine-tuning regarding the environment, energy efficiency, direct and indirect usage of solar energy, limitation in time instead of unnecessary durability, full recycling instead of piling waste, interconnectedness as opposed to linearity and development via trial-and-error processes [64].

Peter Fratzl and Julian Vincent identified the following recurring principles of biology: correlation of form and function, modularity and incremental change, genetic basis, competition and selection, hierarchy and multi-functionality [65, 66].

In 2002 our group investigated the adhesives some diatoms produce to attach to substrates [67]. We found that these adhesives are tough, strong and self-repairing, similar to the properties identified in the abalone glue by Smith and co-workers in 1999 (Fig. 2.11), [68]. The name 'Diatom Tribology' appeared first in one of our conference papers in 2004. In the initial works, species of high relevance to tribology were identified, including *Ellerbeckia arenaria* and *Bacillaria paxillifer* [31, 32]. In a Nano Today article on biotribology inspiring new technologies, diatoms, the switchable cell adhesion molecules selectin and integrin, white blood cells rolling on the endothelium (rolling adhesion) and the dry adhesives of the gecko foot and certain insect attachment pad were introduced as best practice examples from nature [69]. The list of best practice organisms regarding tribology was widened in 2008 [22] by the amazing tribological properties of cartilage (where the friction coefficient can be as low as 0.001, see, e.g., [23–25], the crack redirection properties of the horse hoof that is based on the hoof nanostructure

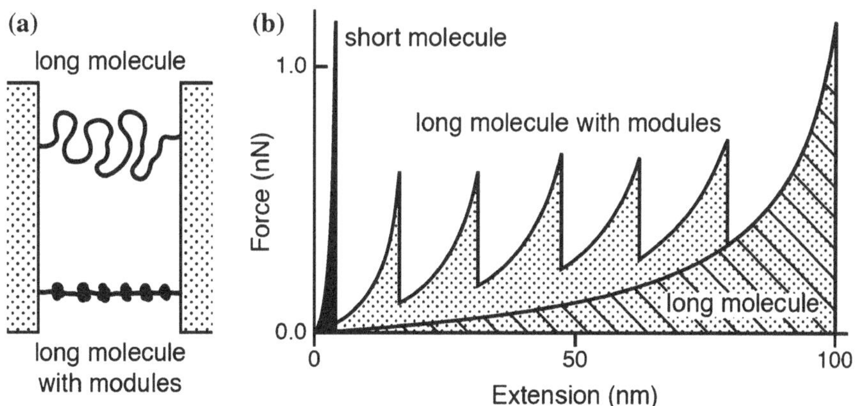

Fig. 2.11 Concept for a self-repairing adhesive [68]. **a** Two ways to attach two particles: with a long molecule or with a long molecule with nodules. **b** When stretched, a *short* molecule could just be extended a bit, and would then break. A *long* molecule would be stretched much more, and finally break. However, a *long* molecule with nodules, with sacrificial bonds that break before the backbone of the molecule breaks, increases the toughness of the adhesive. Such a strategy is applied in the abalone shell, and also in diatom adhesives. Reprinted by permission from Macmillan Publishers Ltd: Nature [68], copyright 1999

(tailored shape of wear particles; [70, 71]) and single huge biomolecules with molecular precision [72].

Gebeshuber, Majlis and Stachelberger investigated in 2009 tribology in biology regarding biomimetic studies across dimensions and across fields [61]. Biotribology is important in addressing tribological challenges at all length scales. The authors apply the Biomimicry Innovation Method to selected current tribological challenges and give ten functions, related biologized questions and nature's best practices concerning optimally designed rigid micromechanical parts (for 3D-MEMS). Regarding pumps for small amounts of liquid (for lab-on-a-chip applications) *Rutilaria philipinnarum*, a fossil colonial diatom that is thought to have lived in inshore marine waters, is identified as best practice organism (for details, see Srajer, Majlis and Gebeshuber [73]).

This type of approach is intensified in a 2011 book chapter [10]. "Biomimetics in Tribology" starts with the historical background (Fig. 2.12) and current developments. The authors present the huge unexplored body of knowledge concerning biological publications dealing with 'wear' and 'adhesives' that make in the text no obvious connection to tribology (Fig. 2.13).

More than 100 references point to a multitude of best practices from nature concerning meliorated technological approaches of various tribological issues. As next step, detailed investigations on the relevant properties of the best practices and extraction of the underlying principles are proposed. Such principles shall then be incorporated into devices, systems and processes and thereby yield biomimetic technology with increased tribological performance.

Fig. 2.12 1771 crash of Nicolas Joseph Cugnot's steam-powered car into a stonewall. Cugnot was the inventor of the very first self-propelled road vehicle, and in fact he was also the first person to get into a motor vehicle accident. Public domain image

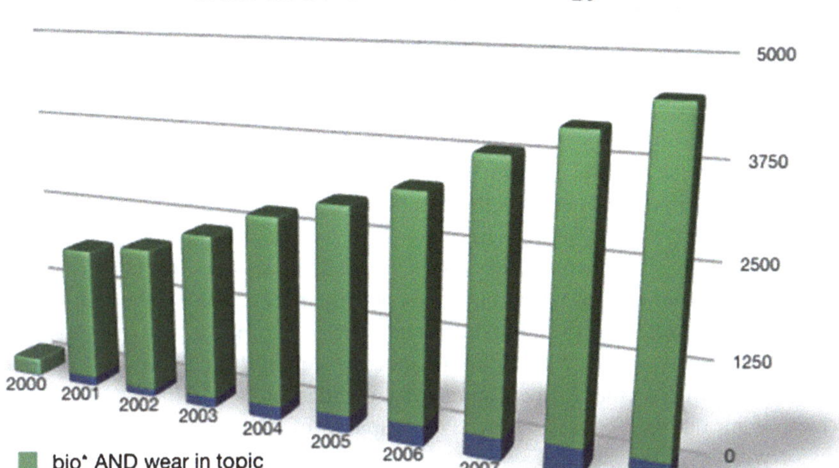

Fig. 2.13 The number of scientific publications in the years 2000–2008 dealing with either wear or adhesives in biology comprise a huge yet unexplored amount of inspiration for technology. *Source* ISI Web of Knowledge, Thomson Reuters, Citation Databases: SCI-EXPANDED (2001-present), CPCI-S (2004-present). http://www.isiknowledge.com (accessed May 5, 2010; Gebeshuber and Majlis [34]). © 2010 W. S. Maney and Son Ltd. Own image reused with permission

Learning from Nature Regarding MEMS Tribology: A Specific Perspective

More often than not, tribological issues prevent successful commercialization of 3D MEMS systems and are the reason why most 3D MEMS are still in prototype stage.

Already in the year 2000 Scherge and Gorb evaluated the adhesion between an insect foot and a variety of surfaces and suggested, inspired by flexible biological micromechanical units found in the attachment pads of the bush cricket, using biological principles to design MEMS [74]. Subsequently they published their book on biological micro— and nanotribology [11]. In 2007, Nano Today invited the author of this chapter to review work on inspiring organisms regarding biotribology. Biotribolological examples with functional units on the micro— and nanometer length range and bioinspired technological development are presented [69].

Engineers and materials scientists can learn by watching, imitating, understanding and generalizing natural approaches to challenges and propose new technology that is recyclable and sustainable, reliable and energy efficient.

As mentioned above, regarding MEMS development, diatoms, the creators of glass castles, can serve as biotribological model systems. Diatoms are unicellular microalgae with a cell wall consisting of a siliceous skeleton enveloped by a thin organic case [3]. Each single diatom cell "lives in a silica box" with an upper and a lower part. When the cell is dividing, a new lower part for the previous upper part and a new upper part for the previous lower part is biomineralized, with the old parts being a bit larger than the newer ones. There are tens of thousands of different diatom species, with a huge variety in shapes and sizes (from some micrometers up to several millimeters). What makes diatoms interesting concerning 3D MEMS development is the fact that they produce the siliceous skeleton just once, and do not repair it. The silica box is hard, tough and rigid, and a multitude of nanostructures, hinges and interlocking devices occur in various species. Some colonial diatom species even act like rubber bands when elongated and subsequently released [32]. Other diatom species exhibit strong, self-repairing underwater adhesives [67]. Some diatom species such as *Pseudonitzschia* sp. and *Bacillaria paxillifer* (the former name of this diatom is *Bacillaria paradoxa,* because if its unusual behavior) are even capable of active movement, with no signs of wear ever detected.

A major issue in MEMS development is adhesion. Understanding this phenomenon on the molecular level not only contributes to improved micro— and nanodevices that do not aggregate or get stuck, but also might aid in the development of tailored man-made adhesives for example in tissue engineering where good adhesive interaction of the implant surface with the surrounding tissue is a necessity and implants should not cause immune reactions via small wear particles [75]. In this regard we can learn from the adaptive adhesion properties of the selectin/integrin complexes exhibited in the interaction of white blood cells with blood vessels.

Biologically inspired technical systems comprise a selectin/integrin inspired technique for cell separation used in lab-on-a-chip applications [76]. This technique involves a microfluidic channel where the silicon surface was functionalized with physisorbed selectin molecules, and thereby captured cells that exhibited integrin molecules. The microfluidic device consisted of arrays of micropillars (made from silicon prepared via deep reactive ion etching) with characteristic spacing somewhat larger than the size of cells (some tens of micrometers in diameter). Control experiments showed no non-specific adhesive interactions between the cells and the microchannels.

Work on the gecko foot by Autumn and colleagues [77, 78] inspired dry adhesives that work via intermolecular van der Waals forces and therefore act between all materials. However, man-made gecko tapes still cannot compete with the adhesive properties of the gecko foot concerning high attachment forces and low detachment forces, rough surface adaptation, self-cleaning ability and durability [79]. The group of Andre K. Geim, the 2010 Physics Nobel Prize winner concerning graphene, is working on improved gecko tapes [80].

In 2009 a novel way to describe the complexity of biological and engineering approaches depending on the number of different base materials was proposed [34, 61, 62]. Approaches in nature and technology can be categorized into material dominated, form dominated and structure dominated ones. When many different

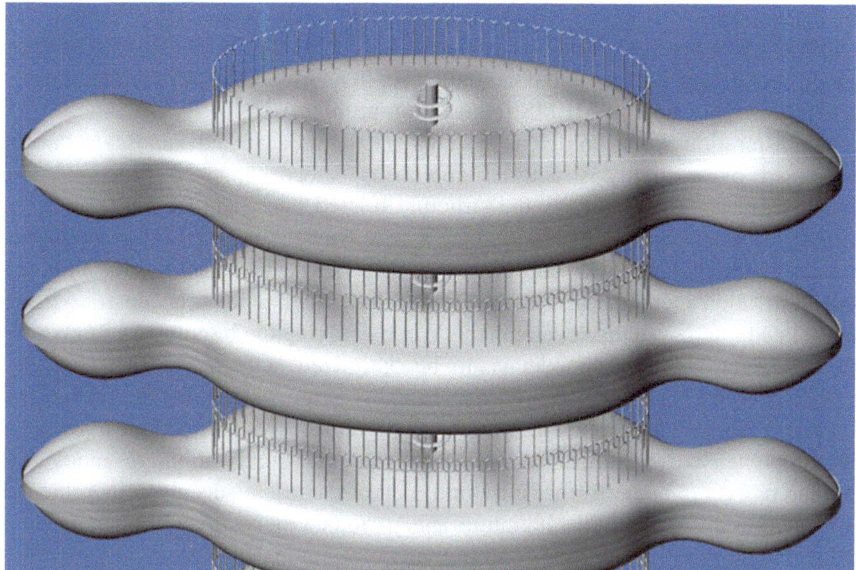

Fig. 2.14 Fluid dynamic simulation of colonial micrometer sized *boxes* (inspired by diatoms that are connected via spacers that ensure minimum and maximum distance) start to exhibit pumping behaviour when water flows along the chain [73]. © 2009 Acta Botanica Croatica. Own image reproduced with permission

materials are or can be used, there is no need to work too much with structures—most of the functions can be accommodated with the materials themselves. This is the approach in most of current technologies on the meso— and macroscale. On the microscale, e.g., for MEMS development, we have to work with just a handful of materials that we can structure, tailor, etch as we want them (e.g., Si, SiO_2, Silicon nitride, GaAs, Silicon carbide, diamond, InP, SiGe, ferroelectric materials and polymers). In MEMS, structures are dominating and accommodate most of the functions. In this regards, MEMS and organisms are highly similar. Also in living nature, just a few base materials are used, and either slightly chemically varied to accommodate specific functions (such as the material in the cornea, in our hairs and in our fingernails, keratin) or the base materials are (nano) structured or hierarchically structured, to accommodate desired functions. Therefore it is concluded that biomimetics is especially promising in MEMS development because of the material constraints in biology and micromachining.

Srajer and coworkers showed in 2009 via microfluidic simulation of a colonial diatom chain that oscillatory movement occurs when water flows along the chain (Fig. 2.14), [73]. The model diatom colony consists of continuously repeated units of ten cells. Undisturbed fluid flow is allowed for between the single cells, and the cells do not actively move—they are solely moved by the water. The computer simulation suggests that a diatom colony subjected to water flow exhibits some kind of oscillatory movement. Such movement might facilitate nutrient uptake of the diatom colony. If in this modelling study one replaces the word "diatom" with "MEMS black box", the result is a microfluidic pump that might be of benefit for lab-on-a-chip applications and potentially also in addressing the problem of how to get rid of excess heat in MEMS (convection problem).

Bioinspiration for tribological systems on the micro- and nanoscale can be addressed by a highly structured approach, categorizing the functions into dynamic, mechanical, surface and structure related ones [81]. Bioinspiration for MEMS tribology is thereby assisted by a new way to analyse best practice biological materials, structures and processes that were established via the biomimicry innovation method, by relating them to four main areas relevant for MEMS and related microsystems development. When applying the biomimicry innovation method regarding a certain device to be optimized, a complete list of related functions is the basis for comprehensive quality improvement for the technology of choice. Table 2.1 presents four representative examples for the four main areas (a comprehensive treatment of MEMS issues via the biomimicry innovation method has yet to be performed). Behind each of the organisms given in Table 2.1 is a huge body of scientific literature and various experts, many of whom have devoted their whole scientific life to studying one single genus or species. Detailed literature searches for archival scientific work and specialists in these fields provide the starting points for collaborative approaches.

It is suggested that the tribology engineers get in contact with the respective biologists working on the organism of interest, perhaps with the additional support of a person who is experienced in biomimetics and in speaking with representatives of both fields. Strong collaboration between biologists and engineers can yield new, sustainable approaches to emerging micro- and nanotechnologies and provide the basis for a new type of scientist/researcher/developer, who understands and feels at home in various fields, and who can transfer concepts and principles across fields. Such people are highly needed in our current time of increasing specialisation.

Summary and Outlook

Biology for tribologists interested in MEMS tribology and materials issues has to be written and categorized in a different way than biology for biologists. Functions are important, as are the relations of structures with functions. Some authors such as Tributsch, Vogel, Alexander and Nachtigall managed to publish works on biology that are highly accessible to engineers [64, 82–89]. These works can serve as a starting point for tribologists who want to learn from nature for improving MEMS tribology and materials issues.

There is a great future ahead of us concerning MEMS. Learning from nature might even yield in the not too distant future sustainable MEMS with expiration date, MEMS that only work as long as we need them, MEMS that are built from substances that are generally available and cheap as well as MEMS that are structured via bottom up technologies based on self-assembly.

Table 2.1 Examples for bioinspiration in MEMS tribology and related fields (from Gebeshuber and Gordon [81])

Area	Function	Biologized question: How does nature...	Nature's best practice	Generated process/product ideas
Dynamic	Movable rigid parts	optimize moveable parts?	The diatoms *Melosira* sp. and *Ellerbeckia arenaria*	3D MEMS with moveable parts
	Pumps	move small amounts of fluids?	The diatoms *Rutilaria grevilleana* and *Rutilaria philipinnarum*	Micropumps for lab-on-a-chip applications
	Energy dissipation in microsystems	dissipate mechanical energy in microorganisms?	The diatom *Solium exsculptum*	3D-MEMS
	Lubrication	prevent wear?	Unknown diatom species	Preventing stiction
Mechanic	Hinges and interlocking devices	mechanically connect hard single cells on the microscale?	Diatoms in chains (*Eunotia sudetica, Bacillaria paxillifer, Ellerbeckia sp.*)	Micromechanical optimization of 3D-MEMS structure
	Click-stop mechanism	unfold microstructures and then irreversibly fix them?	The diatom *Corethron pennatum*	Obtain 3D structures from fabricated 2D structures
	Springs	reversibly store mechanical energy?	*Rutilaria grevilleana* and *R. philipinnarum*; the spasmoneme of *Vorticella convallaria*	Energy storage in MEMS
	Parts connected in a chain with adjustable length	provide stability to chains in turbulent environments?	The diatom *Ellerbeckia arenaria*	MEMS with moveable parts
Surface	Surface texturing	structure surfaces?	The diatom *Solium exsculptum*	MEMS
	Photoprotective coating	protect photo-sensitive plants?	The flowering plants *Begonia sp., Diplazium sp., Phyllagathis rotundifolia*	Coatings of containers for photosensitive reactions
	Photonic components	make structural colours?	Peacock, butterfly scales, iridescent plants, fruits, birds and mammals	Photonic micro- and nanodevices, MEMS, novel lasers
	Selective, switchable adhesion	reversibly adhere to structures?	Immune system, gecko foot, insect attachment pads, plant wax surfaces	Lab-on-a-chip devices, reusable: trap, test and release and start again

(continued)

Table 2.1 (continued)

Area	Function	Biologized question: How does nature...	Nature's best practice	Generated process/product ideas
Structure	Unfoldable microstructures	generate 3D microstructures from rigid parts?	The diatom *Corethron pennatum*	Obtain 3D structures from fabricated 2D structures
	Stability (reinforcement)	mechanically protect viable parts?	The diatom *Solium exsculptum*	Quality assurance of MEMS
	Mechanical fixation	mechanically fix structures on the microscale?	The diatom *Corethron pennatum*	3D-MEMS, lab-on-a-chip
	Pressure resistant containers	deal with high pressures?	*Euglena gracilis* pellicle	Lab-on-a-chip

H. Peter Jost, president of the International Tribology Council, addressed at the 5th World Tribology Congress in Kyoto in September 2009 the situation surrounding the world and green tribology, declaring:

> Green tribology is the science and technology of the tribological aspects of ecological balance and of environmental and biological impacts. Its main objectives are the saving of energy and materials and the enhancement of the environment and the quality of life. [90]

Jost explained in Kyoto that there were now production requirements concerning the supply of energy, mineral resources and food in a way which had not been known before. A focus on tribology might give 'breathing space' while comprehensive solutions to environmental problems were being addressed and suggested that tribology must fall into line with the major politics of world environment and energy.

A smart combination of mechanical, energetic and chemical approaches, combined with optimum designed materials, and minimized stresses to the environment and biology, paves the way towards the future of MEMS tribology and material issues. Successful tribologists are inherently transdisciplinary thinkers—this is needed in our increasingly complex world to successfully contribute to address major global challenges.

Acknowledgments The National University of Malaysia funded part of this work with its leading-edge research project scheme 'Arus Perdana' (grant # UKM-AP-NBT-16-2010) and the Austrian Society for the Advancement of Plant Sciences via our biomimetics pilot project 'BioScreen'. Profs. F. Aumayr, H. Störi and G. Badurek from the Vienna University of Technology are acknowledged for enabling ICG research in the inspiring environment in Malaysia.

References

1. Kinoshita, S.: Structural Colors in the Realm of Nature. World Scientific Publishing Company, Singapore (2008)
2. Gebeshuber, I.C., Lee, D.W.: Nanostructures for coloration (organisms other than animals). In: Bhushan B., Nosonovsky M. (eds.), pp. 1790–1803, Springer Encyclopedia of Nanotechnology. Springer, Berlin (2012)
3. Round, F.E., Crawford, R.M., Mann, D.G.: The diatoms: biology and morphology of the genera. Cambridge University Press, Cambridge (1990)
4. Crawford, R.M., Gebeshuber, I.C.: Harmony of beauty and expediency. Sci First Hand [A good journal for inquisitive people (published by the Siberian Branch of the Russian Academy of Sciences)] **5**(10), 30–36 (2006)

. Gebeshuber, I.C., Crawford, R.M.: Micromechanics in biogenic hydrated silica: Hinges and interlocking devices in diatoms. Proc. IMechE Part J: J. Eng. Tribol. **220**(J8), 787–796 (2006)

6. Gordon, R., Witkowski, A., Gebeshuber, I.C., Allen, C.S.: The diatoms of Antarctica and their potential roles in nanotechnology. In: Chillida, A., Masó, M., Masó M., Chillida, A. (eds.) Antarctica. Time of Change, English edition, pp. 84–95. ActarBirkhauser, Barcelona (2010)
7. Gebeshuber, I.C., Matin, T.R., Halim, L., Ariffin, S.: Nano-scale undergraduate education program. In: Guston, D., Golson, J.G. (eds.) Encyclopedia of nanoscience and society, vol. 2, pp. 528–530. Sage Publications, CA (2010)

8. Gebeshuber, I.C., Majlis, B.Y.: 3D corporate tourism: A concept for innovation in nanomaterials engineering. Int. J. Mat. Eng. Innov. **2**(1), 38–48 (2011)
9. Gebeshuber, I.C., Gruber, P., Drack, M.: Nanobioconvergence. In: Guston, D., Golson, J.G. (eds.) Encyclopedia of Nanoscience and Society, vol. 2, pp. 454–456. Sage Publications, CA (2010)
10. Gebeshuber, I.C., Majlis, B.Y., Stachelberger, H.: Biomimetics in tribology. In: Gruber, P., Bruckner, D., Hellmich, C., Schmiedmayer, H.-B., Stachelberger, H., Gebeshuber, I.C. (eds.) Biomimetics-Materials, Structures and Processes. Examples, Ideas and Case Studies. Ascheron C. (Series ed.) Biological and Medical Physics, Biomedical Engineering, Ch. 3, pp. 25–50. Springer Publishing, Berlin (2011)
11. Scherge, M., Gorb, S.: Biological Micro– and Nanotribology Nature's Solutions. NanoScience and TechnologySpringer, Berlin (2001)
12. Gebeshuber, I.C., Drack, M.: An attempt to reveal synergies between biology and engineering mechanics. Proc. IMechE Part C J. Mech. Eng. Sci. **222**(7), 1281–1287 (2008)
13. Drack, M., Wimmer, R., Hohensinner, H.: Treeplast screw: a device for mounting various items to straw bale constructions. J. Sustain. Prod. Des. **4**(1–4), 33–41 (2006)
14. Johnson, A.T.: Biology for engineers. CRC Press, Boca Raton (2010)
15. Dowson, D.: Engineering at the interface. Proc. Inst. Mech. Eng. Part C: J. Mech. Eng. Sci. **206**, 149–165 (1992)
16. Dowson, D., Unsworth, A., Wright, V.: Analysis of 'boosted lubrication' in human joints. Proc. Inst. Mech. Eng. Part C: J. Mech. Eng. Sci. **12**, 364–369 (1970)
17. Higginson, G.R., Norman, R.: The lubrication of porous elastic solids with reference to the functioning of human joints. Proc. Inst. Mech. Eng. Part C J. Mech. Eng. Sci. **16**(4), 250–257 (1974)
18. Gebeshuber I.C.: Engineering at the interface revisited. Proc. IMechE Part C: J. Mech. Eng. Sci. 50st Anniversary Issue, **223**(C1), 65–101 (2009)
19. Gruber, P., Bruckner, D., Hellmich, C., Schmiedmayer, H.-B., Stachelberger, H., Gebeshuber, I.C. (eds.): Biomimetics–materials, structures and processes. Examples, ideas and case studies, Ascheron C. (Series ed.) Biological and medical physics, biomedical engineering, p. 266. Springer, New York (2011)
20. Gebeshuber, I.C., Kindt, J.H., Thompson, J.B., DelAmo, Y., Brzezinski, M., Stucky, G.D., Morse, D.E., Hansma, P.K.: Atomic force microscopy of diatoms in vivo, Abstr. 15th North American Diatom Symposium, Pingree Park Campus, Colorado State University, vol. 8 (1999)
21. Gebeshuber, I.C., Kindt, J.H., Thompson, J.B., Del Amo, Y., Stachelberger, H., Brzezinski, M., Stucky, G.D., Morse, D.E., Hansma, P.K.: Atomic force microscopy study of living diatoms in ambient conditions. J. Microsc. **212**(Pt3), 292–299 (2003)
22. Gebeshuber, I.C., Drack, M., Scherge, M.: Tribology in biology. Tribol. Surf. Mater. Interfaces **2**(4), 200–212 (2008)
23. Linn, F.C.: Lubrication of animal joints, 1. The Arthrotripsometer. J. Bone Joint Surg. Am. **49A**, 1079–1098 (1967)
24. McCutchen, C.W.: Mechanism of animal joints: Sponge-hydrostatic and weeping bearings. Nature **184**, 1284–1285 (1959)
25. Unsworth, A., Dowson, D., Wright, V.: The frictional behaviour of human synovial joints—Part 1. Natural joints. Trans. Am. Soc. Mech. Eng. Seri. F: J. Lubrication Technol. **97**, 369–376 (1975)
26. Steika, N., Furey, M.J., Veit, H.P.: Biotribology: The wear-resistance of repaired human articular cartilage. ASME Conf. Proc. World Tribol. Congr. **III**(2), WTC2005-63304, 619–620 (2005)
27. Steika, N., Furey, M.J., Veit, H.P., Brittberg, M.: Biotribology: 'In vitro' wear studies of human articular cartilage. Proceedings of NORDTRIB, pp. 773–782 (2004)
28. Tees, D.F., Goetz, D.J.: Leukocyte adhesion: an exquisite balance of hydrodynamic and molecular forces. News Physiol. Sci. **18**, 186–190 (2003)

29. Orsello, C.E., Lauffenburger, D.A., Hammer, D.A.: Molecular properties in cell adhesion: a physical and engineering perspective. Trends Biotechnol. **19**, 310–316 (2001)
30. Gordon R. (ed.): Special issue on diatom nanotechnology. J. Nanosci. Nanotechnol. **5**, 1–178 (2005)
31. Gebeshuber, I.C., Stachelberger, H., Drack, M.: Diatom biotribology. In: Dowson, D., Priest, M., Dalmaz, G., Lubrecht, A.A. (eds.) Life Cycle Tribology. Briscoe, B.J. (Series ed.) Tribology and Interface Engineering. vol. 48, pp. 365–370. Elsevier, NY (2005a)
32. Gebeshuber, I.C., Stachelberger, H., Drack, M.: Diatom bionanotribology–biological surfaces in relative motion: Their design, friction, adhesion, lubrication and wear. J. Nanosci. Nanotechnol. **5**(1), 79–87 (2005b)
33. Macqueen, M.O., Mueller, J., Dee, C.F., Gebeshuber, I.C.: GEMS: A MEMS-based way for the innervation of materials. Adv. Mater. Res. **254**(8), 34–37 (2011)
34. Gebeshuber, I.C., Majlis, B.Y.: New ways of scientific publishing and accessing human knowledge inspired by transdisciplinary approaches. Tribol. Surf. Mater. Interfaces **4**(3), 143–151 (2010)
35. Schmitt, O.H.: Biomimetics in solving engineering problems. Talk given on 26 Apr 1982. http://160.94.102.47/OttoPagesFinalForm/BiomimeticsProblem%20Solving.htm. (1982) (last accessed: 23 Apr 2012)
36. Bar-Cohen, Y.: Biomimetics: Biologically inspired technologies. CRC Press, Boca Raton (2005)
37. Barthlott, W., Neinhuis, C.: The purity of sacred lotus or escape from contamination in biological surfaces. Planta **202**, 1–8 (1997)
38. Sanchez, C., Arribart, H., Giraud-Guille, M.M.: Biomimetism and bioinspiration as tools for the design of innovative materials and systems. Nat. Mater. **4**, 277–288 (2005)
39. Neville, A. (Guest ed.): Special issue on biomimetics in engineering. Proc. IMechE Part C: J. Mech. Eng. Sci. **221**(C10), i, 1141–1230 (2007)
40. Nosonovsky, M., Bhushan, B.: Green tribology. Phil. Trans. Roy. Soc. A **368**(1929), 4675–4890 (2010)
41. Nosonovsky, M., Bhushan, B.: Green Tribology: Biomimetics, energy conservation, and sustainability. Springer, Berlin (2012)
42. Gebeshuber, I.C.: Green nanotribology and sustainable nanotribology in the frame of the global challenges for humankind. In: Nosonovsky, M., Bhushan, B. (eds.) Green Tribology–Biomimetics, Energy Conservation, and Sustainability. Green Energy and Technology, Ch. 5, pp. 105–125. Springer, Berlin (2012)
43. Gebeshuber I.C.: Green nanotribology. Proc. I Mech E Part C: J. Mech. Eng. Sci. **226**(C2), 374–386 (2012)
44. Gebeshuber, I.C., Gruber, P., Drack, M.: A gaze into the crystal ball—biomimetics in the year 2059. Proc. IMechE Part C: J. Mech. Eng. Sci. 50st Anniversary Issue, **223**(C12), 2899–2918 (2009)
45. Fiedeler, U., Nentwich, M., Simkó, M., Gazso, A.: What is accompanying research on nanotechnology? Nanotrust dossier 11. http://epub.oeaw.ac.at/?arp=0x0024e6bf 1–4 Dec (2010) (last accessed 14 Mar 2013)
46. Devenport, W.J., Vogel, C.M., Zsoldos, J.S.: Flow structure produced by the interaction and merger of a pair of co-rotating wing-tip vortices. J. Fluid Mech. **394**, 357–377 (1999)
47. Mattheck, C.: Design in Nature–learning from trees. Springer, Heidelberg (1998)
48. Barth, F.G.: Spider senses—technical perfection and biology. Zoology **105**(4), 271–285 (2002)
49. Barth, F.G.: A spider's world: Senses and behaviour. Springer, Berlin (2002)
50. Makarczuk ,T., Matin, T.R., Karman, S.B., Diah, S.Z.M., Davaji, B., Macqueen, M.O., Mueller, J., Schmid, U., Gebeshuber, I.C.: Biomimetic MEMS to assist, enhance and expand human sensory perceptions: a survey on state-of-the art developments. Proc. SPIE 8066. 80661O(15p), (2011). doi:10.1117/12.886554
51. Lillywhite, P.G.: Single photon signals and transduction in an insect eye. J. Comp. Physiol. A. **122**(2), 189–200 (1977)

52. Crawford, A.C., Fettiplace, R.: The mechanical properties of ciliary bundles of turtle cochlear hair cells. J. Physiol. **364**, 359–379 (1985)
53. Karman, S.B., Macqueen, M.O., Matin, T.R., Diah, S.M., Mueller, J., Yunas, J., Davaji, B., Makarczuk, T., Gebeshuber, I.C.: On the way to the bionic man: A novel approach to MEMS based on biological sensory systems. Adv. Mater. Res. **254**(8), 38–41 (2011)
54. Stachelberger, H., Gruber, P., Gebeshuber, I.C.: Biomimetics—its technological and societal potential. In: Gruber, P., Bruckner, D., Hellmich, C., Schmiedmayer, H.-B., Stachelberger, H., Gebeshuber, I.C. (eds.) Biomimetics—Materials, Structures and Processes. Examples, Ideas and Case Studies. Ascheron, C. (Series ed.) Biological and Medical Physics, Biomedical Engineering, Ch. 1, pp. 1–6. Springer Publishing, Berlin (2011)
55. Jorna R.J.: Sustainable Innovation: The Organisational, Human and Knowledge Dimension. Greenleaf Publishing, Sheffield (2006)
56. Bhushan, B. (ed.): Tribology issues and opportunities in MEMS. Springer, Berlin (1998)
57. Williams, J.A., Le, H.R.: Tribology and MEMS. J. Phys. D Appl. Phys. **39**(12), R201 (2006)
58. Kumar, S.S., Satyanarayana, N.: Nano-Tribology and Materials in MEMS (Eds. Sujeet Kumar Sinha, Nalam Satyanarayana and Seh Chun Lim). Springer, Berlin (2013)
59. Mate, C.M.: Tribology on the small scale: A bottom up approach to friction, lubrication, and wear. Mesoscopic Physics and Nanotechnology SeriesOxford University Press, New York (2008)
60. Rymuza, Z.: Tribology of Miniature Systems. Tribology SeriesElsevier Science Ltd, New York (1989)
61. Gebeshuber, I.C., Majlis, B.Y., Stachelberger, H.: Tribology in biology: Biomimetic studies across dimensions and across fields. Int. J. Mech. Mat. Eng. **4**(3), 321–327 (2009)
62. Gebeshuber, I.C., Stachelberger, H., Ganji, B.A., Fu, D.C., Yunas, J., Majlis, B.Y.: Exploring the innovational potential of biomimetics for novel 3D MEMS. Adv. Mat. Res. **74**, 265–268 (2009)
63. Tiffany, M.A., Gordon, R., Gebeshuber, I.C.: *Hyalodiscopsis plana*, a sublittoral centric marine diatom, and its potential for nanotechnology as a natural zipper-like nanoclasp. Pol. Bot. J. **55**(1), 27–41 (2010)
64. Nachtigall, W.: Vorbild Natur: Bionik-Design für funktionelles Gestalten. Springer, Berlin (1997)
65. Fratzl, P., Weinkamer, R.: Nature's hierarchical materials. Progr. Mat. Sci. **52**(8), 1263–1334 (2007)
66. Vincent, J.F.V.: Deconstructing the design of a biological material. J. Theor. Biol. **236**, 73–78 (2005)
67. Gebeshuber, I.C., Thompson, J.B., Del Amo, Y., Stachelberger, H., Kindt, J.H.: *In vivo* nanoscale atomic force microscopy investigation of diatom adhesion properties. Mat. Sci. Technol. **18**(7), 763–766 (2002)
68. Smith, B.L., Schäffer, T.E., Viani, M., Thompson, J.B., Frederick, N.A., Kindt, J., Belcher, A., Stucky, G.D., Morse, D.E., Hansma, P.K.: Molecular mechanistic origin of the toughness of natural adhesives, fibres and composites. Nature **399**, 761–763 (1999)
69. Gebeshuber, I.C.: Biotribology inspires new technologies. Nano Today **2**(5), 30–37 (2007)
70. Kasapi, M.A., Gosline, J.M.: Design complexity and fracture control in the equine hoof wall. J. Exp. Biol. **200**, 1639–1659 (1997)
71. Kasapi, M.A., Gosline, J.M.: Micromechanics of the equine hoof wall: optimizing crack control and material stiffness through modulation of the properties of keratin. J. Exp. Biol. **202**, 377–391 (1999)
72. Viani, M.B., Pietrasanta, L.I., Thompson, J.B., Chand, A., Gebeshuber, I.C., Kindt, J.H., Richter, M., Hansma, H.G., Hansma, P.K.: Probing protein–protein interactions in real time. Nat. Struct. Biol. **7**(8), 644–647 (2000)
73. Srajer, J., Majlis, B.Y., Gebeshuber, I.C.: Microfluidic simulation of a colonial diatom chain reveals oscillatory movement. Acta Bot. Croat. **68**(2), 431–441 (2009)
74. Scherge, M., Gorb, S.N.: Using biological principles to design MEMS. J. Micromech. Microeng. **10**(3), 359–364 (2000)

75. Freitas, R.A.: Nanomedicine, 15.2.2, Vol. IIA: Biocompatibility, Landes Bioscience Publishing, USA (2003)
76. Chang, W.C., Lee, L.P., Liepmann, D.: Biomimetic technique for adhesion-based collection and separation of cells in a microfluidic channel. Lab Chip **5**(1), 64–73 (2005)
77. Autumn, K., Liang, Y.A., Hsieh, S.T., Zesch, W., Chan, W.P., Kenny, T.W., Fearing, R., Full, R.J.: Adhesive force of a single gecko foot-hair. Nature **405**, 681–685 (2000)
78. Autumn, K., Sitti, M., Liang, Y.A., Peattie, A.M., Hansen, W.R., Sponberg, S., Kenny, T.W., Fearing, R., Israelachvili, J.N., Full, R.J.: Evidence for van der Waals adhesion in gecko setae. Proc. Natl. Acad. Sci. **99**(19), 12252–12256 (2002)
79. Shah, G.J., Sitti, M.: Modeling and design of biomimetic adhesives inspired by gecko foot-hairs. In: IEEE International Conference on Robotics and Biomimetics (ROBIO), p. 873. (2004)
80. Geim, A.K., Dubonos, S.V., Grigorieva, I.V., Novoselov, K.S., Zhukov, A.A., Shapoval, SYu.: Microfabricated adhesive mimicking gecko foot-hair. Nat. Mater. **2**, 461–463 (2003)
81. Gebeshuber, I.C., Gordon, R.: Bioinspiration for tribological systems on the micro- and nanoscale: Dynamic, mechanic, surface and structure related functions", invited article. Micro and Nanosystems **3**(4), 271–276 (2011)
82. Alexander, D.E.: Nature's flyers: birds, insects, and the biomechanics of flight. The Johns Hopkins University Press, Baltimore (2004)
83. Nachtigall, W.: Bionik: Grundlagen und Beispiele für Ingenieure und Naturwissenschaftler. Springer, Heidelberg (2002)
84. Nachtigall, W.: Das große Buch der Bionik. Deutsche Verlagsanstalt, München (2003)
85. Tributsch, H.: How Life Learned to Live. MIT Press, Cambridge (1984)
86. Vogel, S.: Life's devices: the physical world of animals and plants. Princeton University Press, Princeton (1988)
87. Vogel, S.: Life in moving fluids: the physical biology of flow. Princeton University Press, Princeton (1996)
88. Vogel, S.: Cats' paws and catapults: mechanical worlds of nature and people. WW Norton & Co, New York (1998)
89. Vogel, S.: Comparative biomechanics: Life's physical world. Princeton University Press, Princeton (2003)
90. Anonymous.: Summary: World Tribology Congress 2009 (WTC IV) International Tribology Council Information 191, (2010)

Chapter 3
Nanotribology and Wettability of Molecularly Thin Film

Yufei Mo and Liping Wang

Abstract The micro/nano-electromechanical systems (MEMS/NEMS) have received rapid development in the past decades due to their superior performance and low unit cost. However, large surface area-to-volume ratio causes serious adhesive and frictional problems for MEMS operations. Nanotribology is a study of the interaction between contact surfaces at nanoscale, from chemistry and physics to material science and mechanical engineering, and is of extreme technological importance to the application and development of MEMS. This chapter will attempt to cover the range from preparation of organic thin films to instruments and measurement protocols. We will describe this process in steps. The preparation of thin film materials (i.e., ionic liquids, multiply-alkylated cyclopentane or self-assembled molecules) and film deposition are presented. Also, the methods of film evaporation are considered. We examined the relationship of adhesion and lateral force data to their fundamental aspects at molecular level. The main objective will be to provide more thorough examination to the interested reader, and to provide a source to further raise the critical issues concerning the relationship between surface properties and MEMS application. Fluorinated molecules with coplanar structure were successfully self-assembled onto silicon surface. The fluorinated monolayers possessed excellent adhesion-resistance, friction-reduction and anti-wear durability, which were attributed to low interfacial energy of end group and dual layer structure of the films. The spatial distribution of the multi-component film was evaluated by adhesion statistic measurement. Multialkylated cyclopentanes (MACs) are potential lubricants for space and MEMS application due to their extreme low volatility. A series of MACs were synthesized by Dean-Stark trap, autoclave, and phase transfer catalysis methods. Nanoscale dual-layer films consisting of both MACs and self-assembled monolayers (SAMs) were prepared and their morphology, wettability and tribological properties were investigated. Molecularly thin ILs films with different molecular

Y. Mo · L. Wang (✉)
State Key Laboratory of Solid Lubrication, Lanzhou Institute of Chemical Physics,
Chinese Academy of Sciences, Lanzhou 730000, P. R. China
e-mail: lpwang@licp.cas.cn

S. K. Sinha et al. (eds.), *Nano-tribology and Materials in MEMS*,
DOI: 10.1007/978-3-642-36935-3_3,

structures which showed excellent tribological performance were designed, synthesized and prepared successfully on silicon surface by dip-coating method. The influences of anion, cation and post-treatment on wettability and tribological properties of ILs films were investigated systematically. To enhance the wettability and to improve the nanotribology of nano films, surface texture technique is reviewed. Regular and biomimetic surface textures were fabricated by local anodic oxidation (LAO). Dimension of the pillars were precisely controlled by operation parameters such as pulse bias voltage, pulse width and humidity. The H-passivated Si showed higher growth rate and thicker saturated oxide film than common p- or n-type Si under the same oxidation condition. The H-passivated Si employed in LAO process can improve lateral resolution of patterns. The adhesive and friction force of LAO pattern were measured by AFM colloidal probe. The friction forces are closely related to the surface coverage of the nanotexture. The results indicate that nanotextures significantly reduced the friction force, while H-passivated Si showed large friction force, this is because of the less adhesive energy dissipated during sliding on textured surface. The surface nanotextures of biological origins were fully duplicated on surface based on duplication method. The morphology and the size of the surface textures of the replicas are almost in accordance with their biological sources. The wettability of the surfaces was improved with hydrophobicity after duplicating with textures. And the biomimetic textures have shown to improve nanotribological performance.

Contents

Introduction

The microelectromechanical systems (MEMS) have received rapid development in the past decades due to their superior performance and low unit cost. However, large surface area-to-volume ratio causes serious adhesion and frictional problems for MEMS operations. In MEMS devices, various forces associated with the device scale down with the size. When the length of the machine decreases from 1 mm to 1 μm, the area decrease by a factor of million and volume decreases by a factor of a billion. At this scale, mechanical loading is often not the overwhelming force as in macroscale, and surface forces such as van der Waals, electronic and capillary force that are proportional to area, become a thousand times larger than the forces proportional to volume. In addition to the consequence of a large area-to-volume ratio since MEMS devices are designed for small tolerance, physical contact becomes more likely, which makes them particularly vulnerable to adhesion between adjacent components. Slight particulate or chemical contamination present at the interface can lead to failure. Since the start up forces and torques involved in operation available to overcome retarding forces are small in MEMS, the increase in resistive forces such as adhesion force and lateral friction force become a serious tribological concern that limits the durability and reliability in MEMS. A large friction force is required to initiate relative motion between two surfaces, that is large static friction, which has been thoroughly studied in the field of data magnetic storage. The adhesion is generally measured by the amount of force necessary to separate two surfaces in contact. Adhesion, friction, wear can affect MEMS performance and even lead to failure.

Space lubrication and nanotechnology are driven by the trends such as device miniaturization, better integrated functional components and energy saving properties. MEMS as miniaturized devices are operated under very narrow space and small normal load. They cannot be lubricated with lubrication oils, but usually employ thin films whose thickness is well below a few nanometers. The adhesive force comes largely from meniscus force and viscous force, rather than the applied loads. The perfluoropolyether (PFPE) thin films are the most widely used lubricants in data storage devices, but PFPEs usually experience metal catalytic degradation and are normally expensive. The potential of self-assembled monolayers (SAMs), multiplyalkylated cyclopentanes (MACs) and ionic liquids (ILs) as thin lubricant films were exploited by a number of researchers aiming to replace PFPEs. The molecular structure, length of alkyl chains, functional groups, surface microstructure and substrate modification are key factors which affect the wettability and the nanotribological behavior of these thin films.

Tribological Behavior of Perfluorinated Carboxylic Acid and Hydrogenated Carboxylic Acid SAMs

Molecular thin films of several monolayers or less supported by surfaces exhibit considerable departure from bulk behavior, which is mainly due to molecular alignment or ordering. SAMs have been widely investigated during the past decade because of their potential applications in the field of surface modification, boundary lubricant, sensor, photoelectronics, and functional bio-membrane modeling etc. On the basis of the synthetic approaches and the surface chemical reactions, the chemical structures of SAMs can be altered easily both at the individual molecular and at the material levels. The nanotribological properties of SAMs, which are potential lubricants for controlling adhesion and friction, are closely related to their intrinsic chemical composition and structure. For example the friction behaviors of SAMs are termina group and chain length dependent. SAMs with long chains are generally densely packed, while the shorter chain ones are not. With the same terminal group, loosely packed SAMs generally posses higher friction force due to the large energy dissipation during the relative movement, and adhesive force as well due to the liquid-like disordered structure. On the other hand, altering the terminal group, from apolar to polar, could result in the increase of adhesion and friction. This is because SAMs with more relatively strong interaction during the relative movement, and therefore higher adhesion and more energy loss are expected, which leads to a higher friction force. Mix-deposition of molecules with different terminal group or alkyl chain lengths to form mixed SAMs is also extensively studied, which allows an understanding of the relationship between structure and performance of SAMs in wide and depth.

As an example of tribological behavior of perfluorinated and hydrogenated carboxylic acid (FC and HC) SAMs on aluminum surface by chemical vapor deposition were studied [1]. Figure 3.1 shows the mechanism of adsorption of SAMs. The samples were placed in a 100 ml sealed vessel with a glass container filled with 0.2 ml FC or HC precursor. There was no direct contact between the samples and precursor. The vessel was placed in an oven and then nitrogen gas was filled in the oven. The samples were annealed in nitrogen at 200 °C for 3 h, and then cooled in a desiccator. The precursor vaporized and reacted with substrate surface on each sample, resulting in the formation of SAM. Then, the samples were rinsed with chloroform, acetone, ethanol, and deionised water successively to remove other physisorbed ions and molecules. The deposition of SAMs relied on the chemisorption of reactive head groups presented in the adsorbate molecules on the substrate surface in order to anchor them.

X-ray photoelectron spectroscopy (XPS) was used to evaluate the relative atomic composition on the surface of SAMs. The procedure involved the measurement of the Al2p, F1 s and C1 s core level spectra for surfaces of these films. The data of Al2p features from bare surface are shown in Fig. 3.2a and are associated with Al_2O_3 or AlO(OH) (74.7 eV) and Al (72.7 eV). This result indicates that the outmost layer of Al is converted to aluminum oxide under ambient

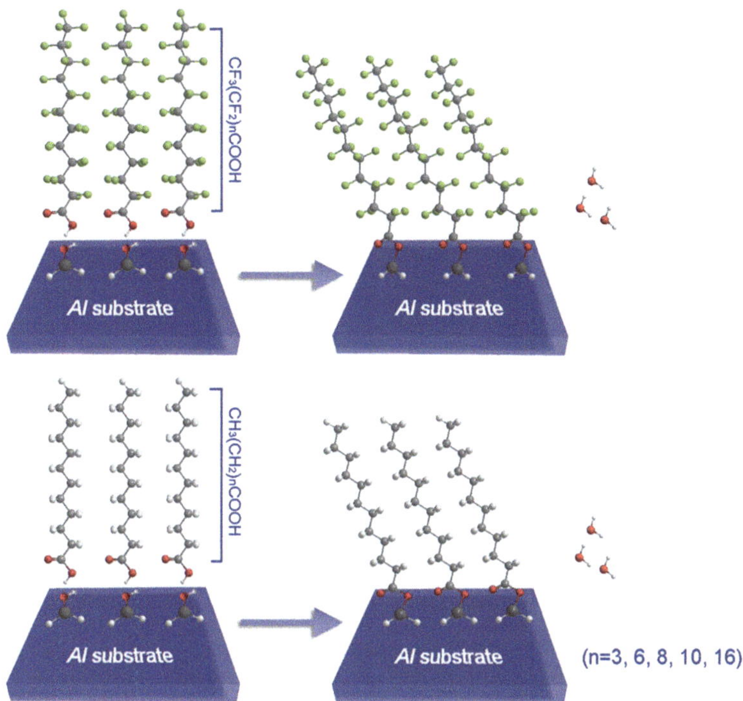

Fig. 3.1 Schematic structure and forming process of perfluorinated carboxylic acid and hydrogenated carboxylic acid molecules chemically adsorbed onto aluminum substrate. Perfluorinated carboxylic acid and hydrogenated carboxylic acid SAMs are of similar chain length and head group monolayers with different backbone groups. Reproduced with permission from Ref. [1]. Copyright (2009) Surface and Interface analysis

condition. A single F1s feature resulting from FC18 adsorption on aluminum oxide is shown in Fig. 3.2b. The F1s feature at 688.7 eV is assigned to $-CF_2-$and $-CF_3$ groups, which indicates the fluorine element on the substrate surface. Figure 3.2c displays the C1s spectra obtained from one set of FC SAMs with various chain lengths (C5–C18) on aluminum oxide surface. The C1s features are assigned to $-CF_3$ group ($\sim$293.5 eV), $-CF_2-$group (C5: 292.1–C18: 291.5 eV) [2, 3], carboxylate group (COO−, 289.2 eV), [4] and a feature (284.8 eV) associated with adventitious carbon possibly from airborne hydrocarbon contamination. It is suggested that the samples were strongly bonded with airborne organics (fatty acid, etc.), which were adsorbed at film defects and imperfections and were not easily removed by vacuum pumping. It is also observed that the intensity of adventitious carbon decreased rapidly with increase of fluorocarbon chain length while that of COO^- and $-CF_3$ remained constant. This is because the long chain FC- SAMs were densely packed and with fewer defects, which prevented airborne organics from adsorbing onto the film. Figure 3.2d shows a similar tendency of increase in C concentration associated with the increase of chain length in

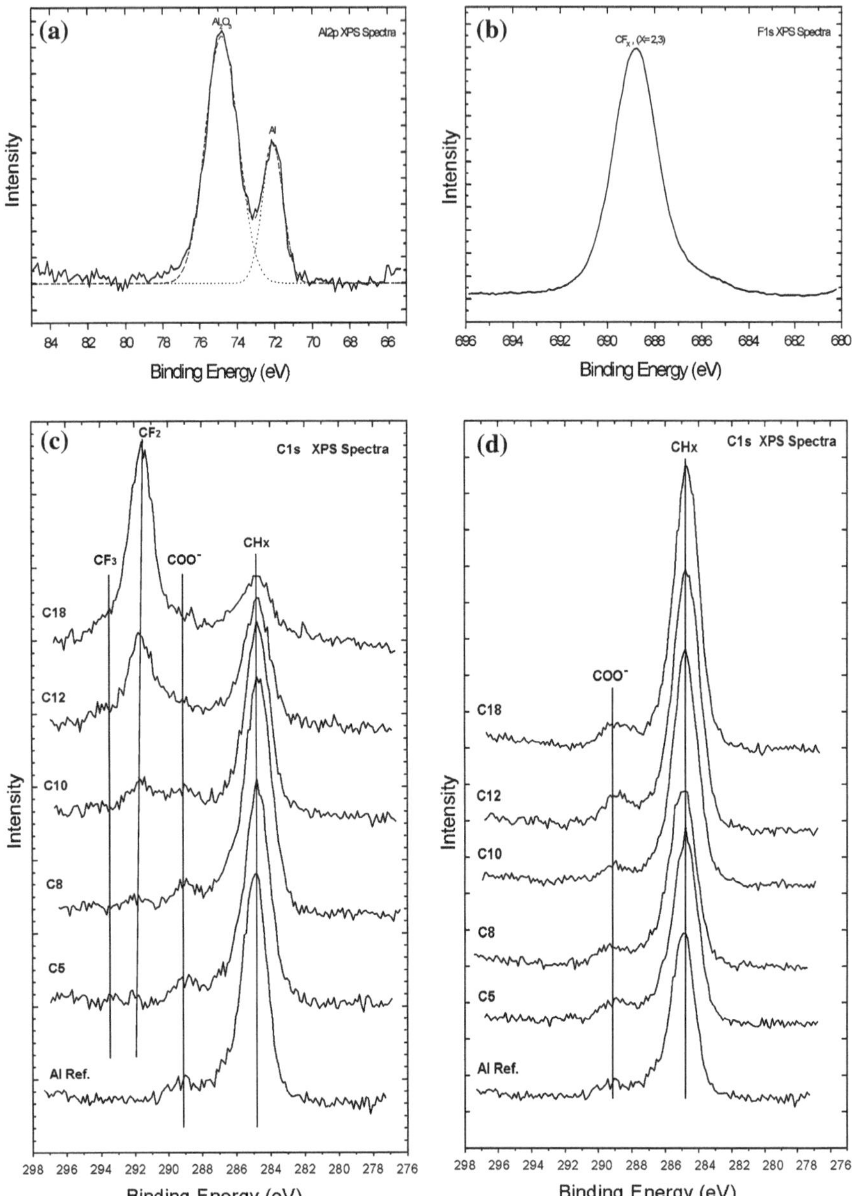

Fig. 3.2 XPS spectra of Al 2p **a** F 1s, **b** C1s, **c** region of the FC5-FC18 SAMs, **d** C 1s region of the HC5-HC18 SAMs. Reproduced with permission from Ref. [1]. Copyright (2009) Surface and Interface analysis

HC-SAMs. In addition, two features arose from C1s, as shown in this figure, the left feature assigned to carboxylate group (COO, 289.2 eV) and the right feature can be attributed to CHx (284.8 eV).

Nanotribological Properties of Monolayers Under Ambient Condition: Effect of Temperature and Humidity

The adhesive force between AFM tip and SAM surfaces under various relative humidity are shown in Fig. 3.3a and b. The bare aluminum surface showed higher adhesive force than SAMs deposited on it. Between FC and HC SAMs, the FC-SAMs showed lower adhesion force than HC-SAMs with same chain length, which indicates that adhesive force is consistent with the difference in surface energy (15 mJ/m^2 for CF_3-terminated compared to 19 mJ/m^2 for CH3-terminated SAMs [5]). The relationship between adhesive forces for bare Al and FC-SAMs with various chain lengths is shown in Fig. 3.3a. It shows that adhesive force

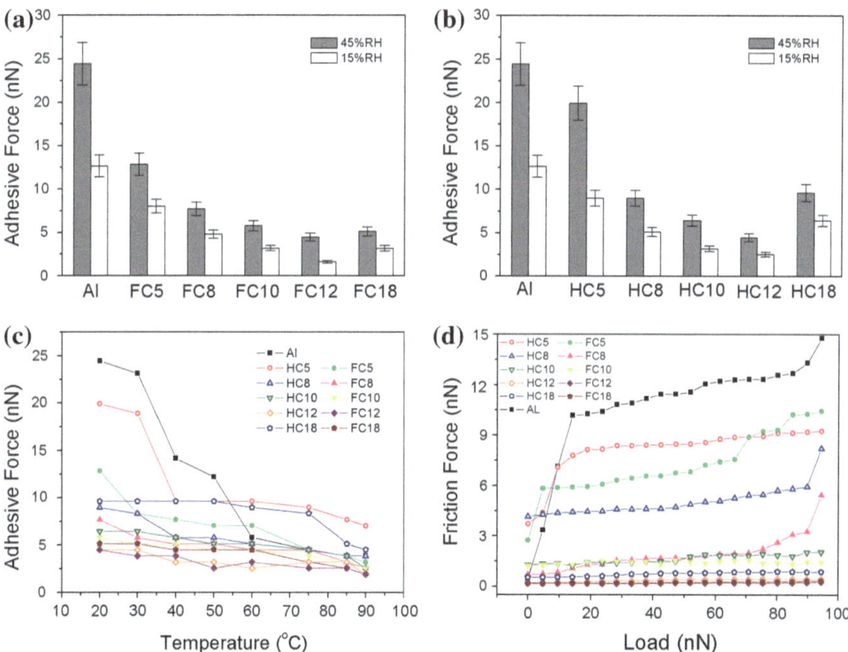

Fig. 3.3 Relative humidity dependence of adhesion for Al substrate and various SAMs. **a** FC-SAMs. **b** HC-SAMs. **c** Dependence on temperature of adhesion for the Al substrate, FC-SAMs (*solid symbols*) and HC-SAMs (*empty symbols*). **d** Plots of friction force versus load for various SAMs on Al substrate, FC-SAMs and HC-SAMs. Reproduced with permission from Ref. [1]. Copyright (2009) Surface and Interface analysis

increased with relative humidity (RH), which is due to water menisci contribution. It is also observed that the adhesive force of FC-SAMs with chain lengths up to ten carbons increased indistinctively and then tended to a stable value. This tendency of the adhesive force agrees well with the change in contact angles which correlates with surface energy [1]. FC12 and FC18 SAMs showed lowest adhesive forces and highest contact angles, which implies that 12-carbon and 18-carbon chains are prone to form more stiff film. In the case of short chain SAMs (n-carbon <10), they formed relatively soft monolayers and tended to disorder under the pressure applied by AFM tip. The pressure induced terminal defects may be sufficient for complete disorder, an effect that will be magnified by the reduced packing density of the short molecules. For SAMs with chain lengths up to ten carbons, low adhesive force may be attributed to high stiffness, which gives rise to a smaller contact area for the same applied load. Figure 3.3c shows the influence of temperature on adhesion. The adhesive force decreased with increase of temperature and then tended to a stable value. The drop in adhesive force is a result of desorption of water molecules and the corresponding decrease of water menisci contribution. The aluminum substrate and short chain SAMs showed more temperature dependence compared with long chain SAMs. The FC and HC SAMs with long chains exhibit temperature independence over the temperature range studied, which is due to the fact that highly hydrophobic nature of these monolayers results in less formation of water menisci. It indicates that long backbone chains and neighboring fluorine atoms provide stronger inter-chain interaction compared to that provided by short backbone chain and hydrogen atoms. SAMs with perfluorinated long chains were densely packed and highly ordered with solid-state-like properties at high temperature due to strong inter-chain van der Waals force.

Figure 3.3d shows the relationship between friction force and external load for bare Al, as well as for FC and HC SAMs with various chain lengths, at RH of 15 % and temperature of 20 °C. The bare aluminum surface without organic film generates the highest friction. This may be attributed to the highest surface energy on the Al_2O_3-covered surface. The highest surface energy can be indicated by the lowest contact angle. It is a general tendency that the friction force decreases as chain length increases and FC-SAMs showed lower friction force than HC-SAMs of corresponding chain length. It is also observed that the friction properties of SAMs do not depend only on the chemical nature of terminal groups. Otherwise, all chain lengths should yield similar friction values. For the formation of SAMs, both surface energy and Inter-chain interactions play important roles and determine quality and character of the SAMs [6]. The decrease of friction is mainly due to SAMs with longer chain, as they generally possess relatively stronger inter-chain interaction, which give rise to a smaller contact area for the same applied load during the sliding. Tribological characterization studies of the SAMs can be summarized as shown in Table 3.1 [1].

Table 3.1 Summary of tribological properties for the FC arid HC SAMs on Al surface

	SAMs property	Adhesive force	Nanofriction	Microfriction
Backbone style	Fluorocarbon backbone	Low	Low	High
	Hydrocarbon backbone	High	High	Low
Chain length	Short (Q < 10)	High	High	–
	Middle (C_n = 10–12)	Low	Low	–
	Long ($C_n > 12$)	Low	Low	–
Terminal group	Methyl	High	High	Low
	Perfluorinated methyl	Low	Low	High

Structural Forces due to Surface Structure: Preparation and Tribological Properties of Perfluorinated Carboxylic Acid Dual-Layer SAMs

SAMs have good anti-rupture properties due to their strong bonding to the substrate surface, and they are expected not to freely migrate on the surface. However, some molecules from SAMs may transfer to the surface of counterpart when external force was applied on the contacting surface [7]. Because of monolayer structure and flexibility, SAMs exhibit poor anti-wear durability [8–10]. To utilize SAMs as lubricants to protect MEMS, it is necessary to consider the molecule layer structure as well as the strongly bonded characteristics of lubricant [11]. Generally, there are two approaches to obtain these more complex structures on surface: one is to synthesize target precursors with functional groups and then assemble them onto surfaces by a one-step method, [12–17] but there are difficulties in purification during the synthesis of more complex molecules. Another is stepwise formation of the film with desired structures based on surface chemical reaction.

Several reports [13–15] have demonstrated that incorporation of amides into hydrocarbon backbones of precursor could improve the stability of SAMs. The reason was that the amide underlayers were capable of being cross-linked by hydrogen bonding. Work has also recently been done on building amide-containing dual layer SAMs on silicon surfaces and found to be very excellent wear-resistant films [16, 17]. Bai et al. [18] designed a perfluorinated dual layer structure which can help to improve the film quality, reduce the friction and significantly enhance their durability, as shown in Fig. 3.4.

A self-assembled dual-layer film was prepared on single-crystal silicon surface by chemisorption of perfluorododecanoic acid (PFDA) molecules on 3-aminopropyltriethoxysilane (APS) SAM with terminal amino group. The dual-layer PFDA–APS film was hydrophobic with the contact angle for water to be about 105° and the overall thickness about 2.5 nm. Atomic force microscopic images revealed that the APS surface was initially characterized with uncontinuous asperities, the surface became relatively smooth and homogeneous after coating with PFDA film by self-assembly. The PFDA–APS film exhibited low

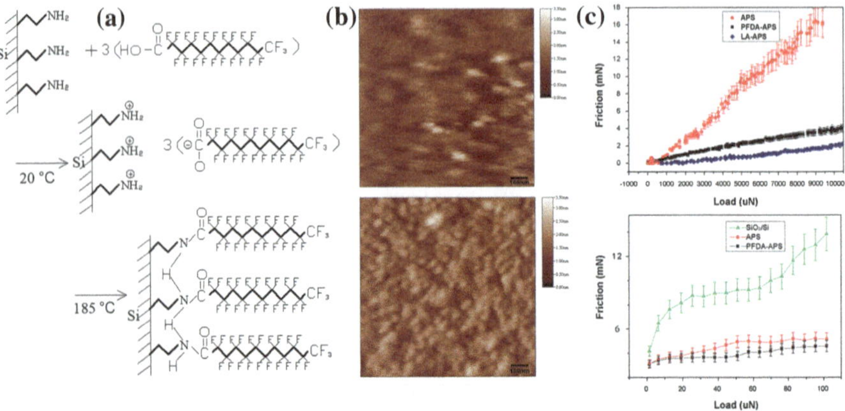

Fig. 3.4 **a** Schematic structure and forming process of PFDA molecules chemically adsorbing onto APS monolayer surface. **b** AFM topographic images of APS monolayer and PFDA–APS dual-layer film on silicon wafer. **c** Microtribological behaviors of the APS monolayer, PFDA–APS dual-layer film and LA–APS dual-layer film surfaces sliding against a steel ball. Reproduced with permission from Ref. [18]. Copyright (2008) Journal of Physical Chemistry C

adhesion and it greatly reduced the friction force at both nano- and microscale. The film exhibited better anti-wear durability than the lauric acid (dodecanoic acid or LA)–APS self-assembled dual-layer film with same chain length and similar structure.

Preparation and Nanotribological Properties of Multi-Component Self-Assembled Dual-Layer Film

Previous results [18–20] indicate that the dual-layer structure can help to improve the film quality and enhance their durability and load bearing capacity. Meanwhile, it is observed that a hydrogenated carboxylic acid dual-layer film exhibits better friction reduction but poorer durability compared to the perfluorinated carboxylic acids dual layer. A lubrication system consisting of dual component self-assembled dual-layer films was designed to minimize friction and a molecular mixture layer to prolong durability. Bai's group [21] reported a novel strategy for a dual-component self-assembled film with control of spatial growth on a large surface area based on a dip-coating nanoparticles method. In selecting among the various SAMs, we particularly focus on the control of both fluorinated and hydrogenated backbone chain molecules because these molecules have strong potential applications in MEMS.

The mechanism of this site-selective growth can be explained as follows. An APS layer was first formed on hydroxylated silicon substrate. Monodisperse Ag

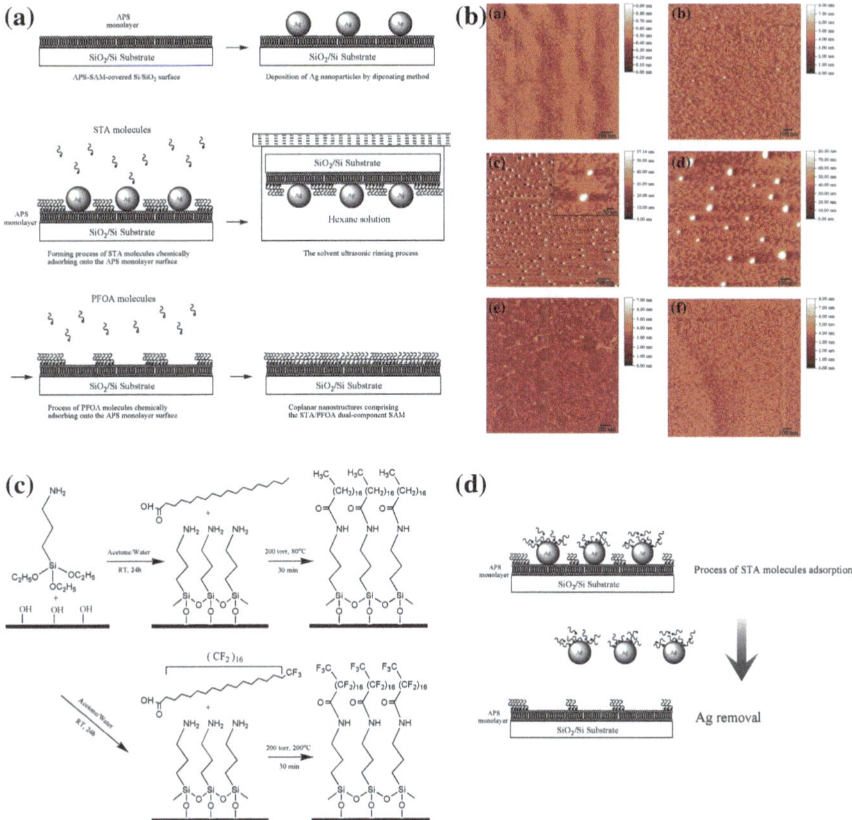

Fig. 3.5 **a** Formation of the coplanar nanostructure STA/PFOA dual-component dual-layer film; **b** AFM images of the film surface in each step; **c** Chemical structure and forming process of STA and PFOA molecules chemically adsorbing onto the APS monolayer surface; **d** Mechanism of STA molecules deposition and Ag nanoparticles removal. Reproduced with permission from Ref. [21]. Copyright (2008) Journal of Physical Chemistry C

nanoparticles capped by long-chain carboxylates played a role in the effective suppression of undesired composite growth on sites. Due to acid amide reaction between stearic acid (STA) and APS molecules, the STA molecules chemisorbed onto the APS-modified surface. The film surface fabricated lunar crater-like pits microstructure and amino-terminated surface exposed in the bottom of the pits after Ag nanoparticles removal. The perfluorooctadecanoic acid (PFOA) molecules absorbed onto the exposed amino terminated surface with acid amide reaction, and the pits of the film were occupied completely by PFOA molecules (Fig. 3.5).

It is important to calculate the spatial distribution of the dual-component film by adhesion statistic measurement, as shown in Fig. 3.6. Adhesive forces of STA and PFOA SAMs were measured as 3.21 and 6.43 nN, respectively. The adhesive

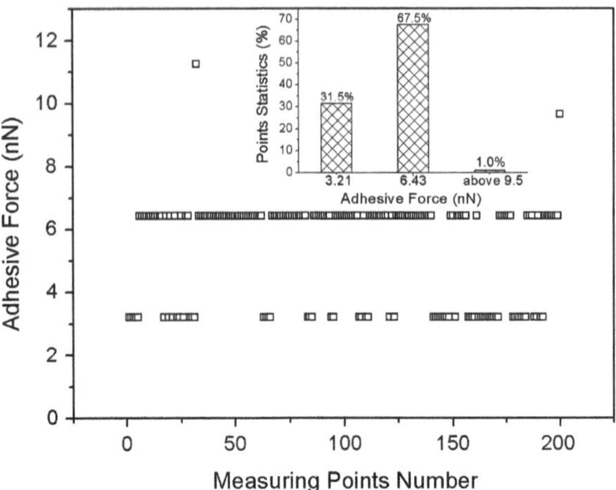

Fig. 3.6 Plot of adhesion force and statistical distribution for the PFOA/STA-APS film (20 °C, 15 % RH). Reproduced with permission from Ref. [21]. Copyright (2008) Journal of Physical Chemistry C

force measurement was typically performed at a rate of 1 Hz along the scan axis and a scan size of 10 × 10 μm during scanning, at least 200 measuring points were carried out for each scan range. From the inset, it can be seen that the adhesive forces of the dual-component layer were calculated statistically as 31.5 and 67.5 %, respectively. The surface coverage of the pits was calculated as a value of about 20 %, which approaches to surface coverage of the pits calculated from the data from the adhesive force measurement. The discrepancy between the surface coverage of pits and the statistical value of PFOA in the adhesion measurement is probably because some STA molecules comprising a SAM exchanged gradually when exposed to the PFOA atmosphere, which results from displacement of SAMs by exchange [22, 23].

Tribolgical Behavior of Multiply-Alkylated Cyclopentane

To utilize SAMs as boundary lubricants, it is necessary to consider the mobile characteristics in addition to the strongly bonded characteristics. The chemically bonded SAMs protect the devices during processing and the early stages of use, while a mobile lubricant is present to replenish the lubricant coating as the SAMs fail. As a result, an ideal boundary lubricant system is pursued.

Multiply-alkylated cyclopentanes (MACs) are composed of one cyclopentane ring with two to five alkyl groups substituted on the ring. They are synthesized by reacting dicyclopentadiene with alcohols of various chain lengths producing a lubricant with a selectable range of physical properties [24]. MAC has excellent viscosity properties, thermal stability and low volatility for use as lubricant and is presently gaining wide acceptance in certain space application [25, 26].

Effect of Wettability on Nanotribology of MACs

Wettability is one of the most important properties of solid surfaces and has attracted much attention since the time of Young in 1805 [27]. It is governed by both chemical composition and topological characteristic of the surface. Controlling wettability is quite important in the study of nanotribological properties.

Wang et al. [28] studied wettability of MACs on silicon substrates that were treated by different cleaning and etching processes. As shown in Fig. 3.7, the wettabilities of MACs on hydroxylated silicon and hydrogenated silicon are better than the wettability on bare silicon without pretreatment, and that outcome is mainly caused by topological structure changes of the surface.

Ma et al. [29] investigated wettability and loading-carrying capacity of MACs on two types of SAMs of decyltrichlorosilane (DTS) and 1H,1H,2H,2H perfluorodecyltrichlorosilane (FDTS). As shown in Fig. 3.8, when MAC was deposited on the DTS-SAM, unlike uniform DTS-SAM, the MAC forms as island-like liquid droplet with a typical diameter of 25 nm and an average height of 3.5 nm was evenly distributed on the DTS-SAM to form dual-layer film with a surface coverage of about 70 %. This research indicate that MAC was deposited on the two SAMs to form dual-layer films with total thickness of 5 nm, the mobile lubricant could markedly decrease friction of DTS-SAM and remarkably promote the load-carrying capacity and durability of both DTS and FDTS SAMs owing to its good self-repairing property.

Distribution and Positioning of Lubricant on a Surface Using the Local Anodic Oxide Method

Local anodic oxidation (LAO) via the atomic force microscope (AFM) is a lithography technique perspective for the fabrication of nanometer-scaled structures and devices. AFM-LAO is based on direct oxidation of the sample by negative voltage applied to the AFM tip with respect to the surface of the sample. The driving force is the faradaic current flows between the tip and sample surface with the aid of the water meniscus. When the faradaic current flows into water bridge, H_2O molecules are decomposed into oxyanions (OH^- and O^-) and protons (H^+). These ions penetrate into the oxide layer because of the high electric field ($E > 107$ V/m), [30] leading to the formation and subsequent growth of SiO_2 on the H-passivated Si surface. The AFM-LAO process can be used not only in fabrication of nanodevices but also in adhesion resistance and friction reduction as in the case of surface texturing. In previous studies [31, 32], AFM-LAO has been demonstrated to be the most promising tool for fabricating nanodots and lines on several types of materials ranging from metals to semiconductors. The LAO process is controlled by several major parameters as follows: pulsed bias voltage, pulsewidth and humidity, as shown in Fig. 3.9.

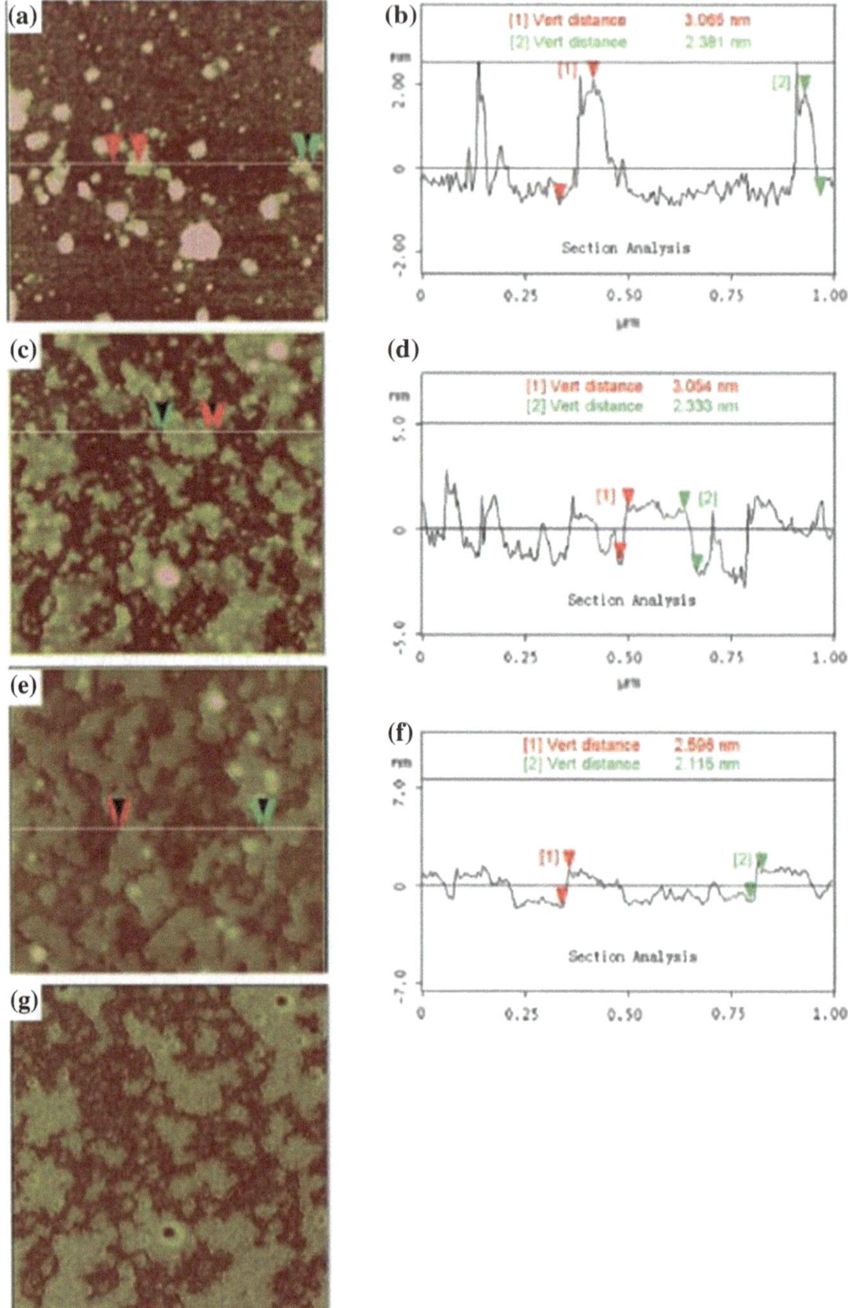

Fig. 3.7 AFM images of MACs films: **a** MACs on bare silicon; **b** line section analysis of (**a**); **c** MACs on hydroxylated silicon; **d** line section analysis of (**c**); **e** MACs on hydrogenated silicon; **f** line section analysis of (**e**); **g** phase map for (**c**). Reproduced with permission from Ref. [28]. Copyright (2010) Tribology Transactions

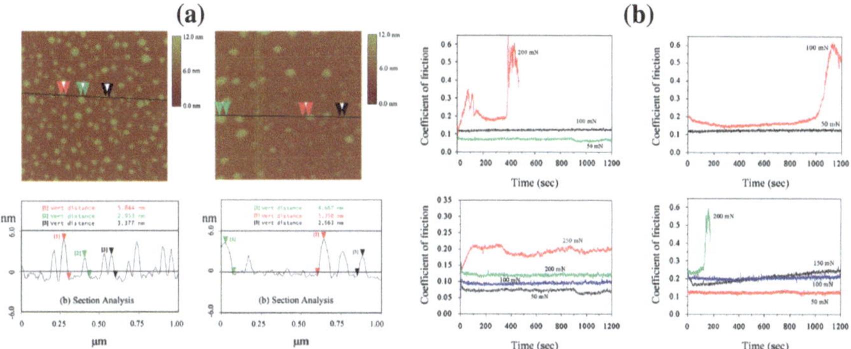

Fig. 3.8 **a** AFM images and section analysis of MAC-DTS dual-layer; **b** friction coefficient and durability of designed SAMs. Reproduced with permission from Ref. [29]. Copyright (2007) Elsevier B. V

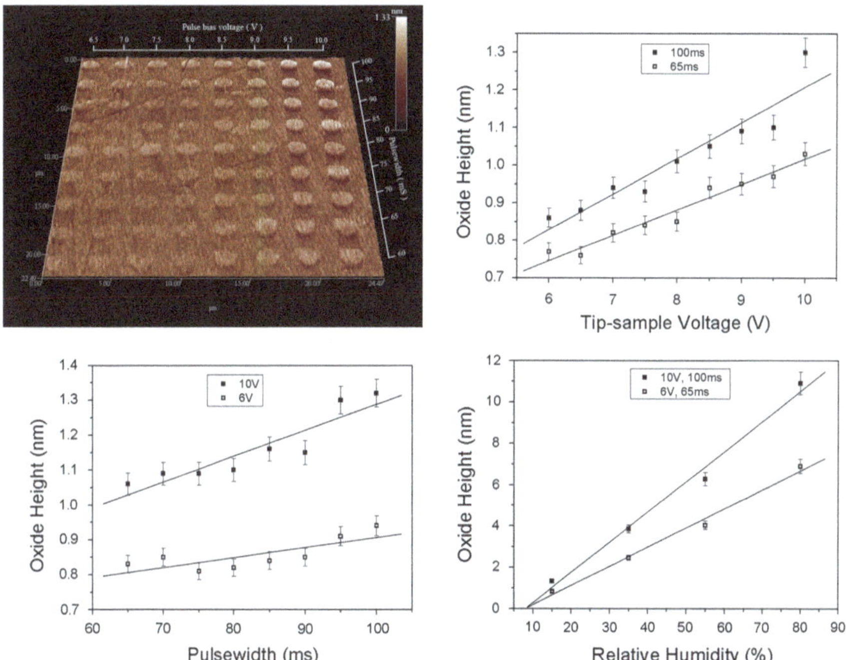

Fig. 3.9 A testing array of Si pillars prepared at different operation parameters, and the oxide height as a function of tip-sample pulse bias voltage, pulsewidth and relative humidity, respectively. Reproduced with permission from Ref. [32]. Copyright (2008) Elsevier B. V

As above mentioned, MAC has excellent viscosity properties, thermal stability, and low volatility for use as a lubricant. Unfortunately, it is difficult to control the size and distribution of lubricants precisely on silicon or DLC. Currently, Bai et al.

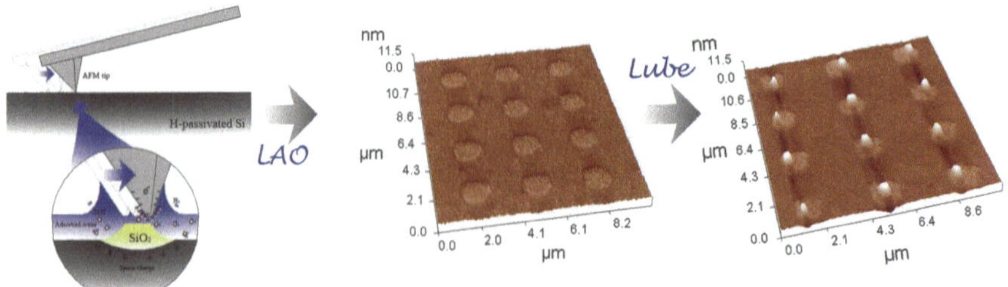

Fig. 3.10 Schematic of distribution and positioning process of MAC on a surface using the local anodic oxide (*LAO*) method. Reproduced with permission from Ref. [33]. Copyright (2009) Langmuir

[33] utilizes an AFM-LAO technique to control the size and distribution of lubricants precisely in an atmospheric environment. The new technique includes two main steps: the production of nanometer-sized nanopatterns using AFM-LAO, followed by the selective adsorption of MAC lubricant onto these patterns using dip-coating method (Figure 3.10).

Ducker [34] first introduced the use of colloidal probe tips by attaching a sphere to the cantilever to measure adhesion. The spherical shape of the tip provides controlled contact pressure, symmetry, and mostly elastic contacts. For adhesion and friction force measurements of the fabricated MAC matrix, the spherical probe tip can fully contact with surface, while sharp tip can only have point contact.

The adhesive force between the colloidal probe and sample surfaces is shown in Fig. 3.11a. A strong adhesive force was observed on the untreated H-passivated silicon surface, on which the adhesive force was about 175 nN. After the patterns were generated, the adhesive force was decreased to 70 nN. This result indicates that the pattern exhibited adhesion resistance. Adhesion is directly related to the bearing ratio, which describes the real area of contact between two solid surfaces. After dip-coating in MAC solution, the adhesive forces of untreated H-passivated

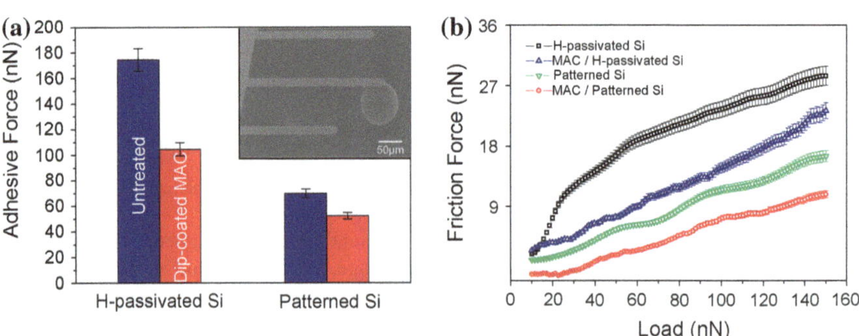

Fig. 3.11 Plots of **a** adhesive forces and **b** friction force between the AFM colloidal probe and the surfaces of samples. The inset shows an SEM image of a typical colloidal probe. Reproduced with permission from Ref. [33]. Copyright (2009) Langmuir

and patterned Si were decreased to 105 and 52 nN, respectively. Such a phenomenon indicates that the MAC layer on the sample surface can obviously lower the interfacial energy and capillarity between the tip and surface. Figure 3.11b presents the plot of friction versus load curves for the bare H-passivated Si and patterned Si and these surfaces treated with MAC. Patterned Si evidently reduced the friction force, and the MAC-cover-patterned Si exhibited the lowest friction force whereas H-passivated Si had a strong friction force. The decrease in friction force is mainly due to a pattern giving rise to a smaller contact area for the same applied load and MAC as the lubricant layer minimizes the shearing strength during sliding.

Tribolgical Behavior of Ionic Liquid Films

Why can ILs be lubricants? ILs have many unique properties, such as negligible volatility at a relatively high temperature, nonflammability, high thermal stability, etc. [35]. These characteristics have attracted great attention and made them available in many potential applications, for example catalysis, electrochemistry, separation science for extraction of heavy metal ions, as solvents for green chemistry, and materials for optoelectronic applications [36–39]. On the other hand, as is well known to tribologists, these characteristics are also just what high performance lubricants demand. Very harsh friction conditions require lubricating oils to have high thermal stability and chemical inertness. The decomposition temperatures of imidazolium ILs are generally above 350 °C, some even as high as 480°C, together with the low temperature fluidity (the glass transition temperature, T_g below 50°C, even 100°C) means that ILs can function in a wide temperature range. In addition, low volatility makes ILs applicable under vacuum, especially for spacecraft application. The above mentioned properties of ILs also make them excellent lubricants. Ye et al. [40, 41] investigated the tribological behavior of two kinds of alkylimidazolium tetrafluoroborate and found them versatile lubricants for the contacts of steel/steel, steel/aluminum, steel/copper, steel/SiO_2, Si_3N_4/SiO_2, steel/Si (100), and steel/sialon ceramics.

Different with large scale mechanical system, MEMS cannot be lubricated with lubrication oils, but can use a thin lubrication film whose thickness is well below a few tens of nanometers. The viscous force comes largely from the viscosity of lubricant films and the meniscus force, rather than the applied loads; this dictates the extent of friction and the mechanisms of lubrication failure. The PFPEs of nanometer thickness are the most widely used lubricants in miniaturized devices, but usually experience metal catalytic degradation and are normally expensive. The potential of ILs in thin film lubrication was exploited by a number of researchers aiming to replace PFPEs [42–46]. The molecular structure, the counter anion, the length of substituted alkyl chains and the functional groups, have key effects on the adhesion and tribological behavior of IL films. The interaction between lubricant and surface cannot only determine the wetting of lubricant but also determine its durability [47].

Effect of Anion and Substrate Modification

The anions have a more complicated effect on tribological properties in that they cannot only change the viscosity but also surface energy. For ILs with same cations, Zhu et al. [48] demonstrated three kinds of 3-butyl-1-methyl imidazolium ILs with anions of hexafluorophosphate, tetrafluoroborate and adipate as ultra-thin film. Mo et al. [49] introduced a series of propylmethylimidazolium (PMIM) base wear resistant ionic liquid with anions of bromide, carbonate, chloride and sulfite. Adhesion and friction measurements at nanoscale were carried out using a colloidal probe. As shown in Fig. 3.12, based on topography analysis, IL films are found to be prone to attach to the silicon substrate surface, leading to more uniform thin films. Bromide and sulfite anions show favorable lubrication as seen from adhesion and friction, which are less than those of carbonate, chloride and uncoated silicon. The wear test of the IL films was evaluated at loads ranging from 60 to 300 mN and sliding frequency in range 1–20 Hz. IL films showed favorable friction reduction and durability. Imidazolium with anions of chloride and carbonate exhibited a low friction coefficient at a normal load of 200 mN. Imidazolium sulfite exhibits low friction and anti-wear durability even at high-frequency sliding (20 Hz).

Effect of Bonding Percentage and Alkyl Chain Length

The lubricant adsorbed onto silicon after the solvent rinsing process, which is termed as bonding lubricant. The bonding percentages of ionic liquid were measured in terms of the thickness of ionic liquid adsorbed onto silicon surface [%bonding $=$ 100 $\times$ (final film thickness/initial film thickness)]. Sinha et al. [50] have also used the similar definition while computing the bonded ratio. To understand the influence of the ratio of bonding to the mobile fraction on the friction in microscale behavior, the mixed IL films were compared with different ratios of bonding to mobile fraction to understand the effect of different bonding percentages. Mo et al. [44] prepared the four kinds of samples (viz. 0, 15, 60 and 85 %) by controlling self-assembled conditions. Fig. 3.13a–e shows the friction coefficients and sliding cycles of 1-alkyl-3-ethylcarboxylic acid imidazolium chloride (AEImi-Cl) ionic liquid with various bonding percentages, as a function of sliding cycles against a steel ball at normal loads ranging between 60 and 500 mN and a sliding velocity of 10 mm/s. Figure 3.13 shows the IL films bonding percentage of 0, 15, 60, and 85 % at a normal load of 60 mN, an average friction coefficient of 0.28, 0.22, 0.18, and 0.16 was recorded, respectively. It was

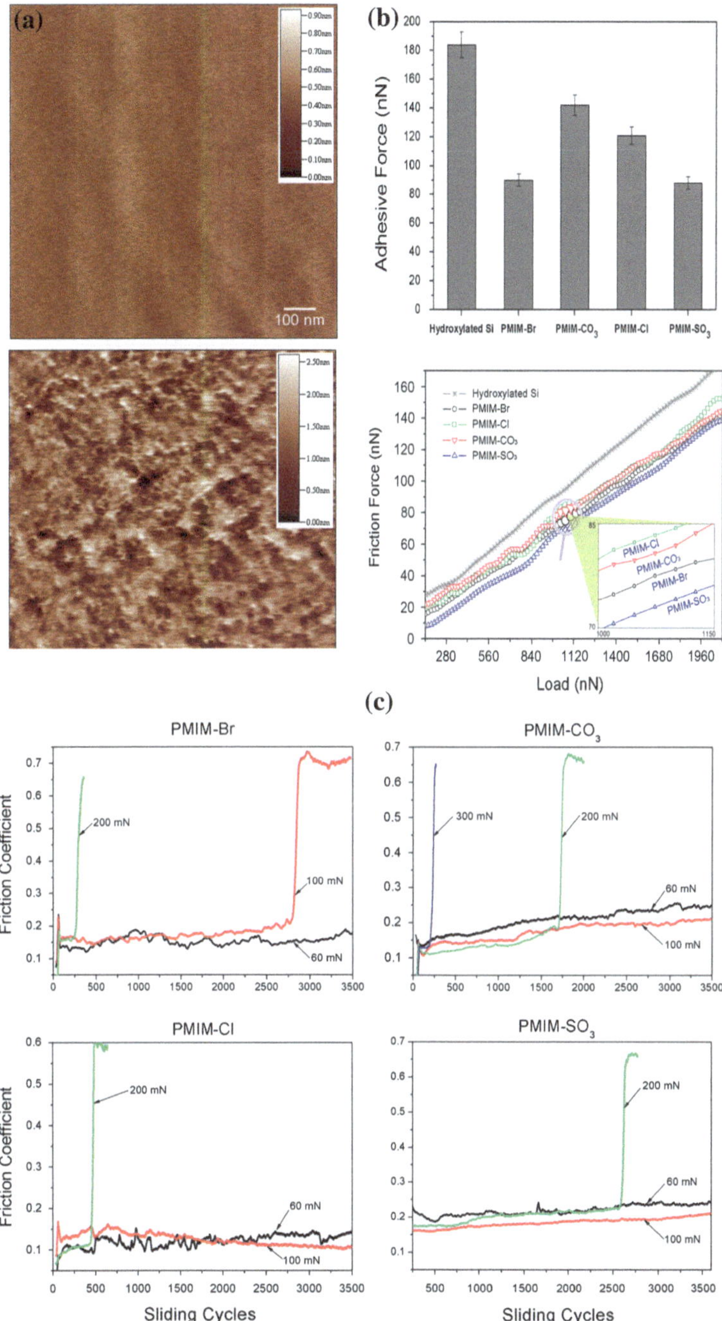

Fig. 3.12 **a** AFM images of hydroxylated Si and IL film surfaces. **b** Adhesive and friction forces between colloidal probe and the surfaces of PMIM-Br, PMIM-CO_3, PMIM-Cl and PMIM-SO_3 IL films. **c** Plots of friction coefficients as function of sliding cycles for PMIM-Br, PMIMOH-CO_3, PMIM-Cl and PMIM-SO_3 film on silicon. Reproduced with permission from Ref. [49]. Copyright (2010) Surface and Interface Analysis

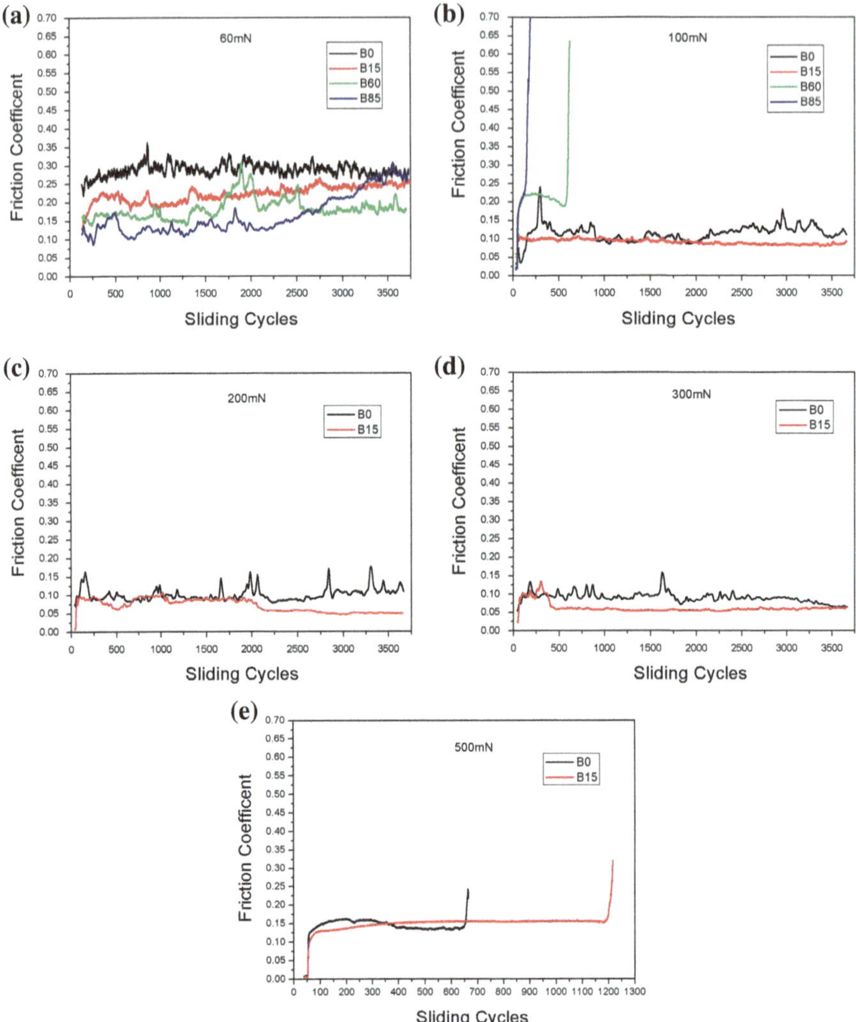

Fig. 3.13 Plots of friction coefficient of AEImi-Cl ionic liquid films with various bonding percentages as function of sliding cycles against steel ball at normal loads of 60, 100, 200, 300, 500 mN with a sliding velocity of 10 mm/s. (The films bonding percentages of 0, 15, 60, and 85 % were denoted as B0, B15, B60, and B85, respectively). Reproduced with permission from Ref. [44]. Copyright (2008) Elsevier B. V

observed that the films with higher bonding percentage exhibited a lower friction coefficient.

The relationship between friction force and external loads for AEImi-Cl ionic liquid with various alkyl chain lengths (viz. C_1, C_4, C_8) is shown in Fig. 3.14. In general, friction is reduced with increase of chain length, and the C_8AEImi-Cl ionic liquid exhibits lowest friction force compared to others. In the formation of

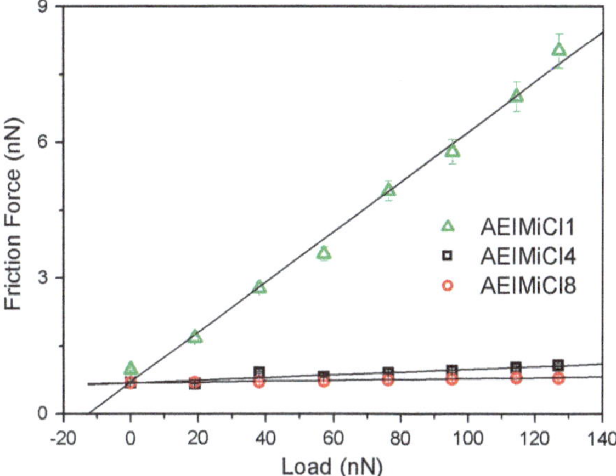

Fig. 3.14 Plots of friction force versus applied loads for AEImi-Cl IL monolayer films with various chain lengths. Reproduced with permission from Ref. [44]. Copyright (2008) Elsevier B. V

the bonding coatings, both the surface energy and inter-chain interactions play important roles and determine quality of the films. Since the AEImi-Cl is the ionic liquid with same terminal group, the nano-friction property is determined by inter-chain interactions.

Effect of Function Group and Annealing Treatment

Zhao et al. [51] successfully prepared four kinds of IL films with different functional cations (1-butyl-3-methylimidazolium hexafluorophosphate, 1-ethanol-3-methylimidazolium hexafluorophosphate, 1-acetic acid -3-methylimidazolium-hexafluorophosphate and 1-phenyl-3-methyl-imidazolium hexafluorophosphate) and characterized their composition and microstructure. The results indicated that IL nanofilms with polar or stiff phenyl cations exhibited relatively higher friction force and better antiwear performance than the ones with apolar alkyl chain structure at micro/nanoscale. The different micro/nanofriction performances of the IL nanofilms were mainly dependent on their different cations which mainly influence their hydrophobic/hydrophilic properties. IL films with more polarized groups generally possessed higher surface energy and a relatively strong interaction during the sliding, and therefore, higher adhesion and more energy loss are expected, which lead to a higher friction force.

Surface morphologies and XPS results [52] indicated that different proportions were formed after post annealing treatment (Fig. 3.15). Annealing treatment of IL film can change the proportions of bonded and mobile molecule in the films. The mobile lubricant fraction present in the partially bonded samples facilitates sliding of the tip on the surfaces; it can rotate with the tip sliding direction easily and

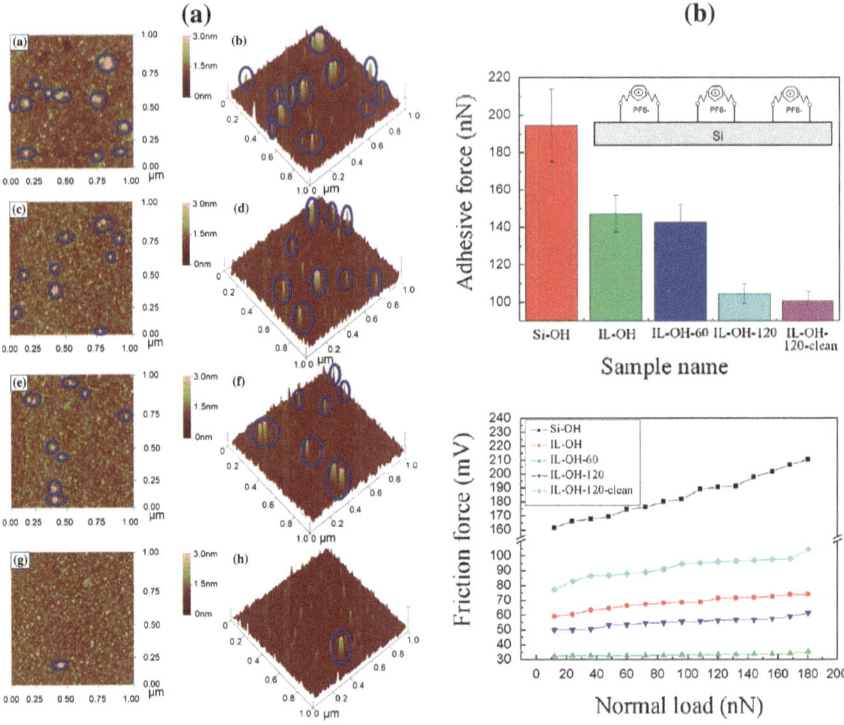

Fig. 3.15 **a** AFM images of IL-OH film at various annealing temperature (from up to down, IL-OH film surfaces are more uniform with reduction in the number of islands); **b** Adhesive and friction force curves of Si, IL-OH films after annealing treatment. Reproduced with permission from Ref. [51]. Copyright (2010) Elsevier B. V

hence the film with higher mobile lubrication fraction exhibits the best nanotribological performance. Annealing treatment significantly changed friction and adhesion performance at nanoscale.

IL Films with Dual-Layer Structure

Choi et al. [53] prepared mixed lubricants with dual-layer structures on a hydrogenated amorphous carbon surface, which consist of alkylsilane SAMs and mobile PFPE lubricants, and found that the friction and durability properties of the mixed lubricants on the carbon surface were mainly dependent on the alkylsilane monolayers. In order to strengthen bonded fraction and further enhance durability of thin IL film, Pu et al. [54] optimized reaction conditions to achieve partial bonding of ILs to silicon substrate by acid-amide reaction between imidazolium-based ILs carrying carboxylic functional groups and amide-containing SAMs as anchor

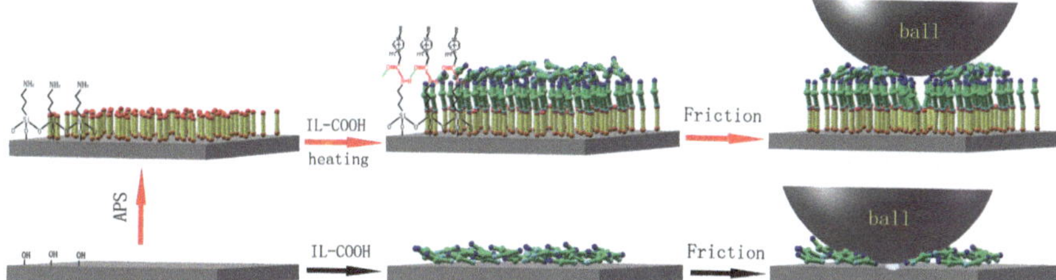

Fig. 3.16 Schematic drawing of the constructing process and frictional mechanism of APS-IL film and IL-COOH film. Reproduced with permission from Ref. [54]. Copyright (2010) Elsevier B. V

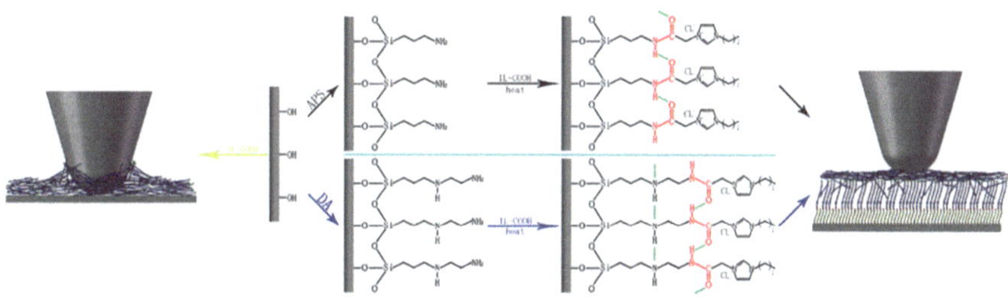

Fig. 3.17 Schematic drawing of the construction process and nanofriction mechanism of APS-IL, DA-IL and IL-COOH films. Reproduced with permission from Ref. [54]. Copyright (2010) Colloids and Surfaces A

layer, and investigated influence of different self-assembled underlayer on the tribological properties of ILs with two-phase structure, aiming to further optimize the nanotribological behaviors of thin IL films and acquire insights into their potential in resolving the tribological problems of MEMS (Figure 3.16).

As shown schematically in Fig. 3.17, a dual-layer film containing both bonded and mobile fractions in IL-COOH layer was constructed on silicon substrates by a two-step process. Two kinds of amino-terminated SAMs which served as anchor layers were formed on hydroxylated silicon surfaces, respectively. Then, the incoming IL-COOH were chemically adsorbed onto amino-terminated SAMs by heat treatment, and formed two-phase structure composed of bonded and mobile IL-COOH molecules.

The formation of chemically bonded phase in IL-COOH layer improves nanotribological properties of the two kinds of dual-layer films as compared with single layer IL-COOH film, which is attributed to synergic effect between mobile phase and steady bonded phase. The protective bonded IL-COOH fraction greatly enhances the stability and antiwear properties of the film, while the mobile IL-COOH fraction serves as lubricant with friction reducing and self-replenishment properties. Generally, the packing density of the underlayer dictates packing density of the overlayer. N-[3-(trimethoxylsilyl)propyl] ethylenediamine (DA) molecules with longer chains as anchors form more densely packed and orderly SAM as compared with APS, thus more IL-COOH molecules are chemically grafted to DA anchor layer, which produce more densely packed bonded phase and reduce meniscus effect resulted from excessive mobile molecules. These characteristics of DA-IL lead to its lowest friction coefficient among studied dual-layer films. The improved durability of DA-IL film is closely related to high load-carrying capacity of more densely packed and ordered bonded phase. Furthermore, the more interlinked hydrogen bonding further strengthens immobile fraction of dual-layer film.

Enhancement of Nanotribology and Wettability by Surface Textures in Adhesion Resistant

Regular Surface Textures

Nature often uses topographic patterning to control interfacial interactions, such as adhesion and release. Examples range from lotus leaf, gecko to jumping spider. Each example demonstrates that additional to chemistry and material properties, geometric structure is also critical for optimizing interfacial design. Although nature has provided guidance, little is known of how topographic patterns can be intelligently used not only to enhance adhesion but also more importantly to tune adhesion. To tune adhesion with patterns, we must understand how material properties and pattern structure interact. Surface textures and chemical modification are commonly used in magnetic disk drives and MEMS to reduce friction and adhesion in order to reduce the possibility of mechanical failure [55–58]. A number of fabrication methods were used to generate micro/nano-hierarchical structures, including laser/plasma/chemical etching [59], soft photolithography [60], sol–gel processing and solution casting [61], electrical deposition [62], dip-pen printing [63], AFM local anodic oxidation [64], and so on.

Zhao et al. [65] prepared hierarchical structures by replication of textured silicon surfaces using polydimethylsiloxane (PDMS) and self-assembly of alkanethiol [$CH_3(CH_2)_9SH$] to create hydrophobic surface and to improve nanotribological properties of MEMS. As shown in Fig. 3.18, the fabrication technique is a low cost, two-step process, which provides flexibility in fabrication of various hierarchical structures. The textured surface with nano-hierarchical structures can be tailored by adjusting the depth and fractional surface coverage of cylinder hole.

For the adhesive force values there is a decrease when the pillar height and fractional surface coverage increases. Adhesive force also decreased greatly after chemical modification. Compared with the nanopatterned Au surface, the Au surface with micro/nanopillar textures greatly improved the adhesive properties and showed lower adhesive forces. Among the Au surfaces with textures, textured surface with the lowest height of 20 nm were fabricated, and chemical

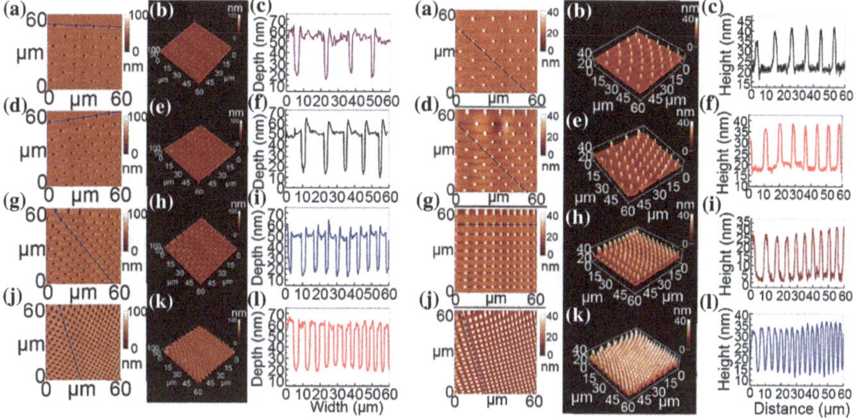

Fig. 3.18 Surface morphologies of template and textured surfaces. Reproduced with permission from Ref. [65]. Copyright (2010) American Chemical Society

modification with ODT-SAMs can lower adhesive force. The results indicate that adhesive force is closely related to the real contact area between the tip and surface, larger area lead to increase adhesive force. With the increase in pillar height and fractional surface coverage, the tip traveling between the pillars results in the decrease of the contact area, responsible for decreased adhesive force. Furthermore, when the solid surfaces were hydrophilic, they would easily form meniscus by the adsorbed water molecules, thus they had larger adhesive force. However, when the surfaces were hydrophobic, they would show lower adhesion (Fig. 3.19).

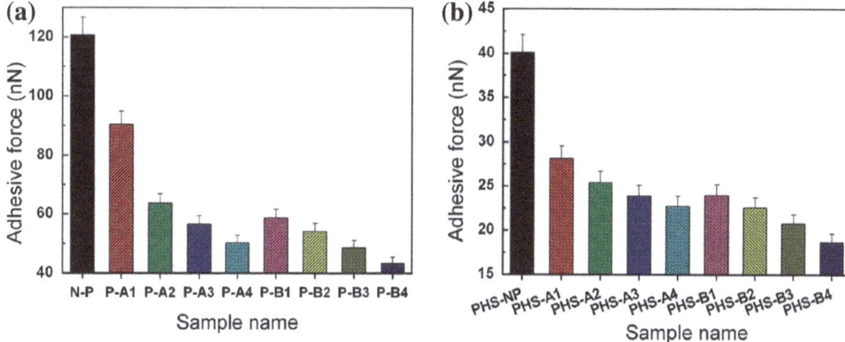

Fig. 3.19 Adhesive forces between AFM tip and Au micro/nano patterned surfaces with different height and surface coverage **a** before and **b** after SAMs chemical modification at room temperature and a relative humidity of 30–40 %. Reproduced with permission from Ref. [65]. Copyright (2010) American Chemical Society

Biomimetic Surface Textures

Functional surfaces with biomimetic micro textures have aroused much interest because of their great advantages in applications such as hydrophobic, anti-adhesion etc. For example, some plant leaves and bodies of animals are known to be hydrophobic in nature because of their intrinsic geometric microstructure. In particular, lotus leaf, on which the water contact angle is larger than 150°, can carry effortlessly the contaminations attached to the leaf when the surface is slightly tilted, which shows self-cleaning function and low hysteresis.

Wang et al. [66] reported three surface micro textures of rice leaf, lotus leaf and snake skin, which were duplicated on surface by combining duplication and electroplating methods. Firstly, a cellulose film is used to replicate the surface micro textures of the biological sample to obtain a negative impression of the biomimetic textures [67, 68]. A metallic layer is electrodeposited on the top of the cellulose film. Then the positive replicas of the original living creature were obtained after removing the cellulose film. Using this method, they successfully duplicated the surface microtextures of the rice leaf, lotus leaf and the snake skin on surface and evaluated wettability of the surfaces (Fig. 3.20). Zhao et al. [69] also used a simple, efficient, and highly reproducible method for producing large-area positive and negative lotus and rice leaf topography on Au surfaces based on PDMS to enhance hydrophobicity.

Mo et al. [70] successfully fabricated biomimetic textures onto silicon surface by local anodic lithography. Furthermore, the dimensions of biomimetic textures can be precisely controlled by controlling pulsed bias voltage, pulse width and RH, as shown in Fig. 3.21.

In this approach, the surfaces of dung beetle and rice leaf were replicated on H-passivated Si surface. The experimental results show that the lowest value of the

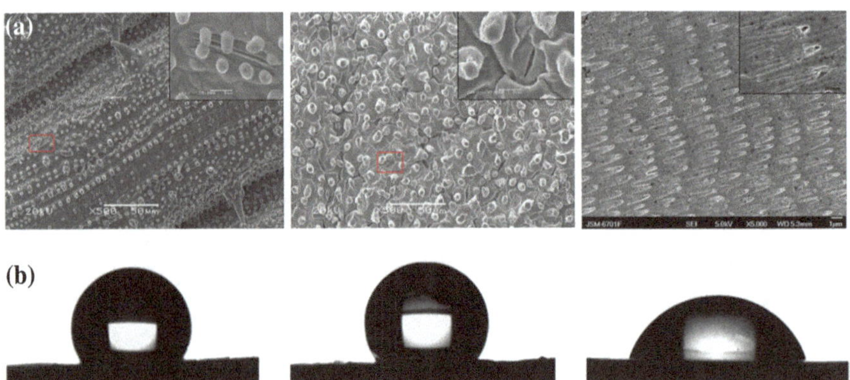

Fig. 3.20 **a** SEM images of surface replica of rice leaf, and on surface. The insets are the high magnification images. **b** Water droplet on surface replica with different textures (157° for rice leaf, 161° for lotus leaf and 65° for snake skin). Reproduced with permission from Ref. [66]. Copyright (2010) Elsevier B. V

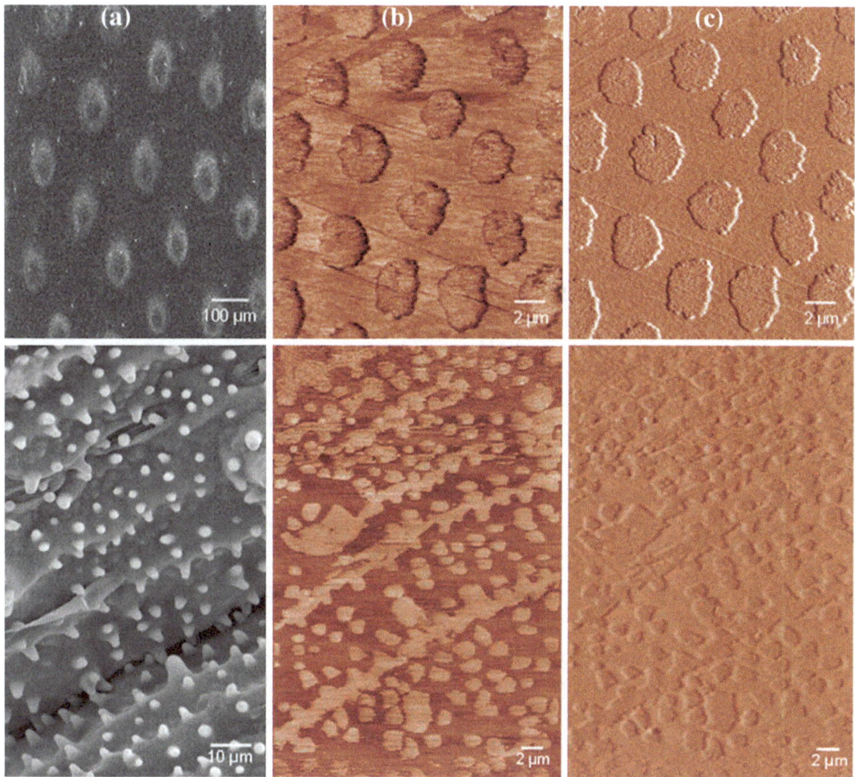

Fig. 3.21 The SEM image **a** of the surface replica of dung beetle and rice **b** Topographic scan of the replica **c** Corresponding frictional force image of (**b**). Reproduced with permission from Ref. [70]. Copyright (2010) Elsevier B. V

Fig. 3.22 Adhesive forces between AFM colloidal probe and surfaces of bare H-passivated Si, rice leaf texture and dung beetle texture. Reproduced with permission from Ref. [70]. Copyright (2010) Elsevier B. V

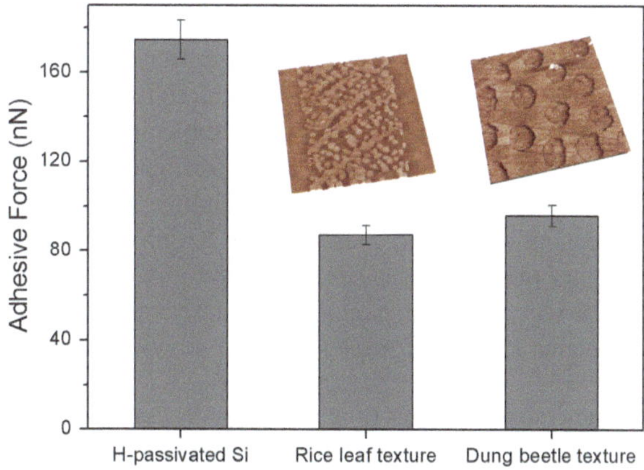

height of the biomimetic nanotexture was about 1 nm. The H-passivated Si treated with biomimetic nanotextures exhibit better adhesive resistance than untreated Si at nanoscale (Fig. 3.22). It is expected that this approach could be extended to duplicate other biological and artificial template surfaces on silicon surface. These surfaces with special nanotextures are of great importance for MEMS practical applications such as microhydromechanics, wettability, and biochips.

Summary and Outlook

Part of the excitement in thin lubricant film is due to the great intellectual simplification associated with the routine way. In this chapter, we have tried to introduce those ideas and concepts. Perfluorinated SAMs have shown remarkably better lubrication and anti-adhesion properties. A comparative research is presented on the surface and nanotribological properties of FC and HC SAMs on aluminum-coated silicon substrate formed by chemical vapor deposition. Furthermore, the influence of environmental conditions, such as RH and temperature, on tribological performance of these SAMs, was investigated. The FC SAMs show obvious environmental independence. In addition, dual-layer film exhibits better anti-wear durability than single monolayer in nanoscale.

MACs and ILs also are potential lubricants for MEMS and space application due to their extreme low volatility. In recent progress we have described important fields of boundary lubrication and the friction of single asperity contact. In ambient conditions, as well as in many tribological applications, surfaces are often covered by a thin film which modifies their tribological properties. Liquid menisci may form, increasing the adhesive force between the contact surfaces. The physical and chemical interactions between the surfaces are affected by the presence of water which can act as a lubricant. The tribological properties of the surface can also be changed directly by covering the surface with a monolayer of organic materials (SAMs with different function group or ILs with different ions). Meanwhile, surface texture is a prospective physical approach to enhance wettability and nanotribology in adhesion resistant conditions.

Summarizing, nanotribology of molecular thin films is a young and emerging field that is maturing fast, as the experiments described show. Due to never ending trend to miniaturization, understanding friction at nanoscale will become of increasing importance, since as the length scale is reduced, friction force become stronger relative to surface force, and thin films may be the only way to lubricate. If a bridge between nano- and macroscopic tribology is found, thin film with a molecular thickness might improve more efficiency and durability of MEMS.

Acknowledgments This work was funded by the National Natural Science Foundation of China (NSFC) under Grant No. 11172300 and 51105352. Prof. Mingwu Bai and Prof. Qunji Xue are greatly acknowledged for their constant support and encouragements to carry out this research work.

The authors want to express their sincere gratitude to Dr. Chongjun Pang, Dr. Jianqi Ma, Dr. Min Zhu, Dr. Wenjie Zhao, Dr. Ying Wang, Dr. Jibin Pu for their great efforts and valuable discussions.

References

1. Mo, Y., Bai, M.: Surf. Interface Anal. **41**, 602 (2009)
2. Hutt, D.A., Leggett, G.J.: Langmuir **13**, 2740 (1997)
3. Wallace, R., Chen, P., Henck, S., Webb, D.: J. Vac. Sci. Technol. A **13**, 1345 (1995)
4. Chance, J.J., Purdy, W.C.: Langmuir **13**, 4487 (1997)
5. Miura, Y.F., Takenaga, M., Koini, T., Graupe, M., Garg, N., Graham, R.L., Lee, T.R.: Langmiur **14**, 5821 (1998)
6. Xiao, X.D., Hu, J., Charych, D.H., Salmeron, M.: Langmuir **12**, 235 (1996)
7. Ren, S., Yang, S., Zhao, Y., Zhou, J., Xu, T., Liu, W.: Tribol. Lett. **13**, 233 (2002)
8. Ruehe, J., Novotny, V.J., Kanazawa, K.K., Clarke, T., Street, G.B.: Langmuir **9**, 2383 (1993)
9. Patton, S.T., William, D.C., Eapen, K.C., Zabinski, J.S.: Tribol. Lett. **9**, 199–209 (2000)
10. Rye, R.R., Nelson, G.C., Dugger, M.T.: Langmuir **13**, 2965–2972 (1997)
11. Hsu, S.M.: Tribol. Int. **37**, 537–545 (2004)
12. Tam-Chang, S.-W., Biebuyck, H.A., Whitesides, G.M., Jeon, N., Nuzzo, R.G.: Langmuir **11**, 4371 (1995)
13. Clegg, R.S., Hutchison, J.E.: J. Am. Chem. Soc. **121**, 5319 (1999)
14. Clegg, R.S., Reed, S.M., Smith, R.K., Barron, B.L., Rear, J.A., Hutchison, J.E.: Langmuir **15**, 8876 (1999)
15. Chambers, R.C., Inman, C.E., Hutchison, J.E.: Langmuir **21**, 4615 (2005)
16. Song, S., Ren, S., Wang, J., Yang, S., Zhang, J.: Langmuir **22**, 6010 (2006)
17. Song, S., Zhou, J., Qu, M., Yang, S., Zhang, J.: Langmuir **24**, 105 (2008)
18. Mo, Y., Zhu, M., Bai, M.: Colloids Surf. A: Physicochem. Eng. Aspects **322**, 170 (2008)
19. Ren, S.L., Yang, S.R., Zhao, Y.P.: Langmuir **19**, 2763 (2003)
20. Ma, J., Pang, C., Mo, Y., Bai, M.: Wear **2007**, 263 (1000)
21. Mo, Y., Bai, M.: J. Phys. Chem. C **112**, 11257 (2008)
22. Fleming, M.S., Walt, D.R.: Langmuir **17**, 4836 (2001)
23. Love, J.C., Estroff, L.A., Kriebel, J.K., Nuzzo, R.G., Whitesides, G.M.: Chem. Rev. **105**, 1103 (2005)
24. Venier, C.G., Casserly, E.W.: US 4,929,782, 1990
25. Venier, C.G., Casserly, E.W.: Lubr. Eng. **47**, 586 (1991)
26. Dube, M.J., Bollea, D., Jones Jr, W.R., Marrcheti, M., Jansen, M.J.: Tribol. Lett. **15**, 3–8 (2003)
27. Young, T.: Philos. Trans. R. Soc. Lond. **95**, 65 (1805)
28. Wang, Y., Mo, Y., Zhu, M., Bai, M.: Tribol. Trans. **53**, 219 (2010)
29. Ma, J., Liu, J., Mo, Y., Bai, M.: Colloids Surf A: Physicochem. Eng. Aspects **301**, 481 (2007)
30. Jian, S.R., Fang, T.H., Chuu, D.S.: J. Phys. D Appl. Phys. **38**, 2432 (2005)
31. Garcia, R., Martinez, R., Martinezz, J.: Chem Soc Rev **35**, 29 (2006)
32. Mo, Y.F., Wang, Y., Bai, M.W.: Phys. E **41**, 146 (2008)
33. Mo, Y., Wang, Y., Pu, J., Bai, M.: Langmuir **25**, 40 (2009)
34. Ducker, W.A., Senden, T.J., Pashley, R.M.: Nature **353**, 239 (1991)
35. Hagiwara, R., Ito, Y., Fluorine, J.: Chem. **105**, 221 (2000)
36. Welton, T.: Chem. Rev. **99**, 2071 (1999)
37. Wasserscheid, P., Keim, W.: Angew. Chem. Int. Ed. **39**, 3772 (2000)
38. Earle, M.J., Seddon, K.R.: Pure Appl. Chem. **72**, 1391 (2000)
39. Nakashima, T., Kawai, T.: Chem. Commun. **12**, 1643 (2005)
40. Ye, C.F., Liu, W.M., Chen, Y.X., Yu, L.G.: Chem. Commun. **21**, 2244 (2001)
41. Liu, W.M., Ye, C.F., Gong, Q.Y., Wang, H.Z., Wang, P.: Tribol. Lett. **13**, 81 (2002)

42. Palacio, M., Bhushan, B.: Adv. Mater. **20**, 1194 (2008)
43. Yu, G., Zhou, F., Liu, W.M., Liang, Y.M., Yan, S.: Wear **2006**, 260 (1076)
44. Mo, Y., Bo, Y., Zhao, W., Bai, M.: Appl. Surf. Sci. **255**, 2276 (2008)
45. Yu, B., Zhou, F., Mu, Z., Liang, Y.M., Liu, W.M.: Tribol. Inter. **39**, 879 (2006)
46. Bhushan, B., Palacio, M., Kinzig, B.: J. Colloid Interface Sci. **317**, 275 (2008)
47. Zhou, F., Liang, Y., Liu, W.: Chem. Soc. Rev. **38**, 2590–2599 (2009)
48. Zhu, M., Yan, J., Mo, Y., Bai, M.: Tribol. Lett. **29**, 177 (2008)
49. Mo, Y., Zhao, W., Zhu, M., Bai, M.: Tribol. Lett. **32**, 143 (2008)
50. Sinha, S.K., Kawaguchi, M., Kato, T., Kenedy, F.E.: Tribol. Int. **36**, 217 (2003)
51. Zhao, W., Wang, Y., Wang, L., Bai, M., Xue, Q.: Colloids Surf. A: Physicochem. Eng. Aspects **361**, 118 (2010)
52. Xiao, X.D., Hu, J., Charych, D.H., Salmeron, M.: Langmuir **12**, 235 (1996)
53. Choi, J., Kawaguchi, M., Kato, T.: Tribol. Lett. **15**, 353 (2003)
54. Pu, J., Huang, D., Wang, L., Xue, Q.: Colloids Surf. A: Physicochem. Eng. Aspects **372**, 155 (2010)
55. Marchetto, D., Rota, A., Calabri, L., Gazzadi, G.C., Menozzi, C., Valeri, S.: Wear **265**, 577 (2008)
56. Pettersson, U., Staffan, J.: Tribol. Int. **36**, 857 (2003)
57. Wakuda, M., Yamauchi, Y., Kanzaki, S., Yasuda, Y.: Wear **254**, 356 (2003)
58. Suh, A.Y., Lee, S.C., Polycarpou, A.A.: Tribol. Lett. **17**, 739 (2004)
59. Laws, G.M., Handugan, A., Eschrich, T., Boland, P., Sinclair, C., Myhajlenko, S., Poweleit, C.D.: J. Vac. Sci. Technol. B **25**, 2059 (2007)
60. Xu, Q.B., Tonks, I., Fuerstman, M.J., Love, J.C., Whitesides, G.M.: Nano. Lett. **4**, 2509 (2004)
61. Bhushan, B., Kocha, K., Jung, Y.C.: Soft Matter **4**, 1799 (2008)
62. Shirtcliffe, N.J., McHale, G., Newton, M.I., Chabrol, G., Perry, C.C.: Adv. Mater. **2004**, 16 (1929)
63. Zheng, Z., Daniel, W.L., Giam, L.R., Huo, F., Senesi, A.J., Zheng, G., Mirkin, C.A.: Angew. Chem. Inter. Ed, **48**(41), 7626
64. Garcia, R., Martinez, R.V., Martinez, J.: Chem. Soc. Rev. **35**, 29 (2006)
65. Zhao, W., Wang, L., Xue, Q.: ACS Appl. Mater Interfaces **2**, 788 (2010)
66. Wang, Y., Mo, Y., Zhu, M., Bai, M.: Surf. Coat. Technol. **203**, 137 (2008)
67. Zhao, X.M., Xia, Y., Whitesides, G.M.: J. Mater. Chem. **1997**, 7 (1069)
68. Sun, M.H., Luo, C.X., Xu, L.P., Ji, H., Ouyang, Q., Yu, D.P., Chen, Y.: Langmuir **21**, 8978 (2005)
69. Zhao, W., Wang, L., Xue, Q.: J. Phys. Chem. C **114**, 11509 (2010)
70. Mo, Y., Bai, M.: J. Colloid Interface Sci. **333**, 304 (2009)

Chapter 4
Mechanical Properties and Deformation Behavior of Ni Nanodot-Patterned Surfaces

Min Zou and Hengyu Wang

Abstract Surface nano-texturing has attracted great attention due to its potential for significantly reducing adhesion and friction in micro-electro-mechanical systems (MEMS) and nano-electro-mechanical systems (NEMS). However, severe deformation of the nano-textures was also observed during tribological testing of nano-textured surfaces (NTSs). Therefore, understanding the mechanical properties and deformation behavior of nano-textures and nano-textured surfaces is of critical importance to the development of durable NTSs for MEMS/NEMS applications. Here, we review our recent work in understanding the mechanical properties and deformation behavior of Ni nanodot-patterned surfaces (NDPSs) on silicon substrates. First, the benefit of nanoscale surface-texturing for MEMS/NEMS application and the size-dependent mechanical properties of nanostructures are introduced. Second, various experimental techniques are described, which include methods of fabricating and characterizing Ni NDPSs as well as studying the mechanical properties and deformation behavior of NDPSs using nanoindentation. Third, methods of determining mechanical properties of Ni nanodots from nanoindentation experiments are presented. Fourth, a multi-asperity contact model for studying the nanoindentation deformation behavior of the Ni NDPSs is described. Fifth, simulation results from the multi-asperity contact model are compared to nanoindentation experiments and validated by the experimental results. Finally, the model is used to study effects of substrate, surface roughness, elastic modulus, and yield strength.

M. Zou (✉) · H. Wang
Department of Mechanical Engineering, University of Arkansas,
Fayetteville, Arkansas 72701, USA
e-mail: mzou@uark.edu

H. Wang
State Key Laboratory of Traction Power, Southwest Jiaotong University,
Chengdu 610031 Sichuan, China

S. K. Sinha et al. (eds.), *Nano-tribology and Materials in MEMS*,
DOI: 10.1007/978-3-642-36935-3_4,

Keywords Nanostructure • Nano-texture • Nano-textured surfaces • Nanodot-patterned surfaces • Anodized aluminum oxide template • Mechanical property • Elastic modulus • Hardness • Deformation

Contents

Introduction

Tribological Issues in Micro/Nano-Systems

Micro-electro-mechanical systems (MEMS) hold great promises to revolutionize nearly every product category. Significant progress has been made in the research and development of MEMS and, recently, in the emerging field of nano-electro-mechanical systems (NEMS) [1]. Commercial applications of MEMS/NEMS could be found in various industries. Examples include integrated accelerometers for air-bag deployment in vehicles and digital micromirror devices for projection displays. However, when a system shrinks to microscale, surface forces such as van der Waals force, capillary force, and electrostatic force become dominant due to the much higher surface-area-to-volume ratio than that of a macroscale system

[2]. The large surface-area-to-volume ratio of surface microstructures makes MEMS/NEMS devices particularly vulnerable to tribological issues, such as adhesion, friction, and wear [3, 4]. For example, adhesion of microstructures to substrates is a fundamental reliability issue in microdevices [3, 5]. Stiction and wear lead to failure of integrated accelerometers in automobile air bags [6] and digital micromirror devices in digital light processing equipment [7, 8]. These tribological issues will escalate as microscale systems are developed into nanoscale systems.

Surface Nano-Texturing to Solve Tribological Issues in MEMS

Various surface engineering techniques have been introduced to alleviate tribological issues in MEMS. These techniques and their pros and cons are described in detail in several recent review articles [3, 9–12]. One general approach is to avoid the formation of water layers on the surfaces of MEMS/NEMS by surface treatments that provide a stable hydrophobic surface, thereby eliminating capillary forces [13]. Another approach is to change the surface interaction forces by using surface roughening or texturing techniques to reduce the real area of contact [14, 15].

Considering the microstructures in MEMS are usually micron-sized, to effectively reduce adhesion and friction, the surface textures must be micron-/submicron-sized or preferably nano-sized. Compared to the micro-roughness produced by the laser and etching methods, nano-surface-texturing with isolated nanoislands has several unique potential advantages for tribological applications that involve ultra smooth surfaces, such as in MEMS/NEMS and computer hard drives. Firstly, the contact areas of nano-textured surfaces (NTSs) are smaller than that of micro-textured surfaces (MTSs) due to the significantly reduced size of the asperities. This reduced contact area can result in more significant reduction of adhesion and friction forces. Secondly, the nanometer-sized asperities are generally much harder than their micron-sized counterparts because of their nearly defect-free nature [16], which can considerably enhance the wear resistance of a surface. Moreover, the presence of nano-sized asperities increases the surface hydrophobicity of hydrophobic surfaces [17], which will lead to the reduction of meniscus-mediated adhesion and friction forces in a humid environment. Therefore, NTSs have greater potential to reduce catastrophic failures in MEMS/NEMS devices involving contacting surfaces.

Recently, using nano-textured surfaces (NTSs) to improve tribological performance of MEMS has started to attract attention of researchers [18–25]. Studies that compare the adhesion and friction performances of a NTS and MTSs provided experimental support for the anticipated benefits of NTSs [26].

Even though quite a few tribological studies on NTSs have shown that significant adhesion and friction reduction can be achieved by using NTSs, severe deformation of the nano-textures was also observed [20, 24, 25, 27]. Therefore, there is a need to develop deformation-resistant and durable NTSs that are suitable for tribological applications in MEMS/NEMS.

Mechanical Properties and Deformation Behavior of NTSs

Understanding the mechanical properties and deformation behavior of nano-textures and nano-textured surfaces is of critical importance to the development of durable NTSs for MEMS/NEMS applications, and to the fundamental understanding of friction, wear, and contact mechanics at the nanoscale. Up to date, many studies have been performed on thin films with micrometer or nanometer thicknesses and have shown that their mechanical properties are size-dependent and sometimes are quite different from those of their bulk counterparts [28–31]. The mechanical properties of nanometer-sized clusters, however, have received much less attention. A few reported such studies, including studies of elastic properties of individual nanometer-sized Au clusters [32, 33] and a study of the hardness of silicon nanospheres [16], have also shown that the mechanical properties of these nanostructures are size-dependent and sometimes they are much stronger and harder than their bulk counterparts. The nanostructures studied, however, are loosely deposited on the substrates. Since for MEMS/NEMS applications, nano-textures or patterned arrays of nano-textures are often engineered to permanently attach to a substrate of possibly a different material. The effects of substrates become significant when the textures are ultra-small. Therefore, understanding the mechanical properties of nano-textures as an integral part of the substrate is becoming increasingly important as the feature sizes of MEMS/NEMS devices continue to shrink. In addition to use materials with proper mechanical properties, NTSs should also have acceptable deformation under the applied load. Therefore, the contact load–deformation behavior of NTSs must also be fully understood.

In this chapter, we review our recent studies on the mechanical properties and deformation behavior of Ni nanodot-patterned surfaces (NDPSs) for potential tribological applications in miniaturized systems [34, 35]. In the following sections, we first describe the fabrication and characterization of Ni NDPSs on silicon substrates. We then present in detail nanoindentation experimental studies [34] to determine the mechanical properties and deformation behavior of the Ni NDPS, followed by numerical simulations [35] of nanoindentation of the Ni NDPS.

Experimental Techniques

Fabrication of Ni NDPS

NDPSs are surfaces covered with orderly distributed arrays of nanodots with approximately the same size and shape. Fabrication of such nanodot arrays has attracted considerable attention recently because these nanostructures show novel physical properties that lead to potential unique applications in electronic, optoelectronics, and magnetic storage [36–38]. Many techniques, such as electron-beam, interference, imprint, and soft lithography, as well as self-assembled anodic aluminum oxide (AAO) template methods, have been used to fabricate NDPS [39–44]. Among the existing techniques, the self-assembled AAO template technique has the advantages of being parallel and scalable; moreover it can be combined with conventional deposition and/or etching processes to pattern surfaces with various nanostructured topologies and materials. For example, using the AAO templates as masks, various metal nanodot arrays with nearly identical dot size, height, and spacing can be fabricated by thermal evaporation of metal sources [45–47]. The size and spacing of the nanodots can be controlled by varying the anodization conditions of the AAO templates while the height of the nano-dots can be partially controlled by the evaporation process parameters. These AAO template techniques are therefore particularly suitable for producing relatively large area, nano-patterned surfaces for tribological applications in miniaturized systems.

Sample processing starts with an AAO template fabricated using a two-step anodization process on a pure (99.999 %) Al foil described in detail elsewhere [48, 49]. First, the Al foil was anodized in 0.3 M oxalic acid at 40 V and 1 °C for 15 h to grow a thick porous oxide layer. The resulting AAO film was then chemically stripped from the Al foil and a secondary anodization in 0.3 M oxalic acid at 40 V and 1 °C for 5 min was carried out. Using this two-step technique, good ordering was obtained over micron-sized regions and resulted in an AAO film approximately 300 nm in thickness with 50 nm pore diameters spaced 100 nm apart. A through-hole mask was prepared by separating the AAO film from the Al foil in a saturated $HgCl_2$ solution and removing the bottom alumina barrier layer in 5 wt % phosphoric acid at 30 °C for 34 min. The remaining AAO layer was then lifted off onto a Si substrate. This resulted in a through-hole AAO mask approximately 3×3 mm in size weakly bonded by van der Waals force to a Si substrate.

Next, a 100 nm layer of Ni was deposited onto the Si through the AAO mask by thermal evaporation using an Edwards E306A system with base vacuum of $\sim 10^{-6}$ Torr. The Ni source was placed directly underneath the sample with a sample-to-source distance of approximately 18 cm. This allowed a significant fraction of the evaporated Ni (with an estimated angular dispersion of $\sim$100/5 mm = 20:1) to travel down the pores of the mask (which had an aspect ratio of 300/50 nm = 6:1). Under these conditions, a 120 nm thick layer of Ni coated on top of the AAO mask results in 75 nm high dots. After evaporation, the AAO film was removed by means of a wet chemical etch (a 1:1 mixture of 6 wt. % phosphoric acid: 1.8 wt. % chromic acid) leaving behind a well-ordered array of conical-shaped Ni dots [20]. The resulting pattern consisted of a well-ordered hexagonal array of conical-shaped Ni nanodots. Alignment markers were fabricated on the Ni NDPS using photolithography. These markers served as references for placing the indents or scratches for the nanomechanical testing and for locating the indents or scratches for scanning electron microscope (SEM) and atomic force microscope (AFM) characterizations after testing.

Characterization of Ni NDPS

Sample topography characterizations before and after indentation testing were performed using a JEOL JSM-880 SEM and a Topometrix Explorer AFM in non-contact mode using an ultra-sharp silicon tip (NSC-35/no Al with specified full cone angle $<30°$ and radius <10 nm) manufactured by MikroMasch. Figure 4.1a illustrates a top-down SEM micrograph of the well-ordered array of Ni nanodots. The hexagonal arrangement of the nanodots results from the hexagonal pore structure in the AAO template mask and exhibits good ordering over micron-sized domains. The inset in the upper right-hand corner of Fig. 4.1a shows a 45° oblique angle view of the Ni NDPS illustrating the conical shape of the nanodots (left) and a view of the same region vertically stretched by 41 % to account for the foreshortened view at 45° (right). Figure 4.1b shows an AFM image of the Ni NDPS. Figure 4.1c shows line cuts A and B [labeled in (b)] plotted on equi-scaled vertical and horizontal axes. A Ni nanodot schematic and foreshortening-corrected image are shown for comparison.

Figure 4.1 illustrates the difficulties in imaging the Ni NDPS with AFM, even with ultra-sharp tips in non-contact. This difficulty arises because the radius of curvature of such ultra-sharp tips ($\sim$10 nm) is not much smaller than the radius of curvature of each nanodot ($\sim$30 nm). As seen in Fig. 4.1c, the line cut of well-ordered nanodots, labeled B, indicates that the tip does not reach the substrate between the nanodots. Only when there is a missing nanodot, as in line cut A, can such a tip actually reach the substrate allowing an accurate measurement of nanodot height. The situation is improved with super-sharp tungsten tips and best for carbon nanotube tips [50], however, these tips do not last long while imaging the Ni NDPS. We find, for routine assessment of Ni nanodot geometry over a wide area, SEM images taken at oblique angles (45°, 60° and even 90°) allow the geometry to be accurately assessed. For example, the right inset image in Fig. 4.1a, with foreshortening correction applied, allows tip height, radius of curvature, and base cone angle to be accurately and easily measured. Based on such images, the height, base diameter, base cone angle, apex radius of curvature and dot-to-dot spacing were found to be 70 nm, 75 nm, 75°, 30 nm, and 100 nm, respectively (all values have 10 % error). Figure 4.1c compares a schematic based on this average nanodot measurement to an AFM nanodot line cut and a foreshortening-corrected typical nanodot. The agreement is good, although the line cut for this fresh-out-of-the-box tip indicates a tip radius of curvature somewhat greater than that specified.

Ni nanodot crystal structure was characterized by plan-view transmission electron microscopy (TEM) and nanobeam diffraction (NBD) using a JEOL 2000FX and JEOL 2010F TEM, respectively. Figure 4.2a shows a bright-field TEM image of a typical area of ordered Ni nanodots. Figure 4.2b shows a higher magnification image of a single Ni nanodot. The dark areas of contrast in the nanodot are duc to grain orientations that satisfy the Bragg condition for strong electron diffraction and give an estimate of approximately 5 nm for the Ni grain

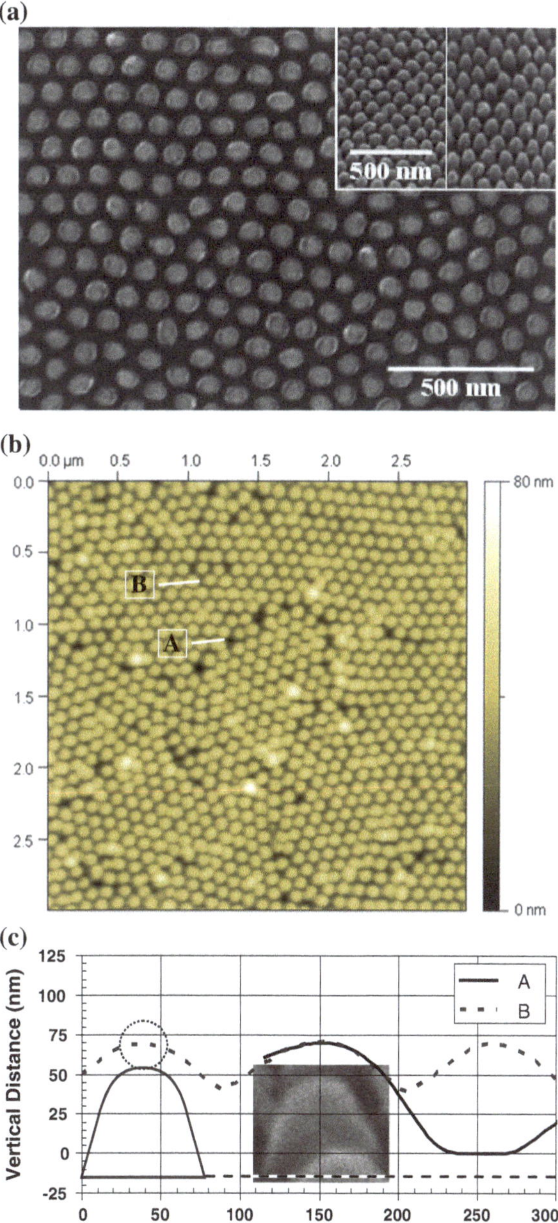

Fig. 4.1 SEM and AFM images and line cuts of the Ni NDPS. **a** SEM top-down view. Inset shows a 45° oblique angle view *(left)* and a vertically stretched view of the Ni nanodots *(right)*. **b** AFM top-down view. **c** *Line cuts A* and *B* [labeled in (**b**)] plotted on equi-scaled vertical and horizontal axes. *A* nanodot schematic and image are shown for comparison (*Source* Data from [34], with kind permission from IOP)

size. Figure 4.2c shows the NBD pattern of the nanodot shown in Fig. 4.2b. The diameter of the beam used for NBD was approximately 70 nm allowing for the diffraction pattern from only one nanodot to be observed. The diffraction pattern

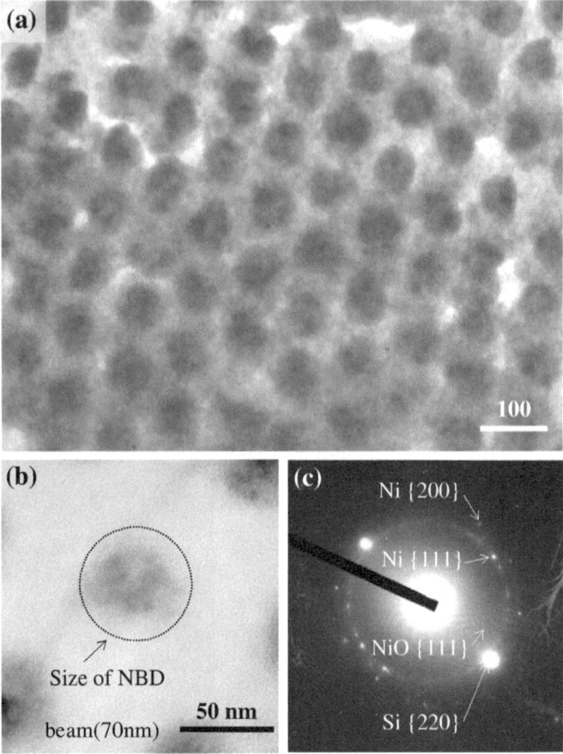

Fig. 4.2 Ni nanodot crystal structure. **a** *bright-field* TEM image of the ordered array of Ni nanodots. **b** higher magnification *bright-field* TEM image of a single Ni nanodot. **c** nanobeam diffraction pattern of the *dot* shown in (**b**) (*Source* Data from [34], with kind permission from IOP)

shows rings of intensity for the {111} and {200} planes of Ni indicating that the Ni nanodots are polycrystalline and intensity spots for the {220} planes of the single-crystal Si substrate. No intensity spots or rings were observed for NiO, so we estimate that the nanodots are covered by a native oxide of thickness <2 nm as observed for uniform thin films [51].

Nanoindentation Experimental Studies

Mechanical property characterization of NDPSs presents tremendous challenge because of the difficulties in preparing the NDPSs, applying ultra low loads on the nanodots, measuring the minute deformations caused by the applied load, and interpreting the nanoindentation data. Nanoindenter and AFM are two of the most popularly used tools for characterizing mechanical properties at the nanometer scale. A state-of-the-art nanoindenter (TriboIndenter from Hysitron, Inc.) was used in our research because the TriboIndenter offers several advantages over an AFM. First, it can apply an indentation load perpendicular to the sample surface, whereas inaccuracy is frequently introduced in an AFM indentation measurement due to the inherent angle between the AFM cantilever and the sample surfaces, and the uncertainties in the AFM cantilever stiffness. Second, a much higher indentation load can be applied by the TriboIndenter, which enables the study of commonly occurring plastic deformations in nanoscale asperities, while these types of studies are more limited with an AFM due to its relatively low cantilever stiffness. Third, since the TriboIndenter does not use a piezoelectric actuation in indentation force generation mechanisms like an AFM does, it can avoid the inherent nonlinearities of the piezoelectric ceramic in an AFM measurement. On the other hand, like the AFM, a TriboIndenter is capable of precisely positioning the tip with nanometer accuracy, and thus enables precise indentations at specific locations and easy locating of the indentations afterward for further investigation.

A Hysitron TriboIndenter was used for the nanoindentation testing. The TriboIndenter has force and displacement sensing capabilities in both vertical and lateral directions. The sensing system of the TriboIndenter consists of two three-plate capacitive force–displacement transducers with integrated electrostatic actuation functionality. The forces are applied electrostatically and the displacements are measured with a differential capacitor technique. The resolution is 0.02 nm with a 0.2 nm noise floor for the vertical displacement, 3 nN with a 100 nN noise floor for the vertical/indentation force, and 500 nN with a 5 μN noise floor for the lateral force. The electrostatic force constants were calibrated following the calibration procedures in the TriboIndenter equipment manual to yield accurate measurements of vertical/indentation and lateral forces. The TriboIndenter also has an optical microscope that allows one to easily locate the alignment markers on the sample's surface and the Ni nanodot-patterned areas for nanoindentation testing.

Conical diamond indenter tips with nominal tip radii of curvature of 1 and 5 μm were used in the nanoindentation studies. Mechanical properties of the Ni NDPS were determined from nanoindentation experiments performed using the 1 μm tip, while load-deformation results were obtained from both 1 and 5 μm tips. All tests were conducted in air at a relative humidity of about 40 %.

Preliminary indentation experiments were performed to determine the appropriate indentation load profiles and the indention loads to be used. Since creep of Ni nanodots was observed under a constant load, a trapezoidal load profile was chosen to minimize this creep effect on their unloading behavior. Our preliminary study showed that creep is mostly complete after 2 s. Therefore a 2-second hold time was chosen for the study. The final indentation load profiles used were as follows: (1) the tip approached the sample from 50 nm above the sample surface in 7 s, (2) the tip indented the Ni nanodots under load control from 0 to a maximum indentation load in 10 s, (3) the tip is held at the maximum indentation loads for 2 s, and (4) the tip load is reduced to 0 μN load in 10 s.

For the tests using the 1 μm radius tip, ten different maximum indentation loads (from 10 to 500 μN) were chosen to induce a maximum indentation approach of up to about 50 nm–enough to study the elastic and severe plastic deformations of the Ni nanodots. Adjacent indents were separated by at least 10 μm to avoid the influence of one indent on another. For the tests with the 5 μm radius tip, seven loads (from 20 to 1000 μN) were chosen, and the minimum spacing between indents was 20 μm. For each load, the indentation tests were repeated three times to ensure reliable and repeatable data was collected. The relatively large conical diamond tips were used to ensure that the tip will indent on the Ni nanodots, not the substrate. A simple calculation based on tip geometry and Ni nanodot location pattern suggests that even if the 1 μm tip does not line up with the top of a Ni nanodot at the beginning of the indentation, it will come into contact with adjacent nanodots when the indenter moves forward by at most 1.6 nm. Therefore, the 1 μm tip will always indent the Ni nanodots, not the substrate. The same is obviously true for the 5 μm tip.

The deformations of the Ni NDPS after nanoindentation were characterized by both SEM and AFM. SEM characterization was performed before the AFM measurements were taken to avoid any damage that may be introduced by the AFM measurement. Location of the indents was facilitated by 8 μm long scratches used for tribological studies [27] near the indents. SEM micrographs provide clear 2-D and 3-D views of the Ni nanodots, while AFM measurements provide more accurate indentation depth information. Figure 4.3 illustrates that with the deformation caused by a scratch nearby, the deformation of Ni nanodots after a 150 μN indentation test can be easily identified from the SEM micrographs and AFM images. As can be seen from Fig. 4.3a and b, SEM micrographs provide top-down and oblique angle views of the shape and spacing of the Ni nanodots. It can also be seen from Fig. 4.3c that AFM images show height changes more clearly as indicated by the bright spots. AFM measurements can provide quantitative indentation deformation values in the height direction. Unfortunately, for loads less than 70 μN, the indentation and scratch marks were hard to locate from SEM and AFM measurements. This is because such loads only produce very small deformations (about 10 nm maximum), which were comparable to the height variation of Ni nanodots, especially at grain boundaries where the height variation of Ni nanodots is the largest. Therefore, the results based on the deformation study are only obtained for indentation loads of 70 μN and above.

Mechanical Properties of Ni NDPS Determined by Nanoindentation Experiments

Elastic Modulus and Critical Shear Stress of the Ni NDPS

As depicted in Fig. 4.4, during a nanoindentation test using a tip with 1 μm radius, as the tip initiates contact with the sample, there are three most likely scenarios with respect to the indenter tip positioning relative to the Ni NDPS, assuming an idealized tip shape and uniform Ni nanodots radii, shape, and spatial distribution. In the first scenario, the indenter tip is centered near the top of a nanodot (point A) and contacts only one nanodot at the beginning of the indentation. Besides the

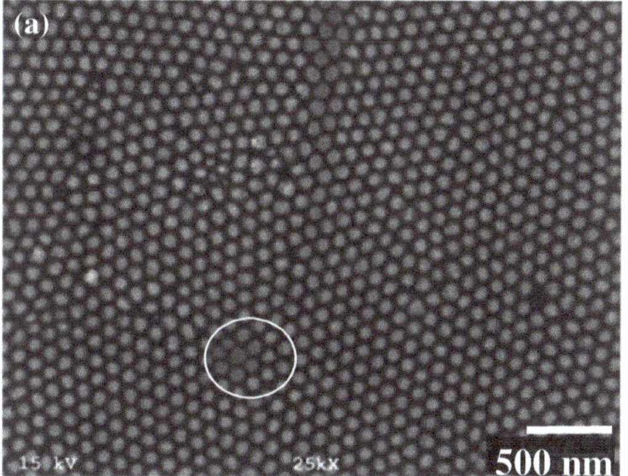

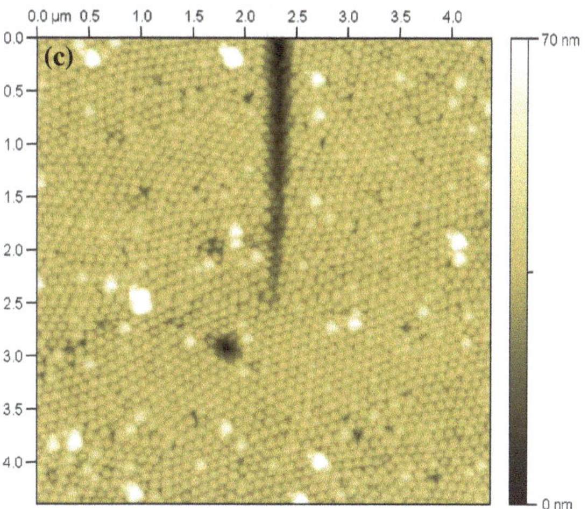

Fig. 4.3 Deformation of the Ni nanodots after a 150 μN indentation test. **a** SEM *top-down* view, **b** SEM 60° oblique *angle view*, and **c** AFM *top-down* view (*Source* Data from [34], with kind permission from IOP)

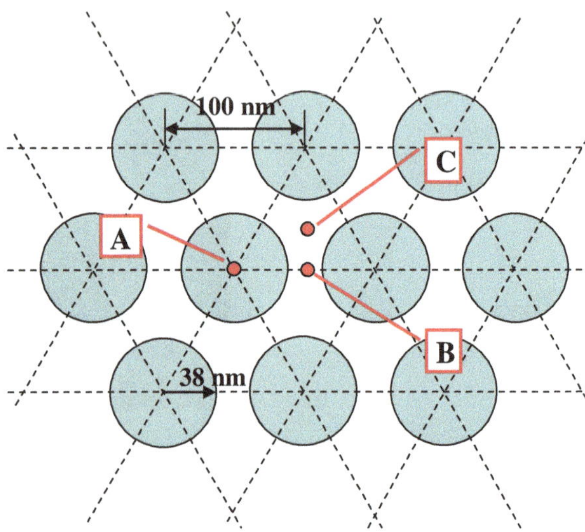

Fig. 4.4 Illustration of *three contact* scenarios between the indenter and the NDPS at different positions (*Source* Data from [34], with kind permission from IOP)

nanodot in contact, there are six nanodots close to point A. The tip will touch the 6 surrounding nanodots after it reaches the indentation depth of 4.9 nm (using 1.0 μm and 30 nm for the radii of curvature of the indenter tip and the Ni nanodot, respectively, and 100 nm dot-to-dot spacing). In the second scenario, the indenter tip is centered near point B. It contacts two nanodots at the beginning of the indentation, and will contact two more nanodots after it reaches the indentation depth of 2.4 nm. In the third scenario, the indenter tip is centered near point C where the tip contacts three nanodots initially, and will contact three more nanodots after an indentation depth of 4.9 nm. In reality, the three scenarios may not happen exactly at the theoretical indentation depth predicted above due to the variations in the Ni nanodots' radii, shape, and spatial distribution. However, we expect that there will be either one, two, or three dots in contact with the tip for indentation depths less than 4.9, 2.4, and 4.9 nm for scenarios 1, 2, and 3, respectively.

The nanoindentation load–displacement curve contains a wealth of information regarding the material deformation process, from which the mechanical properties of materials can be deduced. In the indentation of Ni NDPS, the measured indentation displacement reflects the deformations of nanodot top, δ_{top}, nanodot base, δ_{base}, and the silicon substrate, δ_{sub}.

At very small displacements, the load-deformation relationship of the contact between the top of the Ni nanodot and the indenter tip should approximately follow the classical Hertzian elastic contact theory [52]. However, as the load increases, the loading curve starts to deviate from the Hertzian fit due to additional Ni nanodots coming into contact with the tip or due to plastic deformation of the material. For a single Ni nanodot in contact with the tip, Hertzian contact theory for axisymmetric elastic bodies of parabolic profile yields the following relationship between the indentation load, L, and the deformation of the Ni nanodot top, δ_{top}:

$$L = \frac{4}{3} E^* R^{1/2} \delta_{top}^{3/2} \tag{4.1}$$

where $E^* = \left[(1 - \nu_{tip}^2)/E_{tip} + (1 - \nu_{Ni}^2)/E_{Ni}\right]^{-1}$ is the reduced modulus, and $R = \left[1/R_{tip} + 1/R_{top}\right]^{-1}$ is the reduced radius of curvature. E_{tip} and E_{Ni} are the elastic moduli, ν_{tip} and ν_{Ni} are the Poisson's ratios, and R_{tip} and R_{top} are the radius of curvature of the indenter and the Ni nanodot top, respectively.

It should be noted that Hertzian contact theory deals with two contacting asperities with no center-to-center misalignment. During the indentation of the Ni NDPS, the indenter tip is seldom perfectly aligned with a Ni nanodot. However, since the radius of the indenter tip is much larger than the radius and dot-to-dot spacing of the Ni nanodots, the effect of this center-to-center misalignment can be neglected and Hertzian contact theory can still be applied to the contact between the indenter tip and the nanodot. For multiple nanodots in contact with the tip, the load-displacement relationship in Eq. (4.1) becomes:

$$P = n_0 \frac{4}{3} E^* R^{1/2} \delta_{top}^{3/2} \tag{4.2}$$

where P is the indentation load and n_0 is the number of nanodots in contact. For the three scenarios illustrated in Fig. 4.4 with the indenter tip centered at points A, B, and C, $n_0 = 1, 2, 3$ for each scenario, respectively.

According to SEM and AFM measurements, the base of the Ni nanodot can be viewed as a truncated cone with height, $H = 40$ nm, the bottom radius $R_b = 38$ nm, and the slant angle, $\alpha = 75°$. Assuming a uniform load distribution in the base, the deformation can be calculated as

$$\delta_{base} = \int_0^H \frac{L}{E_{Ni} A(h)} dh = \frac{L}{\pi E_{Ni} \cot\alpha} \left(\frac{1}{R_b - H \cot\alpha} - \frac{1}{R_b} \right) \tag{4.3}$$

where L is the load applied on the nanodot and $A(h)$ denotes the cross sectional area at height h.

The deformation of the substrate under each nanodot consists of two parts. One part is the deformation caused by the distributed load under the nanodot itself and can be determined as [52]:

$$\delta_{sub1} = \frac{2(1 - \nu_{sub}^2)}{E_{sub}} p R_b \tag{4.4}$$

where $p = L/\pi R_b^2$ is the pressure applied by the nanodot. E_{sub} and ν_{sub} are the elastic modulus and Poisson's ratio of the silicon substrate. The other part of deformation comes from the deformation induced by the loads from one or two additional nanodots which are also in contact with the indenter tip. The deformation caused by the pressure on a nanodot that is a distance, D, from the point of interest is [52]:

$$\delta_{sub2} = \frac{4(1-v_{sub}^2)}{\pi E_{sub}} pD\left[\mathrm{E}\left(\frac{R_b}{D}\right) - \left(1 - \frac{R_b^2}{D^2}\right) \cdot \mathrm{K}\left(\frac{R_b}{D}\right)\right] \tag{4.5}$$

where $\mathrm{K}(R_b/D)$ and $\mathrm{E}(R_b/D)$ are complete elliptic integrals of the first and second kind with modulus (R_b/D). Assuming superposition of the deformations, the substrate deformation is

$$\delta_{sub} = \delta_{sub1} + \sum_{i=1}^{N} (\delta_{sub2})_i \tag{4.6}$$

where N is the number of additional nanodots in contact.

In the calculations to determine the deformation components δ_{top}, δ_{base}, and δ_{sub} using Eqs. (4.2), (4.3), and (4.6), the nanodots are assumed to have the same height. Thus, the indenter tip touches the nanodots simultaneously and the indentation load is evenly distributed on each nanodot in contact. The deformations of the nanodot top, the nanodot base and the substrate are calculated under each applied indentation load and the total indentation displacement is evaluated as the summation of the above three deformations.

The elastic modulus of Ni nanodots was obtained by fitting the initial part of the measured load–displacement curves, taking into account the deformations from the Ni nanodot base and the silicon substrate. Figure 4.5a shows three representative load–displacement curves for the Ni NDPS subjected to 70 μN indentation loads while Fig. 4.5b shows the initial portion of one of the indentation curves and the fitted lines for the three contact scenarios. The elastic modulus of the Ni nanodots can be estimated from the fitting of each of the three scenarios to the measured load-displacement curve. Note that there is a small load of about 1 μN at the beginning of the indentation. This pre-load was applied by the nanoindenter when the tip first initiated contact with the sample. To eliminate errors induced by this pre-load, each fitted line starts from the measured pre-load as shown in Fig. 4.5b.

The least-squares method was used to fit the experimental load-displacement curves. Curves were fitted up to an indentation depth small enough (<2.4 nm) such that the assumption of 1, 2, or 3 nandodots in contact with the tip is valid. Assuming 1141 GPa and 0.07 [53] for the elastic modulus and Poisson's ratio of the diamond indenter, respectively, 179 GPa and 0.25 for the elastic modulus and Poisson's ratio of the silicon substrate, respectively, 0.31 for the Poisson's ratio of the Ni nanodots, 1.0 μm and 30 nm for the radii of curvature of the tip and the Ni nanodots, respectively, the elastic modulus of Ni nanodots was estimated from Fig. 4.5b to be 496.8, 172.6, and 104.4 GPa for the first, second, and third scenario, respectively. It was assumed that the second scenario actually happened and the modulus of 172.6 GPa was accepted because the moduli fitted from the first and third scenarios were too far away from the reported modulus of bulk nanocrystalline Ni (186 GPa) or microcrystalline Ni (204 GPa) [54].

The fitting of the load-displacement curves was performed to find out the elastic modulus for all indentation measurements. Out of 28 good indentation curves, 7

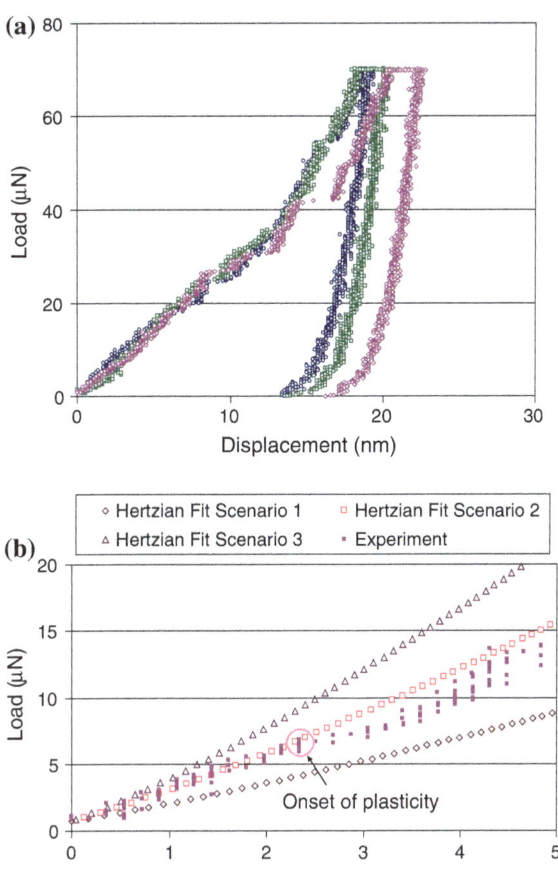

Fig. 4.5 Three load-displacement *curves* of the Ni NDPS subjected to 70 µN indentation load. **a** Three entire indentation curves showing the repeatability of the initial portion of the *indentation curves*. **b** The initial portion of one of the indentation curves with fitted *lines* for three contact scenarios (*Source* Data from [34], with kind permission from IOP)

were found to fit best with the first contact scenario, 13 were best fit with the second contact scenario, and 8 were best fit with the third scenario. The elastic modulus of Ni nanodots from all the indentation data was determined to be 159 ± 22 GPa, which is less than bulk nanocrystalline Ni (186 GPa) and bulk microcrystalline Ni (204 GPa) [54] values. Similarly, a decrease in elastic modulus, compared to bulk Au, was also reported in the nanoindentation of Au nanoislands [32, 33]. Zhou and Huang have shown that electron redistribution and atomic coordination on the surface are competing factors for determining the surface stiffening or softening during elastic deformation [55]. Using a combination of molecular statics and ab initio calculations, they concluded that the surface can be softer or stiffer than the bulk as the length scale decreases. Therefore, in the case of Ni nanodots, due to these surface effects, a smaller elastic modulus than that of bulk microcrystalline Ni is not surprising.

There could be potential error induced when estimating the elastic modulus by fitting the experimentally measured load-displacement curves. According to the calculation results, δ_{top}, δ_{base} and δ_{sub} are approximately 70, 10 and 20 % of the total deformation. Therefore, the uncertainty from the deformation of the top is the main part of the total uncertainty. In the calculation of the deformation of the nanodot top using Eq. (4.2), it was assumed that the Ni nanodots have the same height and radius at top. The measurements of the height and radius show that they both have an uncertainty of about 10 %, which needs to be considered in the uncertainty analysis for the elastic modulus estimated by fitting load-displacement curves. The effect of uncertainty from the nanodot top radius can be directly translated to the final uncertainty through Eq. (4.2). However, the uncertainty due to the nanodot height variation is not explicitly expressed in the equation. If an indentation on Ni nanodots with height variation (real situation) is compared with an indentation on nanodots with the same height (ideal situation), the effect of the nanodot height variation will be reflected in the uncertainty of the indentation curve. More specifically, for a certain indentation load, there is a deviation of the measured indentation depth of the real situation from that of the ideal situation. The effect of the nanodot height variation is reflected in this deviation. Because the uncertainty of the indentation depth includes not only the effect of the uncertainty of the nanodot height but also those of the nanodot shape and spacing, it is a conservative estimation of the uncertainty of the nanodot height variation. According to Eq. (4.2) and the relationship between E^* and E_{Ni}, the uncertainty is estimated to be

$$\Delta E_{Ni} = \left|\frac{\partial E_{Ni}}{\partial E^*}\right| \Delta E^* \tag{4.7}$$

where

$$\Delta E^* = \sqrt{\left(\frac{\partial E^*}{\partial R}\Delta R\right)^2 + \left(\frac{\partial E^*}{\partial \delta}\Delta\delta\right)^2} \tag{4.8}$$

therefore,

$$\frac{\Delta E_{Ni}}{E_{Ni}} = \left|\frac{\partial E_{Ni}}{\partial E^*}\right| \cdot \frac{E^*}{E_{Ni}} \cdot \frac{\Delta E^*}{E^*} = \frac{\left(1-\nu_{Ni}^2\right)E_{tip}^2}{\left[E_{tip} - \left(1-\nu_{tip}^2\right)E^*\right]^2} \cdot \frac{E^*}{E_{Ni}} \cdot \sqrt{\left(\frac{1}{2}\frac{\Delta R}{R}\right)^2 + \left(\frac{3}{2}\frac{\Delta\delta}{\delta}\right)^2} \tag{4.9}$$

In Eq. (4.9), the uncertainty of the reduced radius, $\Delta R/R$, is approximately equal to $\Delta R_{top}/R_{top}$, which is measured to be 10 % as stated earlier; and the uncertainty of the indentation depth, $\Delta\delta/\delta$, obtained from the experimental data was 13.6 %. Therefore, the uncertainty of the elastic modulus of the Ni nanodot $\Delta E_{Ni}/E_{Ni}$ calculated from Eq. (4.9), was 24 %. This number is in good agreement with the uncertainty of a single measurement of 28 % obtained by using the measured standard deviation of 22 GPa and $P_E = tS_E = 45$ GPa, where $t = 2.052$ for a 95 % confidence level. Since we took 28 measurements, the reported average value of 159 GPa has an uncertainty of $P_{\bar{E}} = P_E/\sqrt{28} = 8.5$ GPa, which is 5.4 % of the elastic modulus.

As mentioned earlier, there is a native NiO less than 2 nm thick on the Ni nanodots. We believe that this thin NiO layer has no significant effect on the contact behavior based on a study by Pethica and Tabor [56]. They studied the contact of a tungsten tip with a radius of curvature of 1 μm with a flat single crystalline Ni substrate in an UHV SEM and observed that, under light oxidization conditions, there was no significant change in the contact behavior compared with that of a clean surface.

The point at which the loading curve deviates from the Hertzian fit corresponds to the onset of plasticity of the Ni nanodots. The load at this point is referred to as the critical load here. According to Hertzian elastic contact theory [52], taking into account the number of Ni nanodots in contact with the tip, the maximum elastic shear stress, or the critical shear stress at the onset of plasticity, τ_{max}, can be estimated by:

$$\tau_{\max} = \left(\frac{0.56}{\pi}\right)(P/n_0)^{1/3}\left(\frac{E^*}{R}\right)^{2/3} \tag{4.10}$$

where P is the critical load and n_0 is the number of Ni nanodots in contact with the tip at the critical load. From Fig. 4.5b, we determined the load and displacement at the onset of plasticity to be 6.6 μN and 2.3 nm, respectively. Using Eq. (4.10) the critical shear stress is calculated to be 8.2 GPa. The critical shear stress calculated from all indentation curves was 8.3 ± 1.0 GPa, which is very close to the estimated theoretical shear strength of dislocation-free single crystalline Ni of 7.6 GPa (estimated to be on the order of $G/10$, where G is the shear modulus of Ni (76 GPa)) indicating that these Ni nanodots have unusually high strength.

Hardness of the Ni NDPS

Because the Ni nanodots in this study have a low aspect ratio and are substantially softer than the diamond indenter tip or the silicon substrate, they were squashed during indentation, while the indenter tip and the silicon substrate did not deform plastically. Therefore, the hardness of the Ni nanodots was evaluated using the "mean contact pressure" method by dividing the maximum indentation load with the estimated contact area determined from SEM and AFM analysis of an indentation. Figure 4.6 illustrates the process of determining the measured contact area for two indentations at a small load of 70 μN and a large load of 500 μN. First, SEM and then AFM measurements were taken after indentation tests were performed. The AFM images and SEM micrographs were combined to determine the area of the Ni NDPS that was in contact with the tip during nanoindentation, as shown in Fig. 4.6a and b, as well as d and e for 70 and 500 μN loads, respectively. The SEM images with the contact area circled were then further analyzed to yield

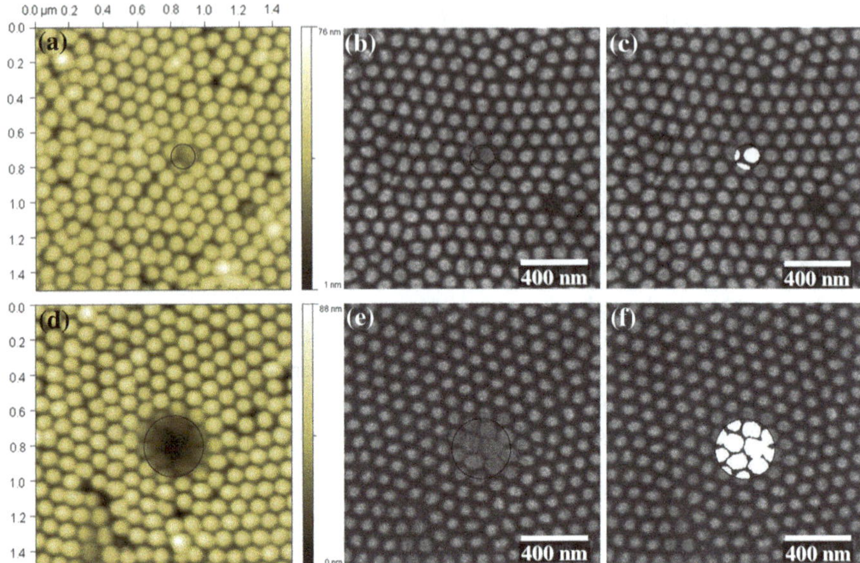

Fig. 4.6 Deformation of the Ni nanodots after 70 and 500 μN indentations. (**a**), (**b**) and (**c**) [(**d**), (**e**) and (**f**)] are an AFM image, an SEM micrograph, and a highlighted SEM micrograph, showing the measured area of contact for the 70 [500] μN indentation (*Source* Data from [34], with kind permission from IOP)

the real area of contact by excluding the areas between the Ni nanodots, as shown in Figs. 4.6c and f. The hardness of the Ni nanodots was then calculated by dividing the indentation load with the real area of contact thus obtained. It should be noted that the contact area determined by SEM or AFM is approximately the same as the contact area under the maximum applied load, and thus can be used for hardness calculations. According to Tabor [57], for spherical indenters (our tip apex can be viewed as spherical), there was elastic recovery of the indented material after the applied load was removed. However, there appeared to be little change in the chordal diameter of the indentation.

Figure 4.7 shows the hardness and real contact area as a function of indentation load based on three measurements under each load. The calculated hardness was 7.7 ± 1.0 GPa, which is in agreement with the hardness for nanocrystalline Ni (8 GPa), but much larger than the hardness of microcrystalline Ni (3 GPa) [54]. The higher value of the hardness is most likely a result of the nanodots having fewer defects due to their much smaller size compared to bulk microcrystalline Ni. The real contact area is proportional to the indentation force, which is understandable because the average pressure between the indenter and the Ni nanodots are expected to be constant throughout an indentation test and equal to the hardness of the Ni nanodots.

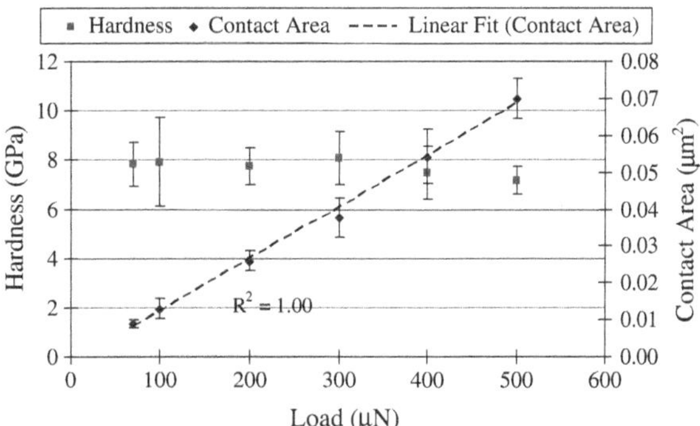

Fig. 4.7 Hardness and measured area of contact as a function of indentation load (*Source* Data from [34], with kind permission from IOP)

Simulations of Nanoindentation on a Ni NDPS

A numerical model was developed to simulate nanoindentation of a Ni NDPS on a deformable Si substrate using spherical diamond tips with different tip radii. The deformation behavior of the Ni NDPS, i.e., the relationship between the indentation load and displacement (normal approach), is modeled and the results are verified by the nanoindentation experiments. The overall structure of the model is presented in "Overall Structure of the Model". To numerically simulate the nanoindentation using the developed model, the deformation of a single nanodot and the deformation of the substrate must be determined for an applied normal load first. The methods used to determine the deformation of a single nanodot and the deformation of the substrate are described in "Deformation of a Single Nanodot" and "Deformation of Substrate", respectively.

Overall Structure of the Model

The contact between a spherical diamond tip and a Ni NDPS can be viewed as a contact between a rigid smooth sphere and a flat surface covered with an array of Ni nanodots. To model the topography of a real Ni NDPS, the nanodots are distributed on the flat surface in a hexagonal arrangement, but the exact sizes and positions of the nanodots are subjected to a random variation mimicking that observed on the real surface (they vary within 10 % of the nominal size and position). The nanodots are assumed to be of the same shape, with the radius of curvature at the top of a nanodot proportional to the height of the nanodot. The substrate below the nanodots is considered to be a deformable elastic half-space (see "Deformation of Substrate" for justification). Because the nanodots do not touch one another, they do not directly interact, but the influence of each nanodot on its neighboring nanodots is taken into account through the deformation of the substrate.

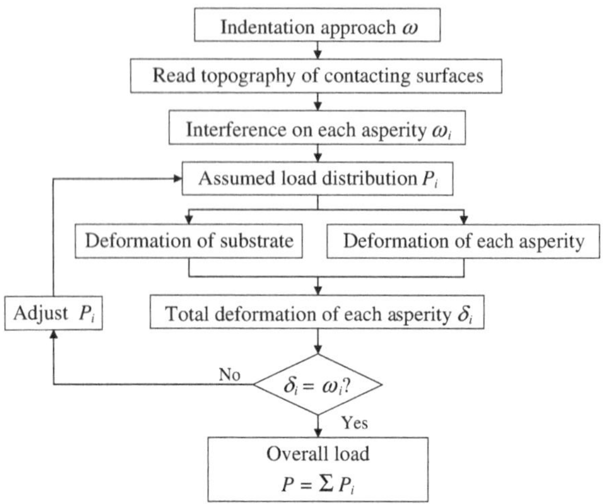

Fig. 4.8 Flow chart of the multi-asperity contact model (*Source* Data from [35])

Numerical algorithms were developed to perform the contact analysis when the spherical indenter tip moves from a reference plane towards the NDPS by a certain distance, ω. In this study, the reference plane was chosen to be the average height of the nanodots. A flow chart of the main contact analysis algorithm is shown in Fig. 4.8. In this algorithm, the radius of the spherical indenter tip and the generated surface topography of the Ni NDPS are first read in. Then the contacting nanodots are identified and the interference for the ith contacting nanodot, ω_i, is calculated according to geometry. From an initially assigned load on each nanodot, P_i, the deformation of the ith nanodot body, $\delta_{\text{dot}i}$, and the deformation of the substrate below this nanodot, $\delta_{\text{sub}i}$, can be calculated. The $\delta_{\text{dot}i}$ is determined by a pre-calculated load-deformation relationship obtained by finite element analysis (FEA) of a single-asperity contact (this relationship is discussed in detail in "Deformation of a Single Nanodot"). The $\delta_{\text{sub}i}$ is determined using the classical elastic theory (see "Deformation of Substrate" for details). The total displacement of each nanodot, δ_i, is then calculated as the summation of the nanodot body deformation and the substrate deformation:

$$\delta_i = \delta_{\text{dot}\,i} + \delta_{\text{sub}\,i} \tag{4.11}$$

If this displacement, δ_i, does not equal the interference, ω_i, a new load distribution will be given and the displacement is calculated again. The iteration continues until δ_i equals ω_i. The new load distribution for the next iteration step is determined by adjusting the load, P_i, on each nanodot with a value that is proportional to the difference between δ_i and ω_i. When a converged solution is achieved, the total contact load is calculated as the summation of the loads on each nanodot:

$$P = \sum\nolimits_{i=1}^{N} P_i \tag{4.12}$$

where N is the number of asperities that are involved in the contact. The relationship between the contact load, P, and the interference ω (normal approach) is then established.

Deformation of a Single Nanodot

The load-deformation relationship of the single-asperity contact between a nanodot and an indenter tip was obtained through FEA using ANSYS. The FEA included the nanodot, the substrate, and the tip. The nanodot material was assumed to be elastic-perfectly plastic, whereas the substrate deforms only elastically. The reason for including the substrate in the FEA was to study the details of the substrate deformation under a nanodot. The indenter tip is assumed to be rigid in this work. To justify this assumption, FEA of the single-asperity contact between a nanodot and a deformable 1 μm radius tip was conducted and compared to the contact between a nanodot and a rigid 1 μm radius tip. It was found there is little difference between the two cases. Figure 4.9 shows a typical mesh generated using ANSYS for a spherical tip with a radius of 1 μm. An axisymmetric 2D model was used. Quadrilateral 8-node elements were used to mesh the nanodot and quadrilateral 4-node elements were used to mesh the substrate and the tip. The inset in Fig. 4.9 is a close-up view of a meshed nanodot and part of the substrate and the tip. The meshes are refined near the contact region between the nanodot and the tip to capture the high stress gradient. The region of contact was meshed by 200 elements. For the smallest contact area calculated in this study, there were at least 64 elements in contact to ensure precision in determining the contact radius and

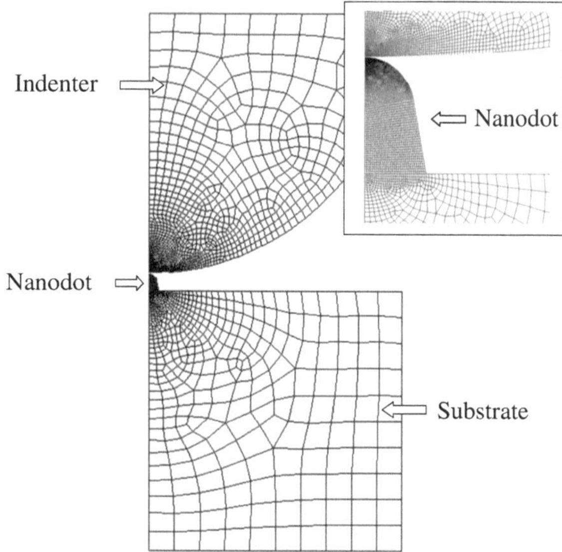

Fig. 4.9 Finite element mesh of the multi-asperity contact model. The inset is a *close-up* view of the mesh on the nanodot and part of the substrate and the indenter (*Source* Data from [35])

stress. The nanodot and the substrate were assumed to be well attached, so that the nodes along the interface between the nanodot and the substrate were set to have the same displacement. Because an axisymmetric model was used, the modeled substrate shown in Fig. 4.9 represents a cylinder, which has a radius and height of 1 μm. The nodes at the base and sidewall of the cylindrical substrate were fixed in all directions. From St. Venant's Principle, this boundary condition is valid because the modeled substrate size is more than 25 times that of the base radius of the nanodot. If the nodes on the current boundary is allowed to move by connecting them to a bulk material that is 6 times larger in size than the current substrate, simulation results showed that there is marginal difference between the two cases: the contact load has less than 1 % difference and the displacement at the positions of the current boundary is less than 0.01 nm in any direction.

A 5 μm radius indenter tip and a flat indenter tip were also simulated. In both cases, the model only included the part of the tip inside a cylindrical volume of 1 μm in height and 1 μm in radius. The elements of the indenter tip were not allowed to have deformation in order to simulate a rigid tip.

To simulate the single asperity contact, a vertical displacement of the indenter was first given. The resulting stress, strain, and the contact area were then computed. After a numerical solution was obtained for the given displacement, the contact load was calculated by integrating the normal stress on the bottom of the nanodot. The contact radius was determined from the position of the node on the edge of the contact area.

The current model was tested by comparing with the Hertzian elastic solution and the FEA conducted by Jackson and Green (JG) [58]. Comparisons were conducted involving the contact between a hemisphere and a rigid flat. The hemisphere was assumed to only deform elastically when compared with the Hertzian solution, and deform elasto-perfectly-plastically when compared with the JG solution. It was found that on average the calculated contact radius from the FEA in this study differed from the Hertzian solution and the JG solution by only 2.8 and 4.3 %, respectively. The average difference on contact load was 3.5 % compared with the Hertzian solution and 4.5 % compared with the JG solution. Therefore, the solutions of our FEA model in this study compares well with the Hertzian and JG solutions.

Deformation of Substrate

To determine the substrate deformation, we used the classical elastic theory of a pressure applied to a circular region on an elastic half-space [52]. Classical elastic theory was used because the FEA results showed that the maximum stress in the substrate was below the yield strength of the Si substrate. For the heaviest indentation modeled (with an interference of 35 nm), FEA showed the maximum von Mises stress as 3.27 GPa, which is lower than the Si yield strength of 4.3 GPa [59]. Moreover, this maximum von Mises stress appears below the outer edge of the nanodot base where a sharp corner exists in the FEA model. This high stress value is partly because of stress concentration caused by the sharp corner, which probably does not exist in reality. Because the thickness and the lateral dimension of the substrate are orders of magnitude larger than the dimension of the bottom of the nanodots, the substrate can be viewed as an elastic half-space.

From our FEA results, it was found that the nanodot-substrate interface had a uniform normal deformation. For example, the simulation of the contact between a nanodot and a 1 μm radius indenter under interference of 35 nm shows that all the points along the interface have normal deformations around 0.94 nm with a standard deviation of 2.7 %. Therefore, it is assumed that the bottom of the nanodot has uniform normal deformation in the current work.

The relationship between the substrate deformation at the bottom of the nanodot, u_y, and the total applied load, L, was given (for a pressure applied to a circular region that causes a uniform deformation of the loaded circle) in Johnson [52] as:

$$u_y = \frac{1 - \nu^2}{2aE} L \tag{4.13}$$

where E and ν are the elastic modulus and Poisson's ratio of the substrate, respectively, and a is the nanodot base radius. The substrate deformation calculated using Eq. (4.13) is in good agreement with the FEA results. Therefore, in the current work, the analytical method is used to calculate the substrate deformation because it is more easily incorporated into the contact model for the nanoindentation of a NDPS.

In the case of contacts that involve multiple nanodots, the substrate deformation under each nanodot consists of two parts: the deformation caused by the load from the nanodot itself and the deformation induced by the neighboring nanodots. Letting δ_{ij} denote the deformation of the substrate under the ith nanodot that is induced by the jth nanodot, Johnson [52] gives:

$$\delta_{ij} = \begin{cases} (1 - \nu^2)P_j / 2aE & (j = i) \\ (1 - \nu^2)P_j \cdot \sin^{-1}(a/d_{ij}) / \pi aE & (j \neq i) \end{cases} \tag{4.14}$$

where P_j is the load applied on the jth nanodot, and d_{ij} is the distance between the center of the ith and the jth nanodot. Assuming the deformations follow the principle of superposition, the total substrate deformation under the ith nanodot, $\delta_{\mathrm{sub}i}$ can be determined:

$$\delta_{\mathrm{sub}i} = \sum\nolimits_{j=1}^{N} \delta_{ij} \tag{4.15}$$

where N is the number of nanodots involved in the contact.

Deformation Behavior Determined by Nanoindentation Simulation

Single-Asperity Contact Results

FEA of a contact between a Ni nanodot and a rigid diamond indenter of 1 μm tip radius was performed to obtain the load-deformation relationship of this single-asperity contact. The material properties used in this study are listed in Table 4.1. The elastic modulus and yield strength used for Ni were obtained from the experimental study of the mechanical properties of the Ni NDPS in "Deformation of Substrate". The yield strength of a Ni nanodot was taken as one third of the hardness, which is 2.57 GPa.

Every combination of elastic modulus (159 or 200 GPa) and yield strength (2.0, 2.57 or 3.0 GPa) was used for Ni in the simulations. The FEA was conducted for interferences up to 35 nm. The load-deformation relation obtained is shown in Fig. 4.10 in a dimensionless form. The dimensionless load, P^*, and the dimensionless deformation, ω^*, were normalized as:

$$P^* = P/P_c \text{ and} \tag{4.16}$$

$$\omega^* = \omega/\omega_c \tag{4.17}$$

where P_c and ω_c are the critical load and critical interference at the yield point for a hemispherical contact. P_c and ω_c are derived semi-analytically by Jackson and Green [58] using the von Mises yield criterion:

$$P_c = \frac{4}{3}\left(\frac{R}{E'}\right)^2\left(\frac{\pi \cdot C \cdot S_y}{2}\right)^3 \tag{4.18}$$

$$\omega_c = \left(\frac{\pi \cdot C \cdot S_y}{2E'}\right)^2 R \tag{4.19}$$

where

$$\frac{1}{E'} = \frac{1 - v_1^2}{E_1} + \frac{1 - v_2^2}{E_2} \tag{4.20}$$

Table 4.1 Material properties of Ni, diamond [53, 59] and Si [59] (*Source* Data from [35])

	Elastic modulus E (GPa)	Yield strength Sy (GPa)	Poisson's ratio v
Ni	159	2.57	0.31
Diamond	1141	35	0.07
Si	179	4.3	0.25

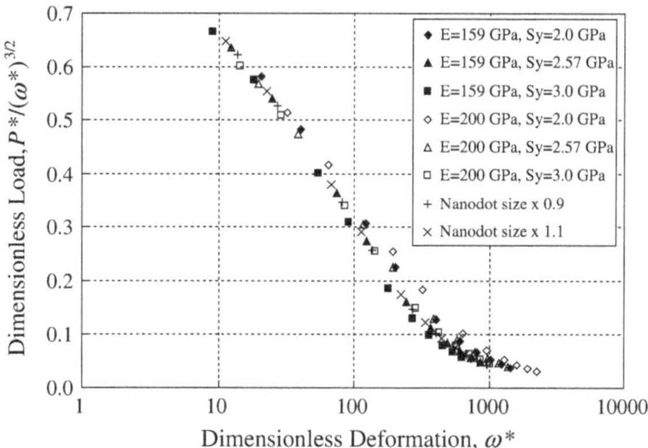

Fig. 4.10 Dimensionless contact load versus deformation for the single-asperity contact between a Ni nanodot and a diamond indenter of 1 μm radius using different material properties of Ni and different sizes of Ni nanodot. The results for different nanodot sizes are calculated using $E = 159$ GPa and $S_y = 2.57$ GPa (*Source* Data from [35])

$$\frac{1}{R} = \frac{1}{R_1} + \frac{1}{R_2} \tag{4.21}$$

and E_1, v_1, R_1 (E_2, v_2, R_2) are the elastic modulus, Poisson's ratio, and radius of the tip (nanodot), respectively. The constant, C, is fitted to the solution by Jackson and Green [58]:

$$C = 1.295 \exp(0.736v) \tag{4.22}$$

In Eq. (4.18), (4.19) and (4.22), S_y and v are the yield strength and Poisson's ratio of the material that yields first. In this study, they are the yield strength and Poisson's ratio of Ni, respectively. In Fig. 4.10, the dimensionless load P^* was also normalized as $P^*/(\omega^*)^{3/2}$ using the load-deformation relationship of $P^*/(\omega^*)^{3/2} = 1$ for Hertzian contact [58].

The effect of the nanodot size variation was also studied. Single-asperity contact of nanodots with 0.9 and 1.1 times that of the original nanodot size (corresponding to nanodots of top radii of 27 nm and 33 nm, respectively) was simulated for all combinations of material properties. Representative results for nanodots with 0.9, 1.0 and 1.1 times that of the original nanodot size using elastic modulus of 159 GPa and yield strength of 2.57 GPa are shown in Fig. 4.10. The results shown in Fig. 4.10 illustrate that, using the dimensionless parameters suggested by JG, the load-deformation curves for the contact between a rigid tip and a nanodot with different material properties and sizes collapse to nearly the same curve. However, there are still small differences between each case, so the curves do not collapse to exactly the same curve.

The dimensionless load-deformation relationship illustrated in Fig. 4.10 represents the relationship between P^* and ω^*, which are combinations of a set of parameters. In order to discuss the dimensional load and deformation of the nanodots, which are important for determining the substrate deformation, the load-deformation relations of selected results are plotted in Fig. 4.11 in a dimensional form. It can be seen that the yield strength has a significant effect on the contact load-deformation relationship, while the elastic modulus has very little effect. The

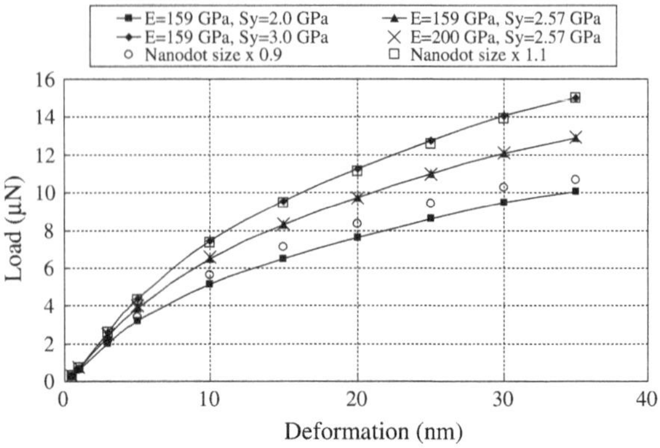

Fig. 4.11 Results of dimensional contact load versus deformation for the single-asperity contact between a Ni nanodot and a diamond indenter of 1 μm radius. The results for different nanodot sizes are obtained using $E = 159$ GPa and $S_y = 2.57$ GPa (*Source* Data from [35])

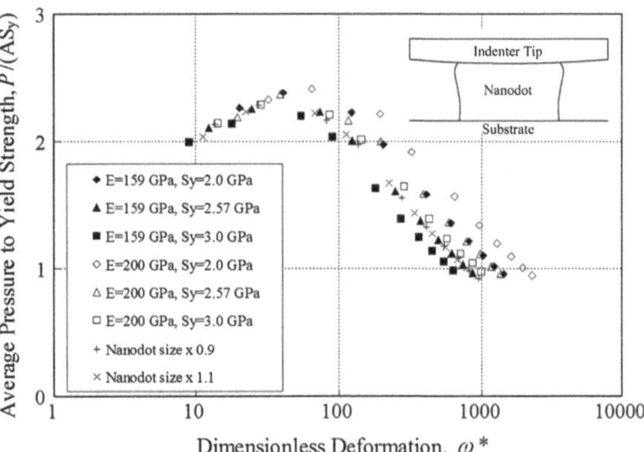

Fig. 4.12 Results of average pressure to yield strength ratio for the single-asperity contact between a Ni nanodot and a diamond indenter of 1 μm radius. The results for different nanodot sizes are obtained using $E = 159$ GPa and $S_y = 2.57$ GPa. The inset is the deformed shape of a nanodot under high pressure (*Source* Data from [35])

reason for the small effect from the elastic modulus is that the contact is mostly plastic and far beyond the elastic regime. Nanodot size variation also showed considerable effects on the contact load-deformation relationship.

Representative results of the average contact pressure to yield strength ratio, $P/(AS_y)$, is plotted in Fig. 4.12, where A denotes the contact area. It can be seen that the average contact pressure first increases as the dimensionless deformation increases. The maximum average contact pressure ratio varies from 2.2 to 2.4. These results are in agreement with the findings by Jackson and Green [58] that the hardness is not a fixed material property that depends only on the yield strength, but it also depends upon the elastic modulus, Poisson's ratio, and the deformation of the asperity.

After the average contact pressure reaches the maximum point, it starts to decrease as the deformation increases because the contact area increases more rapidly than the contact load. The same trend was found by Jackson and Green [58] and Mesarovic [60]. Figure 4.12 shows the average pressure ratio decreases to a value of about 1, which is in agreement with the theoretical prediction by Jackson and Green [58]. In this study, the contact pressure eventually reduces to a value less than the yield stress at large interferences ($\omega^* \sim 1000$). This is because, in the current work, the base of the nanodot is connected to a bulk material such that the base radius is more constrained than the contact radius. At large interference, the nanodot actually deforms to an "hourglass shape" so that the contact area is larger than the minimum cross-sectional area of the nanodot as shown in the inset of Fig. 4.12. Therefore, the average contact pressure is smaller than the pressure on the minimum cross-sectional area at the base of the nanodot.

In order to justify the assumption of a rigid indenter tip, we compared the above single-asperity contact (with a rigid tip) to a contact with a deformable tip. In the simulation of the single-asperity contact with a deformable tip, the tip was allowed to deform elastically and was assigned to the material properties of diamond, shown in Table 4.1. The deformation of the elastic tip was found to contribute less than 1.6 % of the total deformation when the total deformation is larger than 10 nm. The percentage increases when the total deformation is less than 10 nm with a maximum number of only 7.7 % for the smallest simulated total deformation (0.5 nm). The difference between the contact load obtained using a rigid tip and an elastic tip is less than 0.7 % when the total deformation is larger than 10 nm, and the maximum difference is only 6.6 % at the smallest simulated deformation. Thus, considering that many of the nanodots are severely deformed in the nanoindentation of the NDPS, the tip can be assumed to be rigid.

The effects of the indenter tip radius were also studied. A rigid 1 μm radius tip, a rigid 5 μm radius tip, and a rigid flat tip were simulated. The calculated contact loads at all simulated interferences were compared for the three different indenter tips. It was found that the load difference among the results obtained has a maximum of 2.4 % and an average of 1.5 %. Thus the indenter radius has little effect on the single-asperity contact in the range considered. Therefore, in the simulation of the nanoindentation of a Ni NDPS, the load-deformation relationship for the single-asperity contact between a nanodot and a 5 μm radius tip was considered the same as the load-deformation relationship between a nanodot and a 1 μm radius tip.

Multi-Asperity Contact Results

The multi-asperity contact model, outlined in "Overall Structure of the Model", is used to simulate the indentations on a Ni NDPS. Computer programs were written following the previously described procedure in "Overall Structure of the Model" for the numerical modeling of nanoindentation. To mimic the nanodot pattern location arrangement on the Ni NDPS, the height, apex radius, base diameter, and dot-to-dot spacing of the nanodots in the programs were subjected to a random variation of 10 %, the same variations as experimentally measured. The load-deformation relationship of a nanodot was established from the single-asperity contact analysis using FEA in "Overall Structure of the Model". Linear interpolation was used for displacements that fall in between the FEA simulated values.

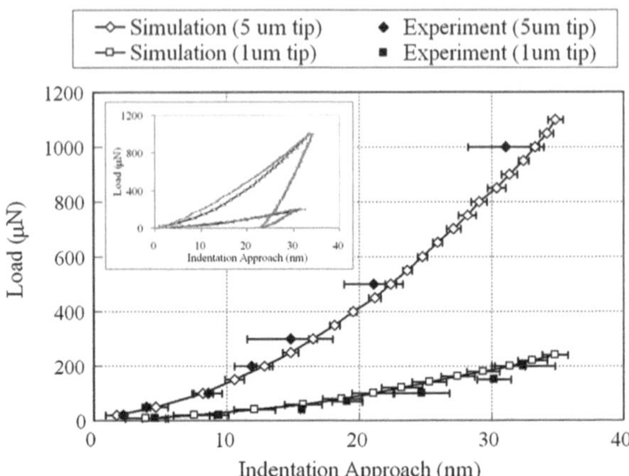

Fig. 4.13 Comparison of the contact load as a function of the indentation approach obtained from the numerical simulation and from the experiments. The inset is an example of the comparison between the indentation curves obtained from the numerical simulation and representative experimental indentation curves (*Source* Data from [35])

The same material properties used in the single asperity simulation were used in the nanoindentation simulation (see Table 4.1). The plane at the average height of the peaks of the nanodots was chosen as the reference plane to determine the normal approach of the indenter tips. For a given normal approach, a contact load was calculated. Contact was simulated for normal approach up to 35 nm and a load-approach curve was obtained. The results of the simulations using 1 and 5 μm radii indenter tips are shown in Fig. 4.13. To account for possible uncertainties in real Ni NDPS topographies, simulations were performed on five different randomly generated NDPSs using the same set of parameters. The simulation data in Fig. 4.13 are the averages of the results from the five generated topographies, with error bars indicating the variation of the simulation results.

Experimental Verification of the Multi-Asperity Contact Model

The results from the nanoindentation experiments are also plotted in Fig. 4.13 to compare with the simulation results based on the model described in "Overall Structure of the Model". An experimental data point in Fig. 4.13 represents the average of the deformations at the maximum indentation load of three different indentations. Because the nanoindentation loading curves at different loads almost overlap, the load-deformation relationship in the nanoindentation experiments can be well described by showing the deformation at the peak load of each indentation curve. It can be seen from Fig. 4.13 that the simulation results shows good agreement with the experimental results, indicating the validity of the multi-asperity contact model developed in this study. The inset of Fig. 4.13 is an example of the comparison between the indentation curves obtained from the numerical simulation and representative experimental indentation curves, which also shows a good agreement between simulation results and experimental results. It should be noted that this was achieved without any fitting parameters.

The variations of the experimental results are indicated by the error bars in Fig. 4.13. These variations exist due to the imperfections of the nanodot array on the Ni NDPS, as shown in Fig. 4.1b. First, there are variations in the nanodots' height, apex radius, and base diameter. Second, there are variations in the arrangement of the nanodots position, for example, missing nanodots in the array may exist. In addition, there may also be variations in the material properties of the nanodots and the relative position of the indenter tip to the nanodots array when the tip lands on the NDPS. Due to these uncertainties, the contact situation is not exactly the same from one indentation to another resulting in differences in the indentation deformation under the same indentation load.

Effect of Substrate

The effects of the substrate on the contact behavior during the nanoindentation of the Ni NDPS were studied using the multi-asperity contact model developed in this work. Indentations using the 5 μm radius tip were taken as an example to show the substrate effects. A nanoindentation load-approach curve of a Ni NDPS on a rigid substrate was also calculated and is shown in Fig. 4.14. It can be seen that the predicted contact load for the rigid and deformable substrate has a maximum difference of 74 % at the 35 nm indentation approach. Although this difference decreases as the indentation approach decreases, there is still a 24 % difference even at a very small indentation approach of 1 nm. This large difference shows that the substrate deformation has a large impact on the indentation behavior of NDPSs.

In order to investigate the details of the effect of substrate, the substrate deformation below each nanodot was investigated numerically for the indentation made using a 5 μm radius tip at an indentation approach of 35 nm using the model

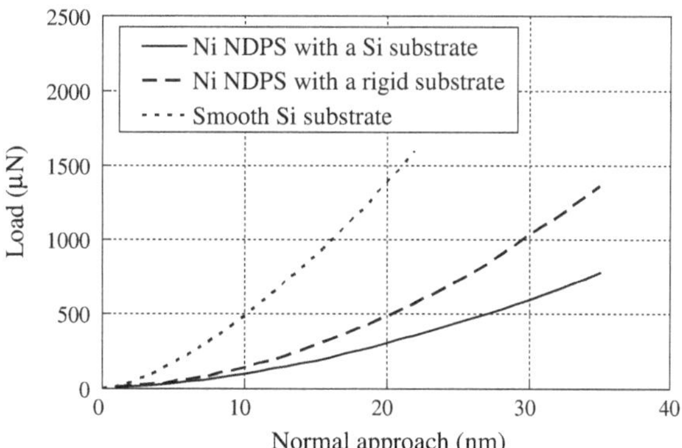

Fig. 4.14 Load-approach relationship for the nanoindentation of a Ni NDPS on a deformable Si substrate using a 5 μm radius tip, the nanoindentation of a Ni NDPS on a rigid substrate using a 5 μm radius tip, and a Hertzian contact between a 5 μm radius tip and a smooth Si substrate (*Source* Data from [35])

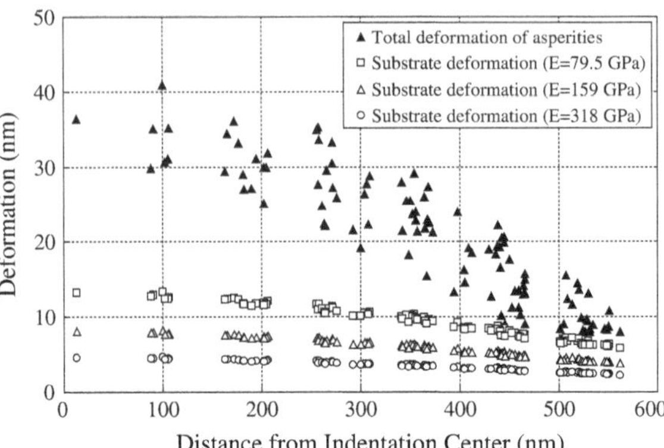

Fig. 4.15 Substrate deformation versus distance from the indentation center for different substrate elastic moduli of the NDPS. Here we used a 5 μm radius tip and an indentation approach of 35 nm (*Source* Data from [35])

described in "Overall Structure of the Model". Simulations were performed using a substrate elastic modulus that is half of, the same as, and two times that of the Ni nanodot elastic modulus (79.5, 159, and 318 GPa, respectively). The total displacement at each nanodot location consists of the deformation from the nanodot body and the deformation from the substrate below the nanodot. Figure 4.15 shows the total displacement at each nanodot and the deformation of the substrate under each nanodot as a function of the distance of the nanodot from the indentation center. It can be seen that, if the substrate is the same material as the nanodots (E = 159 GPa), at least 20 % of the total deformation comes from the substrate for nanodots at the indentation center. This fraction increases to about 60 % at the outer edge of the contact area. If the substrate elastic modulus is half of the elastic modulus of the nanodots, the substrate deformation contributes 40–80 % of the total deformation, as shown in Fig. 4.15. This study shows that the substrate deformation is sensitive to the elastic modulus of the substrate. Therefore, if the substrate material is less stiff than the asperities, the effect of substrate becomes very significant.

The portion of substrate deformation in the total deformation is smaller near the center because the nanodots were more severely squeezed in this area; therefore, the deformation of the nanodot body takes a larger portion of the total deformation. At the edge of the contact area, the substrate effect is very significant. This is because even though the load on each nanodot was very low, the substrate still deforms considerably due to the cumulative effects from the neighboring nanodots.

The large variation of the total deformation shown in Fig. 4.15 is due to the variations in the height, top radius, and position of the nanodots, as well as the relative position of the tip to the nanodot array. These variations were intentionally introduced in the simulation to mimic the real surface topography of the Ni NDPS.

Effect of Roughness

For the Ni NDPS, the array of nanodots forms the "roughness" on the substrate surface. According to many earlier studies, the effect of roughness disappears at high load and the contact between a spherical indenter tip and a rough surface approaches Hertzian contact. Therefore, whether the nanoindentation of the Ni NDPS at high loads approaches Hertzian contact or not needs to be investigated. The model described in "Overall Structure of the Model" was used to study the effects of the roughness on the nanoindentation of the Ni NDPS.

Greenwood et al. [61] suggested a non-dimensional parameter $\alpha = \sigma R/a_0^2$ (where σ is the roughness of a surface, R and a_0 are the effective asperity radius and contact radius if the contact is assumed to be Hertzian contact) to evaluate the roughness effect. When α approaches zero, the contact approaches Hertzian contact between two smooth surfaces. The authors concluded that the difference between rough contact and Hertzian contact is less than 7 % when α is less than 0.05. However, in the nanoindentation of Ni NDPSs, α is at least an order of magnitude larger than 0.05. For example, the minimum value of α is 0.86 for the heaviest indentation at the approach of 35 nm and α increases rapidly when the indentation becomes lighter (for indentation of 20 and 5 nm, α equals 1.5 and 6.0, respectively). Therefore, the effect of roughness can not be neglected during the nanoindentation of Ni NDPSs.

If the roughness caused by the Ni nanodots is neglected, the contact between the tip and the Ni NDPS becomes a Hertzian contact between the tip and the Si substrate. Figure 4.14 shows a load-approach curve of the contact between a 5 μm diamond tip and the Si substrate calculated from the Hertzian solution. It can be seen that, when the indentation approach is larger than 5 nm, the contact load in the Hertzian solution is more than 3 times the load calculated by the current model. Therefore, the effects of roughness on the nanoindentation of the Ni NDPS are significant such that the nanoindentation can not be considered as a Hertzian contact between the diamond indenter tip and the Si substrate.

Effect of Elastic Modulus and Yield Strength of Nanodot

The model described in "Overall Structure of the Model" was used to study the effects of elastic modulus and yield strength of Ni nanodot on the nanoindentation load-approach relationship. Simulations were performed for an elastic modulus of 159 and 200 GPa and yield strengths of 2.0, 2.57, and 3.0 GPa, respectively. The results are shown in Fig. 4.16. For each yield strength, the simulated indentation curves for an elastic modulus of 159 GPa overlap the curves for an elastic modulus of 200 GPa, indicating that the elastic modulus has little effect on the indentation load-approach relationship. In contrast to this, Fig. 4.16 also shows that a variation in yield strength has a large effect on the indentation load-approach relationship.

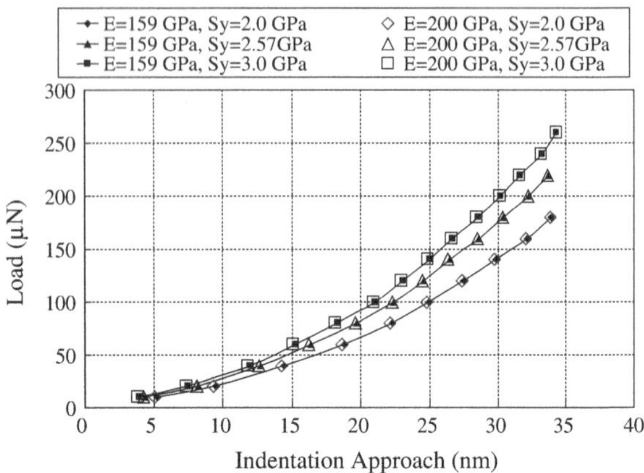

Fig. 4.16 Simulation results of the contact load versus indentation approach for the contact between a 1 μm radius tip and a Ni NDPS using different material properties of Ni (*Source* Data from [35])

Again, this is because severe plastic deformation occurs on many Ni nanodots even at very small indentation approach. Clearly, the yield strength of Ni plays a more important role in the contact of the NDPS than the elastic modulus does.

Conclusions

The mechanical properties of the Ni NDPS on a silicon substrate were studied by nanoindentation. The corresponding deformation behavior was investigated using SEM and AFM. The results showed that the Ni nanodots have a smaller elastic modulus, but a larger hardness than that of bulk microcrystalline Ni reported in the literature. The estimated critical shear stress to initiate plastic deformation in the Ni nanodot was found to be close to the theoretical shear strength in dislocation-free single crystal Ni.

A numerical multi-asperity contact model was developed to simulate the nanoindentation of a Ni NDPS on a deformable Si substrate using diamond indenter tips with radii of 1 and 5 μm. Nanoindentation experiments were conducted to verify the developed model. The simulation and experimental results are in good agreement.

It was found that both the effect of the substrate and the effect of the roughness must be considered in the nanoindentation. The substrate deformation was found to have significant impact on the contact behavior, especially when the substrate material is less stiff than the nanodots. The study of substrate deformation under each nanodot shows that the interaction among asperities through substrate deformation is important. The substrate deformation under a nanodot caused by the neighboring nanodots can be a considerable portion of the total deformation at a nanodot. The yield strength of the Ni nanodot has a significant effect on the load-approach relationship of the nanoindentation. However, the elastic modulus, within the range considered, has very little effect on the nanoindentation because the nano-scale contacts are usually far beyond the elastic regime. In addition, this work shows that for some cases continuum based contact models can be used for even nano-scale contacts.

Since the modeling framework is sufficiently flexible to accommodate different nanodot size and position arrangement, it can also be applied to a surface covered by isolated nanodots with known size and location distributions, but not necessarily following a repetitive pattern.

Acknowledgments Financial support for the work reviewed was provided by the National Science Foundation, Arkansas Biosciences Institute, and the University of Arkansas. Contributions to the work reviewed from our collaborators Drs. Mathew Johnson (University of Oklahoma) and Robert Jackson (Auburn University) are gratefully acknowledged.

References

1. Roukes, M.: Nanoelectromechanical systems face the future. Phys. World **14**(2), 25–31 (2001)
2. Kim, S.H., Asay, D.B., Dugger, M.T.: Nanotribology and MEMS. Nano Today **2**(5), 22–29 (2007)
3. Komvopoulos, K.: Adhesion and friction forces in microelectromechanical systems: mechanisms, measurement, surface modification techniques, and adhesion theory. J. Adhes. Sci. Technol. **17**(4), 477–517 (2003)
4. Van Spengen, W.M., Puers, R., De Wolf, I.: On the Physics of stiction and its impact on the reliability of microstructures. J. Adhes. Sci. Technol. **17**(4), 563–582 (2003)
5. Williams, J.A., Le, H.R.: Tribology and MEMS. J.Phy. D **39**(12), 201–214 (2006). (Applied Physics)
6. Bhushan, B.: Tribology issues and opportunities in MEMS, p. 109. Kluwer Academic, Dordrecht (1998)
7. Douglass, M.R.: "Lifetime estimates and unique failure mechanisms of the digital micromirror device (DMD)," 1998 IEEE International Reliability Physics Symposium Proceedings 36th Annual, pp. 9–16. Anonymous IEEE, New York (1998)
8. Hornbeck, L.J.: The DMDTM projection display chip: A MEMS-based technology. MRS Bull. **26**(4), 325–327 (2001)
9. Bhushan, B.: Adhesion and stiction: Mechanisms, measurement techniques, and methods for reduction. J. Vac. Sci. Technol. B, Microelectron. Nanometer Struct. **21**(6), 2262–2296 (2003)
10. Maboudian, R., Carraro, C.: Surface engineering for reliable operation of MEMS devices. J. Adhes. Sci. Technol. **17**(4), 583–591 (2003)
11. Maboudian, R., Carraro, C.: Surface chemistry and tribology of MEMS. Annu. Rev. Phys. Chem. **55**, 35–54 (2004)
12. Zhao, Y.: Stiction and anti-stiction in MEMS and NEMS. Acta Mechanica Sinica **19**(1), 1–10 (2003). (English Series)
13. Maboudian, R., Howe, R.T.: Critical review: Adhesion in surface micromechanical structures. J. Vac. Sci.Technol. B, Microelectron. Process. Phenom. **15**(1), 1–1 (1997)
14. Ando, Y., Ino, J.: Friction and pull-off force on silicon surface modified by FIB. Sens. Actuators, A **57**(2), 83–89 (1996)
15. Ando, Y.: "The effect of relative humidity on friction and pull-off forces measured on submicron-size asperity arrays," The 2nd International Colloquium on Micro-Tribology, September 15, 1997—September 18, **238**, pp. 12–19. Anonymous Elsevier S.A, Janowice (2000)
16. Gerberich, W.W., Mook, W.M., Perrey, C.R.: Superhard silicon nanospheres. J. Mech. Phys. Solids **51**(6), 979–992 (2003)

17. Choi, C., Kim, J., and Kim, C.:"Nanoturf surfaces for reduction of liquid flow drag in microchannels," 3rd ASME Integrated Nanosystems Conference—Design, Synthesis, and Applications, September 22, 2004—September 24, pp. 47–48. Anonymous American Society of Mechanical Engineers, Pasadena (2004)
18. Song, Y., Premachandran Nair, R., Zou, M.: Adhesion and friction properties of micro/nano-engineered superhydrophobic/hydrophobic surfaces. Thin Solid Films **518**(14), 3801–3807 (2010)
19. Nair, R.P., Zou, M.: Surface-Nano-texturing by aluminum-induced crystallization of amorphous silicon. Surf. Coat. Technol. **203**(5–7), 675–679 (2008)
20. Zou, M., Wang, H., Larson, P.R.: Ni Nanodot-patterned surfaces for adhesion and friction reduction. Tribol. Lett. **24**(2), 137–142 (2006)
21. Zou, M., Cai, L., Wang, H.: Adhesion and friction studies of a nano-textured surface produced by spin coating of colloidal silica nanoparticle solution. Tribol. Lett. **21**(1), 25–30 (2006)
22. Zou, M., Cai, L., Wang, H.: Adhesion and Friction Studies of a Selectively micro/nano-textured surface produced by UV assisted crystallization of amorphous silicon. Tribol. Lett. **20**(1), 43–52 (2005)
23. Morton, B. D., Wang, H., Fleming, R. A.,: Nanoscale surface engineering with deformation-resistant core-shell nanostructures, pp. 1–8 (2011)
24. Yoon, E., Singh, R.A., Kong, H.: Tribological properties of bio-mimetic nano-patterned polymeric surfaces on silicon wafer. Tribol. Lett. **21**(1), 31–37 (2006)
25. Burton, Z., Bhushan, B.: Hydrophobicity, adhesion, and friction properties of nanopatterned polymers and scale dependence for micro- and nanoelectromechanical systems. Nano Lett. **5**(8), 1607–1613 (2005)
26. Zou, M., Seale, W., and Wang, H.: "Comparison of Tribological Performances of Nano- and Micro-Textured Surfaces," Proceedings of the Institution of Mechanical Engineers, Part N (J Nanoeng Nanosystems). **219**(3):103–10 (2005)
27. Wang, H., Premachandran Nair, R., Zou, M.: Friction study of a Ni Nanodot-oatterned surface. Tribol. Lett. **28**(2), 183–189 (2007)
28. Nix, W.D.: Elastic and plastic properties of thin films on substrates: Nanoindentation techniques. Mater. Sci. Eng., A **A234–23**, 37–44 (1997)
29. Shugurov, A., Panin, A., Chun, H.-G.:"Size effects on the mechanical properties of thin metallic films studied by nanoindentation," 8th Korea–Russia International Symposium on Science and Technolog, vol. 3, pp. 168–72. Anonymous IEEE, Piscataway (2004)
30. Son, D., Jeong, J., Kwon, D.: Film-thickness considerations in microcantilever-beam test in measuring mechanical properties of metal thin film. Thin Solid Films **437**(1–2), 182–187 (2003)
31. Kracke, B., Damaschke, B.: Measurement of Nanohardness and Nanoelasticity of thin gold films with scanning force microscope. Appl. Phys. Lett. **77**(3), 361–363 (2000)
32. Schaefer, D.M., Patil, A., Andres, R.P.: Nanoindentation of a supported Au cluster. Appl. Phys. Lett. **63**(11), 1492–1494 (1993)
33. Schaefer, D.M., Patil, A., Andres, R.P.: Elastic properties of individual nanometer-size supported gold clusters. Physical Review B **51**(8), 5322–5332 (1995). (Condensed Matter)
34. Wang, H., Zou, M., Larson, P. R.: Nanomechanical Properties of a Ni Nanodot-Patterned Surface. Nanotechnology, **19**(29), (2008)
35. Wang, H., Zou, M., Jackson, R.L.: Nanoindentation modeling of a Nanodot-patterned surface on a deformable substrate. Int. J. Solids Struct. **47**(22–23), 3203–3213 (2010)
36. Loo, Y., Willett, R.L., Baldwin, K.W.: Additive, nanoscale patterning of metal films with a stamp and a surface chemistry mediated transfer process: applications in plastic electronics. Appl. Phys. Lett. **81**(3), 562–562 (2002)
37. Donthu, S.K., Pan, Z., Shekhawat, G.S.: Near-field scanning optical microscopy of ZnO nanopatterns fabricated by micromolding in capillaries. J. Appl. Phys. **98**(2), 1–5 (2005)
38. Juang, J.Y., Bogy, D.B.: Nanotechnology advances and applications in information storage. Microsyst. Technol. **11**(8–10), 950–957 (2005)

39. Di Fabrizio, E., Cojoc, D., Cabrini, S.:"Nano-optical elements fabricated by e-beam and x-ray lithography," nano- and micro-optics for information systems, August 3,4 2003, **5225**, pp. 113–125. Anonymous SPIE, San Diego (2003)
40. Murillo, R., Van Wolferen, H.A., Abelmann, L.: "Fabrication of patterned magnetic nanodots by laser interference lithography," Proceedings of the 30th International Conference on Micro- and Nano-Engineering, September 19, 2004–September 22. Anonymous Elsevier **78–79**, 260–265 (2005)
41. Kono, Y., Sekiguchi, A., Hirai, Y.: "Study on nano imprint lithography by the pre-exposure process (PEP)," advances in resist technology and processing XXII. Anonymous SPIE—Int. Soc. Opt. Eng. USA **5753**, 912–925 (2005)
42. Yao, J., Yan, X., Lu, G.: Patterning colloidal crystals by lift-up soft lithography. Adv. Mater. **16**(1), 81–84 (2004)
43. Choi, D., Jang, S.G., Yu, H.K.: Two-dimensional polymer Nanopattern by using particle-assisted soft lithography. Chem. Mater. **16**(18), 3410–3413 (2004)
44. Chik, H., Liang, J., Cloutier, S.G.: Periodic array of uniform ZnO Nanorods by second-order self-assembly. Appl. Phys. Lett. **84**(17), 3376–3378 (2004)
45. Masuda, H., Satoh, M.: Fabrication of gold Nanodot array using anodic porous alumina as an evaporation mask. Jan. J. Appl. Phys, Part 2 **35**(1), 126–129 (1996). (Letters)
46. Masuda, H., Yasui, K., Nishio, K.: Fabrication of ordered arrays of multiple Nanodots using anodic porous alumina as an evaporation mask. Adv. Mater. **12**(14), 1031–1033 (2000)
47. Liang, J., Chik, H., Yin, A.: Two-dimensional lateral superlattices of Nanostructures: Nonlithographic formation by anodic membrane template. J. Appl. Phys. **91**(4), 2544–2544 (2002)
48. Masuda, H., Fukuda, K.: Ordered metal Nanohole arrays made by a two-step replication of honeycomb structures of anodic alumina. Science **268**(5216), 1466–1468 (1995)
49. Li, A.P., Muller, F., Birner, A.: Hexagonal pore arrays with a 50–420 Nm interpore distance formed by self-organization in anodic alumina. J. Appl. Phys. **84**(11), 6023–6026 (1998)
50. Rozhok, S., Jung, S., Chandrasekhar, V.: atomic force microscopy of nickel dot arrays with tuning fork and Nanotube probe. J. Vac. Sci.Technol. B: Microelectron. Nanometer Struct. **21**(1), 323–325 (2003)
51. Sandberg, R.L., Allred, D.D., Johnson, J.E.: A Comparison of Uranium Oxide and Nickel as Single-Layer Reflectors from 2.7 to 11.6 Nm. Proc. SPIE Int. Soc. Opt. Eng. **5193**(1), 191–203 (2004)
52. Johnson, K.L.: Conatct mechanics. Cambridge University Press, New York (1987)
53. Oliver, W.C., Pharr, G.M.: An improved technique for determining hardness and elastic modulus using load and displacement sensing indentation experiments. J. Mater. Res. **7**(6), 1564–1583 (1992)
54. Mirshams, R.A., Pothapragada, R.M.: Correlation of Nanoindentation measurements of nickel made using geometrically different indenter tips. Acta Mater. **54**(4), 1123–1134 (2006)
55. Zhou, L.G., Huang, H.: Are surfaces elastically softer or stiffer? Appl. Phys. Lett. **84**(11), 1940–1942 (2004)
56. Pethica, J. B., Tabor, D.:"Contact of characterised metal surfaces at very low loads: deformation and adhesion," Second European Conference on Surface Science, **89**, pp. 182–90. Anonymous Netherlands (1979)
57. Tabor, D.: "A Simple theory of static and dynamic hardness," Proceedings of the Royal Society of London, Series A (Mathematical and Physical Sciences), **192,** pp. 247–274 (1948)
58. Jackson, R.L., Green, I.: A finite element study of elasto-plastic hemispherical contact against a rigid flat. J. Tribol. **127**(2), 343–354 (2005)
59. Ruoff, A.L.: On the yield strength of diamond. J. Appl. Phys. **50**(5), 3354–3356 (1979)
60. Mesarovic, S.D., Fleck, N.A.: Frictionless indentation of dissimilar elastic-plastic spheres. Int. J. Solids Struct. **37**(46–47), 7071–7091 (2000)
61. Greenwood, J.A., Johnson, K.L., Matsubara, E.: A surface roughness parameter in hertz contact. Wear **100**, 47–57 (1984)

Chapter 5
Biomimetic Surfaces for Tribological Applications in Micro/Nano-Devices

R. Arvind Singh, Eui-Sung Yoon, Kahp-Yang Suh and Deok-Ho Kim

Abstract Biological examples both motivate and stimulate new scientific research. Biomimetics is the study and simulation of biological systems for desired functional properties. It involves the transformation of the underlying principles discovered in nature into man-made technology. In recent times, bio-inspiration has given rise to a new promising direction for solving the tribological issues in micro/nano-devices such as Micro/Nano-Electro-Mechanical Systems (MEMS/NEMS). This chapter highlights on the various bio-inspired approaches undertaken based on the 'Lotus Effect' to create biomimetic surfaces as novel tribological solutions for micro/nano-devices. It also presents the recent advancements in this field. Examples presented in the chapter clearly indicate that by learning from natural surfaces, the development of biomimetic surfaces would remarkably enhance the smooth operation, long-term wear durability and reliability of micro/nano-devices.

R. Arvind Singh (✉)
Energy Research Institute @ NTU (ERIAN), Nanyang Technological University (NTU), Singapore, # 06-09, CleanTech One, 1 CleanTech Loop 637141, Singapore
e-mail: arvindsingh@ntu.edu.sg; r.arvindsingh@gmail.com

E.-S. Yoon
Center for BioMicrosystems, Korea Institute of Science and Technology, Hwarangno 14-gil 5, Seongbuk-gu, Seoul 136-791, South Korea
e-mail: esyoon@kist.re.kr

K.-Y. Suh
School of Mechanical and Aerospace Engineering, Seoul National University, Seoul 151-742, South Korea

D.-H. Kim
Department of Bioengineering, University of Washington, Seattle, WA 98195, USA

S. K. Sinha et al. (eds.), *Nano-tribology and Materials in MEMS*,
DOI: 10.1007/978-3-642-36935-3_5, © Springer-Verlag Berlin Heidelberg 2013

Contents

Introduction

With the advent of miniaturized devices, such as micro/nano-electro-mechanical systems (MEMS/NEMS), micro/nano-tribology, the study of surface forces such as adhesion and friction, and wear at small-scales has become popular. Tribology is a multi-disciplinary subject that has grown beyond the fields of engineering. It invokes not only the fundamentals of mechanical engineering and metallurgy, but also the basic principles and concepts from the fields of physics and chemistry, and in recent times it is integrating biology through biomimetics.

Over the last decade, the field of MEMS/NEMS has expanded considerably as these devices have found their niche in industrial, consumer, defense, aerospace, and biomedical applications [1, 2]. However, most of these devices are sensors-based. Actuators-based devices that have elements which undergo relative mechanical motion have not been commercialized yet, mainly due to the tribological issues that include high surface forces and material removal/wear [3]. At the micro/nano-scales at which these miniaturized devices are built, due to the high surface area-to-volume ratio, surface forces such as adhesion and friction become significantly influential and they severely undermine the smooth operation and operating lifetimes of MEMS/NEMS elements that are in relative motion.

It is well-known that adhesion strongly influences friction at small-scales. At micro/nano-scales, friction is in a regime where the contribution from intrinsic adhesion can outweigh that from the asperity deformation [4]. Adhesion arises due to the contribution of various attractive forces such as capillary, electrostatic, van der Waal, and chemical forces under different circumstances [4]. Amongst these forces, the capillary force that arises due to the condensation of water from the environment is the strongest [4]. Thus, in minimizing adhesion, friction also gets minimized and therefore, solutions to reduce adhesion are very important to realize smooth motion between tiny components at small-scales. Hence, there is a need to modify surfaces

Table 5.1 Water contact angle (*WCA*) values and micro/nano-tribological properties of biomimetic surfaces

Material	WCA (deg)	Nano-scale adhesion force (nN)	Nano-scale friction force (nN)	Micro-scale friction coefficient	References
Silicon	22	984	20–28	0.46	[6, 13]
PMMA thin film	69	872	5–8	0.66	[6, 13]
PMMA nano-patterns	91–99	30–40	0.51–1.97	0.24–0.33	[6, 13]
PMMA micro-patterns(replicated surfaces)	91–106	–	–	0.11–0.13	[22–24]
DLC thin film	78	226	5.5–8.3	0.19	[25]
DLC nano-dot surfaces	95	67	0.98–1.96	0.13	[25]
Silicon micro-pillars	59	–	–	0.16	[28]
Silicon micro-pillars coated with DLC	145	–	–	0.12	[28]
Silicon micro-pillars coated with ZDOL	105	–	–	0.08	[28]

in order to achieve increased hydrophobicity that will drastically reduce adhesion arising due to the capillary force, which in turn would also reduce friction.

Traditionally, silicon is the widely used material for the fabrication of MEMS/NEMS devices. However, silicon does not have good tribological properties, owing to its hydrophilic nature (Table 5.1) and brittleness [5]. It shows high adhesion and friction at nano-scale, owing to the high capillary force that forms meniscus/water-bridges with the counterface (Table 5.1) [5, 6]. At micro-scale, silicon shows high friction value (Table 5.1) and undergoes severe wear generating large amounts of wear debris [5]. Thus, the inherent hydrophilic and brittle nature of silicon makes it unsuitable for tribological applications in MEMS/NEMS. Hence, most of the tribological solutions for MEMS/NEMS devices are focused towards enhancing the micro/nano-tribological properties of silicon. In the past, tribologists have investigated lubricant coatings such as self-assembled monolayers (SAMs), diamond-like carbon (DLC) coatings and perfluoropolyether (PFPE) coatings in order to enhance the tribological performance of silicon surfaces through chemical modification approach [7–9]. In recent years, bio-inspired topographical modifications of silicon surfaces have emerged as the novel route to effectively enhance the tribological properties of the material.

In this chapter, we present a brief overview on the bio-inspired approaches undertaken based on the 'Lotus Effect' to create biomimetic surfaces as the new solutions to the tribological issues found in miniaturized devices. Recent developments that have advanced these solutions have also been highlighted.

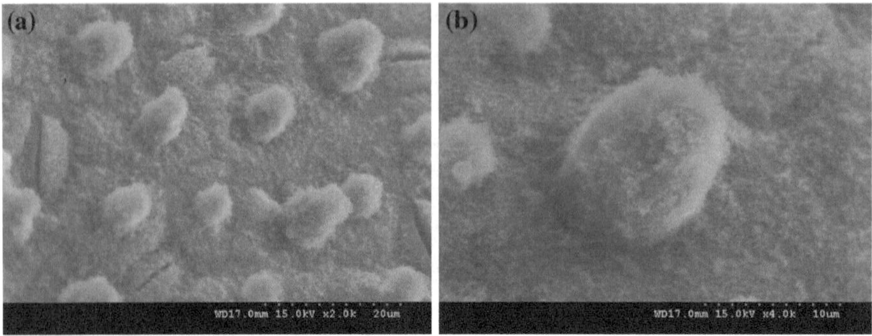

Fig. 5.1 SEM images of a real lotus leaf surface: **a** micro-scale protuberances on the leaf surface and **b** wax present on a protuberance

Lotus Effect

The unique ability of lotus leaf surface to avoid getting wet is popularly known as the "Lotus Effect". Lotus (*Nelumbo nucifera*) leaves have superhydrophobic property (water contact angle $\sim 162^{\circ}$ [10]). Their leaves have micro-scale protuberances and waxy nano-crystals covering the protuberances (Fig. 5.1). These features give rise to the unique ability of lotus leaves to avoid getting wet by the surrounding water [10]. Such an effect arises due to heterogeneous/composite wetting of water, wherein water drop sits on top of the protuberances with air trapped in between them, reducing the solid contact area with water drop [11]. Inspired by the protuberances that lower surface energy and reduce the contact area, scientists have topographically modified silicon surfaces that mimic the protuberances on lotus leaves in order to achieve hydrophobicity and reduction in contact area, so as to decrease adhesion and friction. It is noteworthy that although in nature, lotus leaves do not have any tribological activity related to them [12], the underlying principles of the features present on their leaves inspire novel tribological solutions.

Biomimetic Surfaces

Polymeric Nano-Patterns

Inspired by the intriguing features on lotus leaf surfaces, researchers have fabricated nano-scale polymeric patterns on silicon surfaces that mimic the protuberances on lotus leaves [13]. Investigation of micro/nano-scale tribological properties of the surfaces showed that they reduced adhesion and friction forces significantly due to their hydrophobic nature and reduced contact area. These surfaces were recognized as prospective tribological candidates for MEMS/NEMS [14]. This was

Fig. 5.2 Morphology of the biomimetic polymeric nano-patterns fabricated using capillary force lithography, at the holding times of: **a** 15 min, **b** 30 min and **c** 60 min [16]

the pioneering bio-inspired work that opened a new direction towards enhancing micro/nano-scale tribological properties by creating biomimetic surfaces.

The nano-scale polymeric patterns were created on poly(methyl meth-acrylate) (PMMA) polymer spin coated on silicon wafers by using capillary force lithography [13]. This technique principally utilizes the competition between capillary and hydrodynamic forces in the course of pattern formation. The nano-patterns were fabricated on PMMA films, as this is a polymer often found in MEMS devices [15]. Polyurethane acrylate (PUA) stamps were used to fabricate these nano-patterns. The processing sequence of pattern formation can be found in Ref. [13]. Figure 5.2 shows scanning electron microscopy (SEM) images of the nano-patterns fabricated at different holding times [16]. Water contact angle measurements showed that these surfaces were hydrophobic in nature when compared to silicon wafer and PMMA thin film (Table 5.1) [13]. Investigation on nano-scale adhesion and friction properties using atomic force microscopy (AFM) showed that these samples exhibited superior tribological properties when compared to silicon wafer and PMMA thin film [5, 13]. Their adhesion force and friction force were significantly lower than those of silicon and PMMA thin film (Table 5.1) [5, 13]. The drastic reduction in adhesion and friction forces observed in these nano-patterns was attributed to their hydrophobic nature (hydrophobicity reduces inherent adhesion and thereby reduces friction [4]) and the reduction in contact area [5, 13]. According to the friction law given by Bowden and Tabor, the friction force directly depends on the real area of contact, for a single asperity contact. ($F_f = \tau A_r$, where τ is the shear strength, an interfacial property, and A_r the real area of contact) [17]. Thus, any reduction in real area of contact directly reduces friction between surfaces.

Approximate calculations of the contact area in the case of the nano-patterns, based on simple geometry showed that the contact area reduced by almost an order of magnitude [13], which is quite significant to reduce both adhesion and friction forces in the patterned samples. Amongst the nano-patterns, the ones fabricated at the highest holding time of 60 min showed higher adhesion and friction properties when compared to the rest. These fully grown patterns undergo elastic deformation upon their contact with the AFM tip, thereby increasing the contact area, and in turn increasing the adhesion and friction values [13]. Figure 5.3 shows AFM traces during the friction measurements on the nano-patterns formed at the holding time of 60 min. The gap/lag between the trace and retrace curves shown by the arrows is

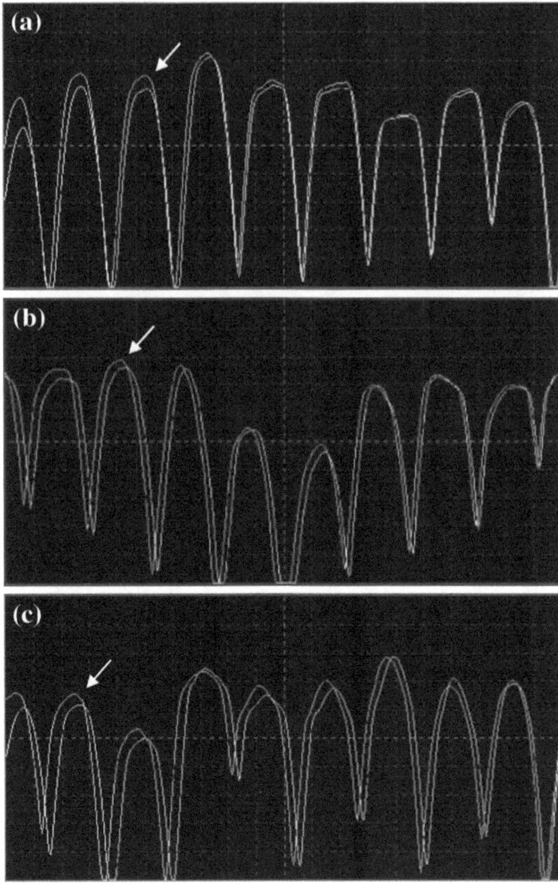

Fig. 5.3 AFM traces during the friction measurements on the PMMA polymeric nano-patterns formed at the holding time of 60 min, at the applied normal load of: **a** 20 nN, **b** 40 nN and **c** 80 nN. The gap/lag between the trace and retrace curves shown by the *arrows* is indicative of the occurrence of elastic deformation of the nano-pillars. Note the increase in the gap/lag with the increase in the applied normal load

indicative of the occurrence of elastic deformation. Understandably, the gap/lag increases with the increase in the applied normal load. Using the capillary force lithography technique which is simple, cost effective, and less time consuming, patterns with different sizes and shapes can be formed by varying the parameters of time and temperature [18]. The variation in size and shape of patterns influence their adhesion and friction properties at micro/nano-scale [18–20]. In case, if the polymeric material from which the patterns are formed is hydrophilic, then the patterns can be made hydrophobic by coating them with a hydrophobic film such as perfluorodecyltriethoxysilane (PFDTES) self-assembled monolayer, so as to improve their tribological properties [20]. The PMMA polymer nano-patterns which showed excellent tribological properties at nano-scale [13] also showed lower coefficient of friction values at micro-scale when compared to those of silicon and PMMA thin film (Table 5.1) [5, 13]. However, this was observed only for a short duration of time ($\sim$5 min), after which the nano-patterns exhibited wear [13, 21].

Replicated Surfaces of Water-Repellent Leaves

The fact that the PMMA nano-patterns underwent wear at micro-scale revealed that they do not have good load-bearing capacity at that scale. With the view that the solution towards having patterned surfaces with good load-bearing capacity at micro-scale may lie in creating surfaces with micron-sized patterns, direct replication of the surfaces of natural lotus (*Nelumbo nucifera*) and colocasia (*Colocasia esculenta*) leaves were conducted, and the replicated surfaces were investigated for their micro-frictional behavior [22–24]. Similar to lotus, the colocasia plant also has water-repellent leaves (water contact angle $\sim 164^{o}$), due to micro-scale protuberances and wax on its leaves [10]. The replication of the leaves was done by using the same capillarity-based lithographic technique that was used to fabricate the PMMA nano-patterns, using the real leaves as the natural templates. Two different molds (PUA and PDMS molds) and molding techniques were used to replicate the surfaces of real leaves. The replication procedures can be found in Refs. [22, 23]. Indeed, this is the first biomimetic approach of creating effective tribological surfaces by the direct replication of natural surfaces.

Figure 5.4a shows a SEM image of a lotus-like surface (replicated surface) fabricated using a PUA mold [22]. These surfaces are hydrophobic in nature (Table 5.1) [22]. Results from micro-friction tests showed that these surfaces exhibited values of coefficient of friction that were ~ 4 times lower than that of silicon wafer and ~ 6 times lower than that of PMMA thin film (Table 5.1) [22]. The superior micro-friction behavior of these replicated surfaces is attributed to their hydrophobic nature and the reduced contact area projected by them. Observations of the surfaces after the friction tests by using SEM showed no wear of the micro-patterns (i.e. no complete removal of the patterns from the surface). However, roughening and plastic deformation at their tips were seen [22]. In another work, replication of the surfaces of lotus and colocasia leaves were done by using poly(dimethyl siloxane) (PDMS) molds [23, 24]. The colocasia leaf surface was replicated both in its fresh and also in its dried conditions, as the leaf upon drying exhibits a significant change in its morphology. The 'bumpy' protuberances in its fresh condition become 'depressions' upon drying, leaving behind a bump at the center and a ridge that surrounds each bump. Therefore, both fresh and dry colocasia leaves were replicated and tested. Figure 5.4b–d shows the SEM images of the replicated surfaces. Measurement of the water contact angles showed that the replicated surfaces were hydrophobic in nature (Table 5.1) [23, 24]. Micro-friction tests conducted for these surfaces showed that the values of the coefficient of friction of the surfaces were ~ 3.5 times lower than that of silicon wafer and ~ 5 times lower than that of PMMA thin film (Table 5.1) [23, 24]. No wear was observed on these surfaces after the tests, but roughening and plastic deformation at their tips was seen [23, 24]. From Fig. 5.4, it could be also seen that the real leaves surfaces have been replicated on a smoother scale, even though the real leaves have hierarchical roughness due to the presence of micro-scale protuberances and nano-scale wax on them [10]. It is quite possible to reproduce the

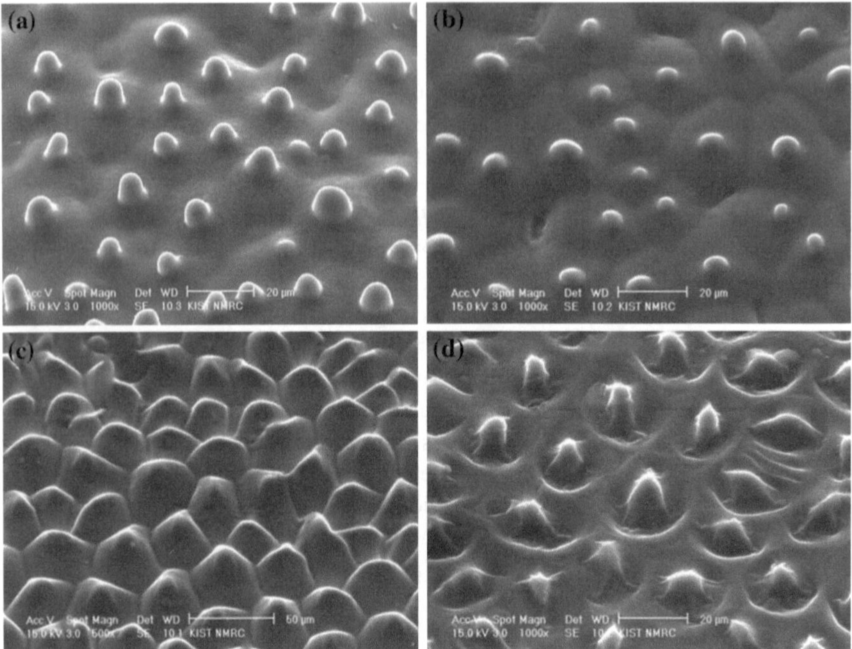

Fig. 5.4 Replicated surfaces of real leaves of water-repellant plants: **a** lotus-like (*fresh*) surface using PUA mould [22], **b** lotus-like (*fresh*) surface, **c** colocasia-like (*fresh*) surface and **d** colocasia-like (*dry*) surface. The replicated surfaces shown in **b–d** were fabricated using PDMS molds [23, 24]

surface topographies of real leaves with high accuracy, but the process would be relatively time-consuming. Most importantly, for the enhancement of micro-frictional property through the reduction in contact area, a simple replication of the surface structures of natural leaves in a fast and efficient route would suffice. This is clearly demonstrated from these works [22–24].

DLC Nano-Dot Surfaces

DLC nano-dot surfaces created by depositing diamond-like carbon (DLC) films on nano-sized nickel (Ni) dots fabricated on silicon wafers by annealing Ni thin films, is another biomimetic surface that has good micro/nano-tribological properties [25]. Figure 5.5a shows an AFM image of the DLC nano-dot surface with features resembling the protuberances on lotus leaves, at nano-scale. The surface is hydrophobic in nature (Table 5.1) [25]. At the nano-scale, the adhesion force of the nano-dot surface was lower by an order of magnitude when compared to those of the silicon wafer and DLC coated silicon wafer (Table 5.1) [6, 25]. Further, the

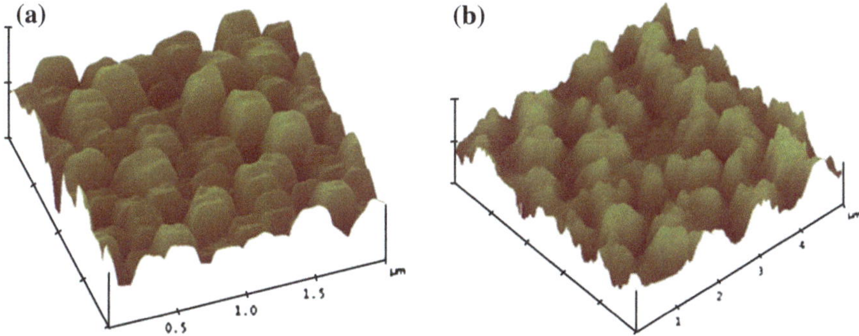

Fig. 5.5 AFM images of: **a** DLC nano-dot surface (scan area: 2 × 2 μm, Z scale: 100 nm) [25] and **b** SU8 polymer film exposed to oxygen plasma and coated with PFPE nanolubricant (scan area: 5 × 5 μm, Z scale: 1 μm) [29]

nano-dot surface showed friction force values that were considerably lower than those of the silicon wafer and DLC coated silicon wafer (Table 5.1) [6, 25]. At the micro-scale, the friction coefficient of the nano-dot surface was ~3.5 and 1.5 times lower than those of silicon wafer and DLC coated silicon wafer, respectively (Table 5.1) [25]. Unlike the other two surfaces, the nano-dot surface did not show any wear at the micro-scale [25]. The reduction of the surface forces in the nano-dot surface is due to its hydrophobic property and the reduced contact area when compared to the flat silicon and DLC coated silicon surfaces.

Silicon Micro-Patterns with Lubrication

Interestingly, direct experiments on real lotus leaf surfaces have shown that the wax on their surface can act as a lubricant and can reduce friction force at micro-scale [26]. The friction coefficient of the leaves in their fresh condition (~0.01) was found to be significantly lower than in their dried conditions (~0.15), owing to the presence of wax in its fresh condition, which acts like a lubricant at the surface and assists in lowering the friction property [26]. Taking the inspiration from the surface features of lotus leaves i.e. the protuberances that can reduce surface energy and contact area, and wax that can further reduce surface energy [27] and can also lubricate/reduce interfacial shear strength, researchers fabricated cylindrical micro-pillars on silicon wafers using photolithography and coated them with DLC and ZDOL lubricants [28]. Results showed that the surfaces were hydrophobic in nature (Table 5.1) [28]. It was found that the silicon micro-pillars without lubricant coatings showed coefficient of friction value that was lower by ~2.8 times than that of the silicon wafer, due to the reduction in contact area by the micro-pillars (Table 5.1) [28]. Upon coating with the lubricants, the silicon micro-pillars showed coefficient of friction value that was lower by ~4 to 6 times than that of the silicon wafer (Table 5.1) [28]. The lubricants on the top of the micro-pillars further reduce the friction property in the similar fashion as the wax that acts as a lubricant on the lotus leaf surfaces.

Recent Advancements

As seen in the previous section, biomimetic polymeric micro/nano-patterns are very effective in reducing the surface forces at small-scales due to their hydrophobic nature/reduced surface energy and reduction in contact area. However, while the reduction in contact area decreases the friction force, on the other hand it also increases the contact pressure, due to which these polymeric micro/nano-patterns showed low wear durability at micro-scale (occurrence of severe wear/plastic deformation in $\leq$15 min at the load of 3 mN and the speed of 1 mm/s) [13, 22–24]. It is important to note that the wear durability of a tribological surface is an important factor as it defines the useful operating lifetime. Hence, modifications of surfaces for tribological applications should not only result in the reduction of surface forces, but should also effectively increase their wear durability. In order to reduce the surface forces and simultaneously increase the wear durability, surfaces should be topographically and chemically modified by the application of nanolubricants. This is demonstrated by the two recent works highlighted below.

Textured Polymeric Surfaces with Lubrication

Bio-inspired textural and chemical modifications were carried out on SU8 polymer thin films (thickness $\sim$1.8 μm) spin coated on silicon wafers [29]. The polymer films were exposed to oxygen plasma (power: 150 W) using a reactive ion etching machine for a duration of 1 min. After the oxygen plasma treatment, the polymer films were coated with perfluoropolyether (PFPE) nanolubricant, using a dip coating technique. The samples were dipped into a beaker containing the solution of PFPE (0.5 weight % in H-Galden solvent) for a duration of 1 min. Figure 5.5b shows an AFM image of SU8 polymer film treated with oxygen plasma and coated with PFPE (SU8 TF/O_2/PFPE), which has nano-scale texture on it, resembling the protuberances on a lotus leaf, at nano-scale. Analyzing the AFM image for the roughness showed that the SU8 TF/O_2/PFPE surface had a roughness value which was higher by three orders of magnitude (R_a $\sim$178 nm) when compared to that of the polymer film (R_a $\sim$0.23 nm). The increase in the roughness of the polymer film when exposed to oxygen plasma is due to the etching effect of the polymeric material by the energetic plasma [30]. Water contact angle measurements showed that the polymer film was semi-hydrophobic (WCA $\sim$68°) and when coated with PFPE nanolubricant (SU8 TF/PFPE) it became near-hydrophobic (WCA $\sim$80°). When the polymer film was exposed to oxygen plasma its surface became extremely hydrophilic (WCA $\sim$5°) and upon coating the plasma treated sample with PFPE nanolubricant the surface

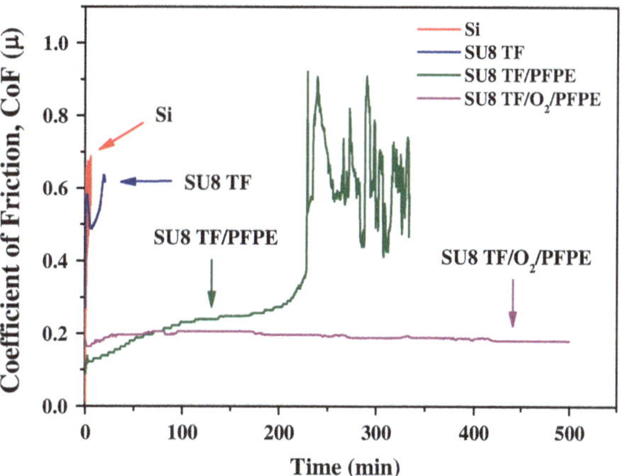

Fig. 5.6 Coefficient of friction as a function of time (applied normal load: 0.3 N, speed: 42 mm/s), of silicon (Si), SU8 polymer film (SU8 TF), SU8 polymer film coated with PFPE (SU8 TF/PFPE), and the SU8 polymer film treated with oxygen plasma and subsequently coated with PFPE i.e. SU8 TF/O_2/PFPE surface [29]

exhibited hydrophobicity (SU8 TF/O_2/PFPE, WCA ~105°). The hydrophilic nature of the plasma treated sample is due to the generation of active/polar carboxyl groups (-COOH) [28]. PFPE nanolubricant when coated onto any surface increases the water contact angle due to the presence of fluorine atoms in its molecular structure [9, 31]. The SU8 TF/O_2/PFPE surface showed hydrophobicity owing to its nano-scale roughness and the presence of PFPE material.

Figure 5.6 shows the coefficient of friction as a function of time (applied normal load: 0.3 N, speed: 42 mm/s), of silicon (Si), SU8 polymer film (SU8 TF), SU8 polymer film coated with PFPE (SU8 TF/PFPE), and the SU8 polymer film treated with oxygen plasma and subsequently coated with PFPE i.e. SU8 TF/O_2/PFPE surface. Experiments were stopped when the surfaces of the test materials showed evidences of surface failure (coefficient of friction >0.3 and/or traces of wear at the surface). Figure 5.7 shows the optical images of the surfaces of Si, SU8 TF, SU8 TF/PFPE and SU8 TF/O_2/PFPE, taken after the friction tests. Si and SU8 TF showed high friction coefficients of ~0.7 and 0.6, respectively. Both these materials had low wear durability <0.5 min and exhibited severe wear. The plasma treated surface showed high friction coefficient of ~0.7 and low wear durability (t < 0.5 min), because of its hydrophilic property arising due to the presence of carboxyl functional groups. SU8 TF/PFPE showed low friction coefficient of ~0.22, owing to its higher water contact angle value/low surface energy and the presence of PFPE nanolubricant. SU8 TF/O_2/PFPE showed low friction coefficient of ~0.17, due to its hydrophobic nature/low surface energy, lower contact area (because of the texture/high surface roughness), and the lubrication effect of PFPE. Further, SU8 TF/O_2/PFPE showed high wear durability of >500 min (t > 8 h) that was >2 times than that of the SU8 TF/PFPE (t ~ 216 min i.e. 3.6 h), which is mainly due to the chemical bonding of the PFPE nanolubricant on its surface. For SU8 TF/O_2/PFPE, the tests were stopped at ~500 min due to long test duration (>8 h). It is to be noted that the molecules of PFPE nanolubricant have hydroxyl (-OH) groups at both their terminal ends (PFPE: OH-CH_2-CF_2-[(O-CF_2-CF_2)$_p$-(O-CF_2)$_q$]-OCF_2-CH_2-OH), and the plasma treated SU8 film surface (SU8

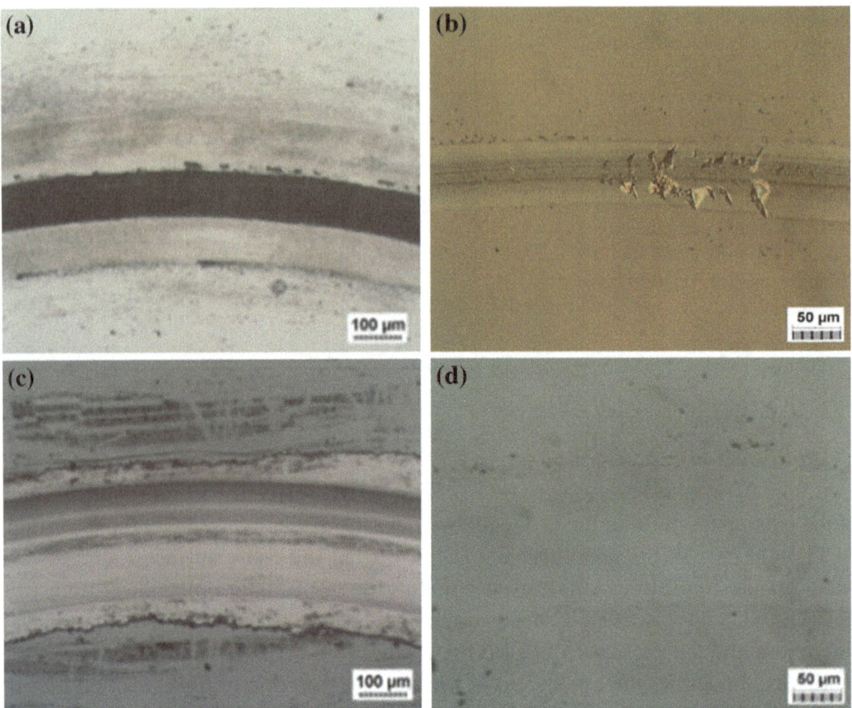

Fig. 5.7 Optical images of surfaces of the test materials taken after the friction tests: **a** Si (at ~5 min), **b** SU8 TF (at ~20 min), **c** SU8 TF/PFPE (at ~6 h) and **d** SU8 TF/O_2/PFPE (at ~8.3 h). Si and SU8 TF show severe wear within a short duration of time. SU8 TF/O_2/PFPE shows no recognizable wear on its surface even after a long duration of time [29]

TF/O_2) has active/polar groups (-COOH) [30]. As both the PFPE molecules and the plasma treated SU8 polymer thin film surface have polar reactive chemical groups, when SU8 TF/O_2 is coated with PFPE, strong chemical bonds form between them [32, 33]. Due to this reason, PFPE molecules get anchored strongly, thereby resisting their removal under shear loading, which increases their wear durability. Thus, it was seen that the texturally (O_2 plasma treated) and chemically modified (PFPE coated) SU8 polymer thin film (SU8 TF/O_2/PFPE, bio-inspired surface) showed the best tribological results with low friction property and high wear durability (>1,000 times higher than that of Si), indicating that it can perform as a good tribological candidate for silicon based miniaturized devices.

Polymeric Micro-Patterns with Lubrication

In a recent work, biomimetic polymeric micro-patterns have been made significantly durable by chemically anchoring PFPE nanolubricant [34]. This could be the first report on enhancing the wear durability of any polymeric patterns, to date.

Fig. 5.8 Biomimetic polymeric micro-patterns fabricated by nano imprint lithography, which were made significantly durable by chemically anchoring PFPE nanolubricant [34]

In this work, SU8 micro-patterns in the form of micro-bumps with dimensions: diameter (top) ~6 μm, diameter (bottom) ~10 μm, height ~1.5 μm and distance between any two adjacent micro-bumps ~10 μm, were fabricated on SU8 thin films (thickness ~3 μm) spin coated on silicon wafers, using nano imprint lithography. Figure 5.8 shows a representative SEM image of the fabricated SU8 micro-patterns. The height of ~1.5 μm was selected for the micro-patterns as the clearance/gap between the elements of MEMS devices is usually about a few microns [1], and also for the reason that patterns with high aspect ratio tend to undergo deformation when subjected to external loads [13, 19].

The PFPE nanolubricant was applied to the fabricated SU8 micro-patterns in two different ways [34]: (1) directly and (2) after argon/oxygen plasma treatment, for a duration of 1 min. Water contact angle measurements showed that the micro-patterns were hydrophobic (WCA ~105°), owing to the heterogeneous/composite wetting of water, similar to that of the surface of a lotus leaf. The micro-patterns when coated with PFPE, the WCA value increased to ~108°. The micro-patterns when treated with argon and oxygen plasmas became hydrophilic with WCA values of ~61 and 5°, respectively. The hydrophilic property of the plasma treated surfaces is due to the activation of the polymeric surfaces by the creation of polar functional groups: (1) argon plasma generates carbonyl groups (C–O and C–O–O groups) and (2) oxygen plasma generates carboxyl groups (COOH groups). Further, when the argon and oxygen plasma treated micro-patterns were coated with PFPE their WCA values increased to ~83 and 91°, respectively.

Friction tests conducted at 0.3 N and 15 mm/s, as a function of time, revealed that SU8 micro-patterns showed high value of coefficient of friction ($\mu \geq 0.3$) and low wear durability (t < 0.5 min). In a short duration of time, plastic deformation of micro-bumps was observed. The micro-patterns coated with PFPE and those treated with argon/oxygen plasmas and coated with PFPE showed lower values of coefficient of friction of <0.22. These surfaces showed high wear durability of >500 min (t > 8 h), which was >1,000 times higher than those of the silicon wafer and the SU8 micro-patterns. After the tests, no recognizable wear was observed on these surfaces. Further, friction tests were conducted on these surfaces at the higher conditions of load and speed of 1 N and 42 mm/s, respectively. At these conditions, although all these surfaces showed lower values of coefficient of friction of <0.2, clear difference was observed in their wear durability. The micro-patterns coated with PFPE had the lowest wear durability of $\sim$14 min. The micro-patterns treated with argon plasma and coated with PFPE showed increased durability of $\sim$32 min, and those treated with oxygen plasma and coated with PFPE had the highest wear durability of $\sim$60 min.

In the case of micro-patterns coated with PFPE, the nanolubricant was applied over the regular SU8 material, whereas in those treated with argon/oxygen plasmas; the nanolubricant was coated over chemically activated SU8 material. As mentioned earlier, the molecules of PFPE nanolubricant have hydroxyl group at both their terminal ends, and the plasma treated surfaces have polar functional groups, due to which strong chemical bonds form between the two. As a result, the nanolubricant gets chemically anchored onto the plasma treated surfaces giving rise to their increased wear durability. The difference in the wear durability amongst the argon and oxygen plasma treated surfaces coated with PFPE could be attributed to the type of bonding that would occur between PFPE molecules and chemically activated surfaces. Hydroxyl groups form covalent bonds with carboxyl groups [32, 33], whereas with carbonyl groups they form hydrogen bonds [35, 36]. Covalent bonds are stronger than hydrogen bonds [37], and when nanolubricants are covalently bonded onto surfaces, the surfaces exhibit significant improvement in their wear resistance capacity/wear durability [38]. Hence, it was found that the micro-patterns treated with oxygen plasma and coated with PFPE showed higher wear durability than those treated with argon plasma and coated with PFPE. Polymeric micro-patterns with enhanced wear durability are the best candidates for MEMS/Bio-MEMS devices. Also, in other applications such as optoelectronics and photonics, polymeric patterns are usually employed, and in addition to their regular function if an application requires them to perform reliable mechanical motion, micro-patterns such as the ones in this work [34] will be the attractive candidates for such applications.

Summary

Inspired by nature and encouraged by the early success, scientists and researchers are exploring biomimetic approaches as never before. By integrating biomimetics into tribology, they are exploring nature-based solutions for tribological issues that confound the smooth operation and operating lifetimes of micro/nano-devices. Examples given in this chapter highlight that the biomimetic/bio-mimetically engineered surfaces reduce adhesion and friction forces through the reduction in surface energy and contact area, and by the lubrication effect. For their real-time application in miniaturized devices, it is also important that these surfaces should have high wear durability/wear resistance. Recent advancements in the field have shown that the solution to enhance the wear durability of biomimetic surfaces lies in the pre-functionalization of their surfaces to chemically bind the nanolubricant molecules. It is also seen that polymer science is closely connected with most of these biomimetic approaches as it offers a wide range of materials and easy routes of fabrication to mimic biological surfaces.

Acknowledgments Dr. R. Arvind Singh wishes to acknowledge the financial support given to a part of the work by the National Research Foundation (NRF), Singapore (Award no. NRF-CRP 2-2007-04; Project Title: Biologically-Inspired Design, Nano-Fabrication and Nano-Lubrication of MEMS, NEMS and Micro-Mechanical Devices). Dr. D. -H. Kim thanks Department of Bioengineering at the University of Washington for the new faculty startup fund.

References

1. Adams, T.M., Layton, R.A.: Introductory MEMS: Fabrication and Applications. Springer, Berlin (2010)
2. Kim, D.H., Lee, H.J., Lee, Y.K., Nam, J.M., Levchenko, A.: Biomimetic nanopatterns as enabling tools for analysis and control of live cells. Adv. Mater. **22**, 4551–4566 (2010)
3. Kim, S.H., Asay, D.B., Dugger, M.T.: Nanotribology and MEMS. Nanotoday **2**, 22–29 (2007)
4. Maboudian, R., Howe, R.T.: Critical review: Adhesion in surface micromechanical structures. J. Vac. Sci. Technol. B **15**, 1–20 (1997)
5. Yoon, E.S., Arvind Singh, R., Oh, H.J., Kong, H.: The effect of contact area on nano/micro-scale friction. Wear **259**, 1424–1431 (2005)
6. Arvind Singh, R., Yoon, E.S.: Biomimetics in tribology—recent developments. J. Korean Phys Soc **52**, 661–673 (2008)
7. Arvind Singh, R., Yoon, E.S., Kim, H.J., Kong, H., Park, S.J., Lee, K.R.: Friction behaviour of diamond-like carbon films with varying mechanical properties. Surf. Coat. Technol. **201**, 4348–4351 (2006)
8. Arvind Singh, R., Yoon, E.S., Han, H.G., Kong, H.: Friction behavior of chemical vapor deposited self-assembled monolayers on silicon wafer. Wear **262**, 130–137 (2007)
9. Satyanarayana, N., Sinha, S.K.: Tribology of PFPE overcoated self-assembled monolayers deposited on Si surface. J. Phys. D Appl. Phys. **38**, 3512–3522 (2005)
10. Neinhuis, C., Barthlott, W.: Characterization and distribution of water-repellent, self-cleaning plant surfaces. Ann. Bot. **79**, 667–677 (1997)
11. Gould, P.: Smart, clean surfaces. Mater. Today **6**(11), 44–48 (2003)
12. Arvind Singh, R., Yoon, E.S., Jackson, R.L.: Biomimetics: The science of imitating nature. Tribol. Lubr. Technol. **65**(2), 40–47 (2009)
13. Yoon, E.S., Singh, R.A., Kong, H., Kim, B., Kim, D.H., Jeong, H.E., Suh, K.Y.: Tribological properties of bio-mimetic nano-patterned polymeric surfaces on silicon wafer. Tribol. Lett. **21**, 31–37 (2006)
14. Gnecco, E., Meyer, E.: Fundamentals of Friction and Wear on the Nanoscale, NanoScience and Technology Series. Springer, Berlin (2007)
15. Bhushan, B.: Springer Handbook of Nanotechnology. Springer, Berlin (2004)
16. Yoon, E.-S., Singh, R.A., Kong, H., Kim, B.K., Kim, D.H., Suh, K.Y., Jeong, H.E:. Tribological properties of nano/micro-patterned PMMA surfaces on silicon wafer, WTC2005-63964, 3rd World Tribology Congress, Washington DC, USA, (2005)
17. Bowden, F.P., Tabor, D.: The Friction and Lubrication of Solids. Clarendon Press, Oxford (1950)
18. Suh, K.Y., Jeong, H.E., Kim, D.-H., Arvind Singh, R., Yoon, E.S.: Capillarity-assisted fabrication of nanostructures using a less permeable mold for nanotribological applications. J. Appl. Phys. **100**(034303), 1–5 (2006)

19. Arvind Singh, R., Yoon, E.S., Kim, H.J., Kong, H., Suh, K.Y., Jeong, H.E.: Topographical modification of polymeric surfaces by capillarity directed soft-lithographic technique for improved tribological properties. ASIATRIB 2006, Kanazawa, Japan, pp. 117–118, 16–19 Oct 2006
20. Burton, Z., Bhushan, B.: Hydrophobicity adhesion, and friction properties of nanopatterned polymers and scale dependence for micro- and nanoelectromechanical systems. Nano Lett. **5**, 1607–1613 (2005)
21. Arvind Singh, R., Yoon, E.S.: Friction of chemically and topographically modified Si (100) surfaces. Wear **263**, 912–919 (2007)
22. Arvind Singh, R., Yoon, E.S., Kim, H.J., Kong, H., Park, S., Jeong, H.E., Suh, K.Y.: Enhanced tribological properties of lotus leaf-like surfaces fabricated by capillary force lithography. Surf. Eng. **23**, 161–164 (2007)
23. Arvind Singh, R., Yoon, E.S., Kim, H.J., Kim, J., Jeong, H.E., Suh, K.Y.: Replication of surfaces of natural leaves for enhanced micro-scale tribological property. Mat. Sci. Eng. C: Biomimetic Supramolecular Syst **27**, 875–879 (2007)
24. Arvind Singh, R., Kim, H.J., Kong, H., Yoon, E.S., Jeong, H.E., Suh, K.Y.: A biomimetic approach for effective reduction in micro-scale friction by direct replication of topography of natural water-repellent surfaces. J. Mech. Sci. Technol. **21**, 624–629 (2007)
25. Arvind Singh, R., Na, K., Yi, J.W., Lee, K.R., Yoon, E.S.: DLC nano-dot surfaces for MEMS tribology. Appl. Surf. Sci. **257**, 3153–3157 (2011)
26. Kim, H.J., Arvind Singh, R., Yoon, E.S., Kong, H.: Surface characteristics and micro-scale friction property of natural surfaces. Proceedings of the KSTLE Fall Conference 2005, pp. 115–119. Seoul, South Korea, (2005)
27. Cheng, Y.T., Rodak, D.E., Wong, C.A., Hayden, C.A.: Effects of micro- and nano-structures on the self-cleaning behaviour of lotus leaves. Nanotechnology **17**, 1359–1362 (2006)
28. Arvind Singh, R., Pham, D.C., Kim, J., Yang, S., Yoon, E.S.: Bio-inspired dual surface modification to improve tribological properties at small-scale. Appl. Surf. Sci. **255**, 4821–4828 (2009)
29. Arvind Singh, R., Satyanarayana, N., Sinha, S.K.: Bio-inspired advanced materials for reducing friction and wear in MEMS devices. International Conference on the Advancement of Materials and Nanotechnology (ICAMN), p. 29. Kuala Lumpur, Malaysia, 1 Nov (2010)
30. Walther, F., Davydovskaya, P., Zurcher, S., Kaiser, M., Herberg, H., Gigler, A.M., Stark, R.W.: Stability of the hydrophilic behavior of oxygen plasma activated SU-8. J. Micromech. Microeng. **17**, 524–531 (2007)
31. Satyanarayana, N., Gosvami, N., Sinha, S.K., Srinivasan, M.P.: Friction, adhesion, and wear durability of an ultra-thin perfluoropolyether-coated 3-glycidoxypropyltrimethoxy silane self-assembled monolayer on a Si surface. Phil. Mag. **87**, 3209–3227 (2007)
32. Wu, C.S., Liao, H.T.: Study on the preparation and characterization of biodegraded polyactide/multiwalled carbon nanotubes nanocomposites. Polymer **48**, 4449–4458 (2007)
33. Kumeta, K., Nagashima, I., Matsui, S., Mizoguchi, K.: Crosslinking reaction of poly(vinyl alcohol) with poly(acrylic acid) (PAA) by heat treatment: Effect of neutralization of PAA. J. Appl. Polym. Sci. **90**, 2420–2427 (2003)
34. Arvind Singh, R., Siyuan, L., Satyanarayana, N., Kustandi, T. S., Sinha, S.K.: Bio-inspired polymeric patterns with enhanced wear durability for microsystems applications. Mat. Sci. Eng. C Accepted, July 2011
35. Su, Y.C., Kuo, S.W., Yei, D.R., Xu, H., Chang, F.C.: Thermal properties and hydrogen bonding in polymer blend of polybenzoxazine/poly(N-vinyl-2-pyrrolidone). Polymer **44**, 2187–2191 (2003)
36. Manjunath, M.S., Sivagurunathan, P., Sannappa, J.: Studies of hydrogen bonding between N, N-Dimethylacetamide and primary alcohols. E-J. Chem. **6**, 143–146 (2009)
37. Israelachvili, J.N.: Intermolecular and Surface Forces, p. 125. Academic Press, USA (1991)
38. Yang, G.H., Dai, S., Cheng, G., Zhang, P., Du, Z.: Evaluation of nano-frictional and mechanical properties of a novel Langmuir-Blodgett monolayer/self-assembly monolayer composite structure. J. Mat. Res. **22**, 2429–2436 (2007)

Chapter 6
Molecular Simulation of Polymer Nanotribology

Y. K. Yew, Z. C. Su, Sujeet K. Sinha and V. B. C. Tan

Abstract In order to improve the tribological complication between miniature moving components, such as those used in micro-electromechanical systems, polymer-based coatings have been found to be an effective solution. We have examined the effectiveness of polymer-based coating by using molecular mechanics simulation, and explained the frictional mechanism of polymer at nanometer length scale. Results from the simulations show that the frictional force of the sliding surfaces is strongly dependent on surface structures and external load, respectively. The frictional contact between two sliding polymers can be attributed to the interplay between interfacial adhesion and deformation of the polymers. This study also showed the promising future of molecular simulation in complementing the laboratory experiments as well as predicting the physical phenomena in the nanotribology of polymers.

Contents

Y. K. Yew · Z. C. Su · V. B. C. Tan (✉)
Department of Mechanical Engineering, National University of Singapore,
9 Engineering Drive 1, Singapore 117576, Singapore
e-mail: mpetanbc@nus.edu.sg

S. K. Sinha
Department of Mechanical Engineering, Indian Institute of Technology,
Kanpur 208016, India
e-mail: sujeet@iitk.ac.in

S. K. Sinha et al. (eds.), *Nano-tribology and Materials in MEMS*,
DOI: 10.1007/978-3-642-36935-3_6, © Springer-Verlag Berlin Heidelberg 2013

Introduction

Tribology, or surface engineering, is gaining more importance in research than ever because of the extensive use of micro-electromechanical systems in engineering applications from commercially available electronic devices, automotive systems, space technology to biological equipments. Micro-electromechanical systems (MEMS) are integrated micro-devices or systems composed of micromechanical and micro-electrical structures, usually a combination of micromachinery and sensing/controlling circuitry.

Due to the integration of micromechanical devices, many MEMS devices suffer from a range of reliability issues depending on design functionality, fabrication process, and operational environment. These reliability issues have gained the attention of the MEMS industry, because poor reliability poses a major hindrance to their commercialization [1, 2]. Being small, MEMS devices have a relatively large surface to volume ratio, making them particularly vulnerable to stiction and friction failure—a key reliability issue for MEMS products [3–6]. Micromachinery such as micro-motors, micro-gears, comb drives, etc., which are designed to be in constant motion to perform certain tasks, operates in contact mode. The contact between two moving surfaces or between a moving surface and an underlying substrate greatly influence the friction and wear performance of the MEMS. Stiction, a phenomenon in which the adhesion force exceeds restoring force leading to permanent surface attachment, can occur in a wide range of MEMS devices, e.g. micro-mirror devices, RF or mirror-based optical microswitches, etc. Either intentional or accidental, stiction is a major issue affecting the reliable operations of most MEMS devices. Therefore, tribological issues like adhesion, friction and wear of contacting surfaces in moving components are hampering the lifespan, operation efficiency, power output and steady-state speed of dynamic micromechanical devices.

One way of alleviating tribological problems is to modify the surface topographies or properties of the underlying material using special coatings or solid lubricants with low surface adhesion, low friction and good wear durability. The advantage of surface modification over other methods is its ease of incorporation with existing fabrication process of MEMS systems, because they can be applied at the very last stage of the manufacturing process without impacting the fabrication flow. Due to the low dimensions of the spatial features of MEMS systems, the lubricant and wear resistance layers must be only a few nanometers thick. Such thin surface coatings have been devised for MEMS application. The real challenge of surface coating is in the suitability of the coating material for specific applications. In general, the material should be mechanically durable and strongly bonded to the device surfaces to minimize wear. However, it is also desirable to have low surface energy to minimize adhesion and low interfacial shear strength to minimize friction.

There are a few types of commonly used coatings, namely, the hard coatings like silicon nitride and diamond-like carbon (DLC) [7], superhydrophobic organic coatings like self-assembled monolayers (SAM) [8, 9], polymer coatings like polydimethylsiloxane (PDMS) and polymethylmethacrylate (PMMA) and perfluoropolyether (PFPE) [10, 11], as well as hybrid coatings like polyethylene-aminopropyltrimethoxysilane self-assembled monolayer (PE-APTMS SAM) and polydimethylsiloxane self-assembled monolayer (PDMS-SAM) surface tethered networks [12, 13]. Studies of these various coatings have demonstrated that microfabrication-compatible surface treatments with ultrathin coatings can be exploited to improve the functionality and reliability of existing MEMS devices. However, further basic research of surface interactions is necessary before the adhesion and friction behavior of the polymer coatings can be thoroughly understood.

Nanotribology and MEMS

Feynman postulated that more interesting mechanisms could be discovered as length scale decreases because not all things simply scale down in proportion. In order to illustrate this point, he has cited the effect of materials sticking together by the molecular (van der Waals) attractions as an example [14]. In fact, studies have proven the preciseness of Feynman's hypothesis, in that adhesive forces can dominate over applied forces when the length scales of the devices are scaled down to the micro- and nanoscale [15, 16].

The main reason for the change in tribological behavior is due to the high surface to volume ratio increase, rendering increased significance to surface forces. Appreciation of the major role that adhesion plays in generating friction as two contacting bodies move in relative motion has shown that assumptions made for the macroscopic level will not always hold at the microscopic level. Factors that are negligible at macro scale but important at micro scale include the effects of van der Waals forces between contacting and noncontacting surfaces, contribution of capillary forces from liquid surface tensions [5], and the true contact area from microscopic surface roughness [17].

Therefore, the conventional methods used for the macroscopic engineering systems are insufficient to solve the tribological problems associated with MEMS technologies. It is critical to have a good understanding of the adhesion, friction and wear of the materials involved, without which the reliability and quality of MEMS devices could be compromised. The emerging field dealing with adhesion, friction, lubrication and wear at nanometer length scale is often referred to as 'nanotribology'. Recent developments in this field have begun to lend valuable insights into surface interactions and characteristics of material properties at the scale relevant to MEMS technologies.

The first important step towards the efforts to determine the atomic scale origins of tribology phenomena was taken when a surface force apparatus (SFA) was modified to measure the forces between contacting surfaces between two atomically smooth mica sheets sliding against each other [18]. These were followed by the invention of atomic force microscopy (AFM) [19]. In 1987, Mate et al. demonstrated the first friction measurement technique, known as friction force microscopy (FFM) or lateral force microscopy (LFM) using AFM and studied the fundamental forces at atomic scale [20]. The moving of a sharp tip over a sample surface closely resembles single asperity contact. Other techniques such as the quartz crystal microbalance, are also providing exciting new insights into the origin of friction [21, 22]. These breakthroughs have established new milestones for nanotribology research, which can be used for quantifying adhesion, friction, wear resistance, and nanohardness of well-defined contacts, thereby facilitating fundamental understanding of tribological phenomena at the nanometer length scale (a few hundred nanometers or less). Further studies with these experimental methods have shown that tribological behaviors deviate considerably from the established Amonton's law, since frictional forces have been found to be proportional to the true contact area on the nanometer scale, which is not necessarily proportional to the loading force [23, 24, 25, 26]. Therefore, there is an obligation to clarify the atomic basis of many well-established tribological laws and empirical observations.

To obtain a comprehensive picture of the fundamental frictional processes at the atomic or nanometer scale, it is important to fully understand the phenomenological events at the contact interfaces or subsurface regions during and after frictional sliding. In general, physical, chemical, electrochemical, and mechanical properties of materials change at the interfacial contact depending on the conditions of contacting asperities. Hence, the in-depth understanding of the evolution of interfacial surface when two surfaces interact and slide against each other is paramount to the design of reliable and robust MEMS devices.

Currently, there is no experimental technique to obtain data on the evolution of material surface properties or subsurface properties of bulk systems. Atomic or molecular level of modeling and simulation has proven to be an effective tool in providing information at the nanometer length scale and is being utilized to analyze experimental results. With the advancement in computer technology over the past decades, computer simulations have greater capacity to include more features to predict physical phenomena. Several of these simulation techniques have been adapted for studying tribological processes. The flexibility of controlling the parameters independently in simulations, for example, the geometry, sliding conditions and interactions between atoms, expands our understanding of the physical mechanism and energy-conversion processes during sliding contact at the molecular level. Simulation techniques can be employed to provide information down to the details of each atom, be it displacement or forces acting on the individual atom, supplementing experimental studies to present a more complete picture of the sliding process.

The Understanding of the Tribology

The widely accepted law of friction had been proposed by Amontons centuries ago [27]. The first Amontons' law of friction states that the friction force is proportional to the normal load; and his second law of friction states that the friction force is independent of the contact area. The Amontons' frictional laws can be summarized in a simple equation as

$$F = \mu L \tag{6.1}$$

where L represents the external loading force and μ the friction coefficient. These laws describing the friction force F between sliding bodies against each other are essentially phenomenological laws based on experimental observations. The basic idea was that the friction force originated from the mechanical interlocking of the asperities from the surface roughness of two contacting bodies. In other words, the friction force depends only on the applied load and the coefficient of friction will depend only on the actual combination of surfaces in contact. There are arguments from more recent research work made against the physical basis of this law, notably in publications on the energy dissipation of the dynamic friction [28], exclusion of surface force [16], true contact area [29, 30].

Friction force was later considered to originate from two different mechanisms, i.e. adhesive friction and deformation friction [27]. Therefore, the sum of friction force is expressed as

$$F = F_{adh} + F_{def} \tag{6.2}$$

where F_{adh} and F_{def} are the adhesion force and deformation force respectively. Adhesion force is needed to shear the contacting asperities; and the deformation force is needed to plow the contacting asperities of the harder surface through the softer surface. The adhesion force to shear all the junctions and sustain the sliding is directly proportional to the real area of contact,

$$F_{adh} = \tau A_r \tag{6.3}$$

where τ is the shear strength and A_r is the real area of contact. If the real area of contact is proportional to the normal load and the shear strength of contacting asperities is independent of load, Eq. [6.3] reduces to Amontons' law of Eq. [6.1]. The shear strength τ is usually assumed to be a constant and experiments performed with friction force microscopy have reinforced this assumption [23, 25, 26]. However, more recent experiments [31, 32] showed that shear strength, τ, can be linearly dependent on the contact pressure when the interactions across the sliding interface are relatively weak (Briscoe and Evans [33]).

At the microscopic viewpoint, contact occurs through a large number of contacting asperities, and the true area engaged in sliding may be many orders of magnitudes smaller than the apparent contact area [17, 30]. For calculation of the real area of contact, one can refer to Greenwood and Williamson's work on elastic deformation of multi-asperities contact [34]. The model is an extension from the continuum Hertzian theory of elastic contact for a sphere on a flat surface (single asperities contact) to multi asperities contact that illustrates the dependence of contact area on the surface roughness and elastic modulus. The Hertzian contact assumes that no adhesive forces act between the bodies. If attractive forces are present, the contact mechanics models that are usually used are the Johnson-Kendall-Roberts (JKR) and Derjaguin-Muller-Toporov (DMT) models depending on the adhesive interaction stresses [35, 36, 37]. The JKR model is applicable for strong adhesion coupled with highly compliant materials, whereas the DMT model is more suitable for low adhesion coupled with rigid materials. Tabor have introduced a nondimensional parameter that is useful in determining the validity of these two different models [37].

Besides the development of these general models, various theoretical and analytical models had been proposed. These include the models proposed by Tomlinson [38], Frenkel-Kontorova [39, 40], Rice-Ruina [41], Sokoloff [42], Hirano [28, 43], Persson [44], as well as the model from Baumberger et al. [45, 46]. The key assumptions in these models are homogeneity, isotropy, linearity, and elasticity of the materials [47]. These idealized assumptions simplified complex phenomena in order to offer some basic mechanistic explanations.

More recently, atomistic simulations have systematically studied how friction depends on the atomic structure of the sliding materials and the interaction potentials between the atoms. Molecular simulation provides a compromise between analytical models and experiments [47, 48].

Müser and Robbins [49] modeled pinning between commensurate and incommensurate surfaces using Molecular Dynamics (MD) simulations. The MD models of bare surfaces sliding against each other were always found to be pinned for commensurate surface when the total surface potential grew with increased surface area. Incommensurate sliding surfaces were completely unpinned and exhibited frictionless sliding, unless one of the surfaces is very compliant. In their more recent research, Luan and Robbins [50] demonstrated that the atomic-scale surface roughness, which is produced by the discreteness of atoms on the surface, plays a pivotal role at the nanoscale and that it leads to critical deviations from continuum level contact mechanics. The authors also showed that friction force has a non-linear dependence with normal load for amorphous and incommensurate surfaces.

However, contradicting results were reported by Gao et al. [51] using MD analysis. They have shown that even at the molecular scale, friction force depends linearly on normal load as Amontons' law suggests in the case of dry friction and non-adhering material. They also argued that contact area is not useful in describing friction at the nano-, micro- or macroscale. In a more reconciling finding, Mo et al. [30] used the MD simulations to prove that friction force scales linearly with the area of contact and Amontons' law still applies for non-adhesive and high roughness surfaces at the nanoscale.

Despite the tremendous insights brought about by MD simulations of frictional sliding, the origins of the observed differences in the dependence of friction force on load and contact area are yet to be determined for both simulations and experiments.

Nanotribology of Polymer

Recently, the focus of tribology has been on polymer-based coatings, be they synthetic or naturally occurring macromolecules at material interfaces. This was driven by the improvement in mechanical and tribological properties of coated surfaces, especially in dry friction conditions, where lubricants cannot be used. The needs for lightweight, non-corrosive and easily manufactured materials in advance MEMS applications have propelled polymeric materials into the tribological field of research. However, very little information is known concerning the friction and wear of polymer interfacial sliding.

The tribology of polymers is different from tribology of metals and ceramic materials. The difference in applications of polymers in frictional contacts in comparison to metals and ceramic materials relates mainly to the chemical and physical structures as well as to the surface and bulk properties. The polymers show very low surface free energy and also have viscoelastic properties. These will bring about significant tribological differences when adhesive and mechanical components of friction force are considered. Also polymers can be easily modified both on surface and in bulk. Therefore they are often and easily used as a background material to produce many composites with easily varied physicochemical properties. This makes polymers very promising materials with the ability to control their frictional and wear behaviors of sliding contacts. Such "elasticity" of physical and chemical structures facilitates the production of interesting components for various tribological systems involving polymers, metals, ceramic materials as well as polymer-on-polymer sliding and even rolling tribological contacts.

The first MD simulation of self-assembled monolayers (SAM) friction were performed by Glosli and McClelland [52], who simulated sliding friction between two ordered monolayers of alkane. They found that energy dissipation occurred through two different mechanisms, i.e. discontinuous "plucking" mechanism and continuous "viscous" mechanism, depending on the interfacial interaction strength. Tupper and Brenner [53] have studied friction in self-assembled monolayers of alkanethiols on gold. They concluded that friction force is proportional to the applied load. They also demonstrated that structure of the monolayer film changes during compression, but no changes in the friction coefficient is observed as the structure changed.

In regard to the friction of monolayer films, Koike and Yoneya [54] studied the friction in Langmuir–Blodgett organic film. They found that the large differences in friction forces for different types of monolayers were due to van der Waals interaction. Steven et al. [55] simulated a layer of hexadecane film confined between two plates with nanometer scale separations using MD simulations. They found that the film tends to crystallize at large loads. The hexadecane system also showed the tendency to adopt a linear configuration to be aligned with the shearing direction.

In more recent publication, Yew et al. [56] discussed the shear mechanism in polymer-on-polymer sliding. They showed in their studies that there are two distinct regions of shearing, i.e. a localized severely deformed band at the interface and a uniformly sheared bulk that reaches a finite steady-state value of shear strain at the subsurface of the polymers. They also found that friction strongly depends on interfacial adhesion, surface roughness and shear modulus of the sliding system.

Polymer Simulation

Molecular mechanics simulations to investigate the tribology of polymer-on-polymer sliding are reported in this section. The computational models are parameterized for real material and found to reproduce the mechanical properties of polyethylene (PE) within certain accuracy. PE is widely used in engineering and commercial applications. It is recently being used as a low-friction material in coatings and lubricating fluids for a growing number of mechanical devices where adhesion, friction and wear are important factors for determining performance and lifetime [57]. PE is preferred by the industries over other polymers as it is more economical to produce than other polymeric materials. By manipulating their molecular weight, co-monomer contents, branching characteristics and polymerization processes, different types of PE can be fabricated. Different types of PE have different properties and are used in different applications. As the different types of PE are usually characterized by their density and molecular weight, these two parameters are varied in the simulation models.

Potential Model

The phenomenological behaviors of polymeric materials pose a challenge for multiscale simulations as such mechanisms span several length and time scales ranging from electron interactions to entire macroscopic structures and from femtoseconds to years of reaction period. However, a full atomistic molecular simulation with detailed chemical interactions is computationally intractable for studying dense systems consisting of many long polymer chains over a long relaxation period. To reduce the complexity of the molecular model and yet be able to span entire hierarchy of length and time scales, a commonly adopted approach is to describe polymer chains as interactions of groups of atoms rather than discreet atoms [58, 59, 60, 61–63]. Variations of the scheme are mainly distinguished by the method of discretizing polymer chains, distributing the mass center of interaction of the new coarse-grained unit and incorporating different interatomic force fields.

Table 6.1 Parameter values for interatomic potentials of polymeric system

Parameters	Approximated value for CH_2
σ	5Å
ε	30 meV
R_0	1.5 σ
k	30 ε/σ^2
r_{cutoff}	2.2 σ

Coarse-graining refers to the mapping of a group of atoms to a bead-spring model, .i.e., the bead-spring model represents a polymer chain as a collection of beads connected by elastic springs. Each bead typically represents one or a few monomer units while the springs reproduce the Gaussian distribution of separations between monomers connected through a number of chemical bonds. As some of the information on physical and chemical properties will be lost through coarse-graining, the choice of the force field governing the interactions of the bead-spring model is crucial in determining the practicability and accuracy of the simulations. Despite lacking fine chemical details, simplified general models are capable of predicting many physical properties and phenomena of polymer networks, solutions and melts [59, 60, 63, 64, 65–67]. Such results help to identify key parameters (e.g. chain length, architecture, and stiffness) that are mainly responsible for particular properties.

The coarse-grained model introduced by Kremer and Grest [58] has been applied for various polymer studies with great success. It allows treatment of longer length and time scales which would save at least an order of magnitude of computational time compared to a corresponding fully atomistic model to get the same results. The same force field model is employed in our present tribological simulations. In our simulations, all the beads in the polymer system interact through a truncated Lennar-Jones (LJ) potential,

$$V_{LJ}(r_{ij}) = 4\varepsilon\left(\left(\frac{\sigma}{r_{ij}}\right)^{12} - \left(\frac{\sigma}{r_{ij}}\right)^{6}\right), \quad r_{ij} < r_{cutoff} \tag{6.4}$$

where r_{ij} is the distance between two interacting beads i and j, ε and σ are the energy and length at which $V_{LJ} = 0$. A repulsive cutoff radius of $r_{cutoff} = 2^{1/6}\sigma$ and geometric mean combination rules are employed [68]. The parameter values used in the simulations are tabulated in Table 6.1.

For chemically bonded beads, an additional potential is needed. The Finite Extensible Non-Linear Elastic (FENE) potential which reproduces a simple harmonic potential for small extensions is added for the beads connected along the sequence of the chains,

$$V_{FENE}(r_{ij}) = \begin{cases} -0.5kR_0^2 \ln\left[1 - \left(\frac{r_{ij}}{R_0}\right)^2\right], & r_{ij} \leq R_0 \\ \infty, & r_{ij} > R_0 \end{cases} \tag{6.5}$$

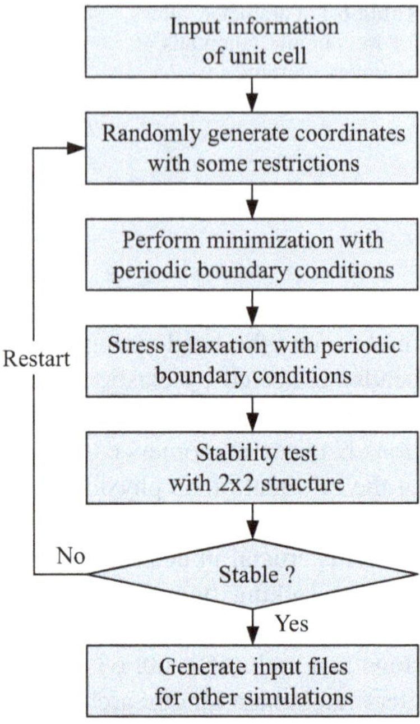

Fig. 6.1 Flowchart of unit cell generation for polymers

Here $k = 30\varepsilon/\sigma^2$ and $R_0 = 1.5\sigma$ are the spring constant and maximum bond extension respectively. The chosen spring constant was strong enough to maintain the maximum extensions of bonds to be less than 1.2σ. As a result, bond crossings are energetically infeasible. Mapping of the potential with these specific parameter values to typical hydrocarbons yields values between 25 and 45 meV for ε and between 0.5 and 1.3 nm for σ [58].

The influence of adhesion at the contacting surfaces can be controlled by varying the LJ potential cutoff distance for the intermolecular interaction. A value of $r_{cutoff} = 2^{1/6}\sigma$ gives a purely repulsive potential representing zero adhesive interaction. The larger the cutoff distance, the longer the attractive tail of the LJ potential and the stronger the adhesive effects. As the focus of the simulations was on interfacial friction, a cutoff radius of $r_{cutoff} = 2.2\sigma$ was applied for non-bonded beads in all simulations. This cutoff was large enough for the attractive portion of the potential to be active without sacrificing computational efficiency.

Once the potential form is determined, we proceed to generate the polymer unit cell. The flow of the procedures is shown in Fig. 6.1. The input information is the number of chains, number of beads per chain, density, equilibrated bond length, and restriction distance. The restriction distance is defined as the minimum allowable distance between any two beads. With this input, the coordinates of the beads are generated randomly. On top of that, the bonds cannot cross each other, and the distance between any two non-bonded beads should be larger than the restriction distance. The minimization is performed with periodic boundary

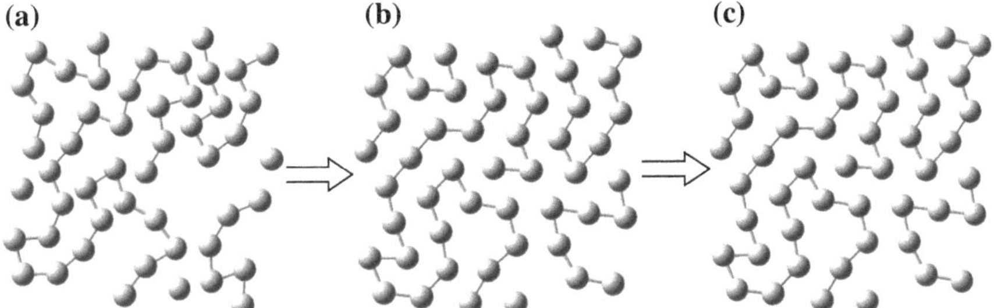

Fig. 6.2 An example of unit cell generation for polymer. **a** Initial structure. **b** After equilibration. **c** After stress relaxation

conditions. Usually, the structure obtained has non-zero internal virial stresses. These stresses can be relaxed by gradually adjusting the dimensional length of the cell in its respective direction. Energy minimization iterations will be carried out with every single step of dimensional change until the virial stress becomes zero or insignificant before going into next procedure. The next and final procedure is the stability check of the unit cell. The unit cell is considered stable if all the eigenvalues of the stiffness matrix are positive. Unfortunately, the calculations of stiffness matrix or Hessian matrix are computationally intensive and require a lot of memory allocation for larger unit cell. One alternative and easier way is to perform energy minimization on a configuration of 2×2 units of cell or even larger configuration. If the minimization does not cause any change to the atomic arrangement, the unit cell can be considered to be stable.

An example of such a checking process of the unit cell is shown in Fig. 6.2 using a two dimensional example. As can be seen, the energy minimization brings about significant changes to the initial structure whereas the configurations before and after stress relaxation are almost the same. Although changes in the dimensions of the cell are only slight during stress relaxation, the initial strain when free surfaces are introduced might be very high if the stress is not relaxed at the cell level.

Polymer Sliding

The simulation model replicates a symmetric system of a polymer surface sliding on another polymer substrate [69]. It comprises an upper slider block and a lower substrate block of bead-springs as depicted in Fig. 6.3. Periodic boundary conditions were applied on the vertical surfaces while the contacting surfaces of the two blocks were stress free prior to contact. A layer of beads at the bottom of the substrate block was fixed while a layer of beads at the top of the slider block was prescribed to firstly displace downwards toward the slider block to attain specific contact pressures.

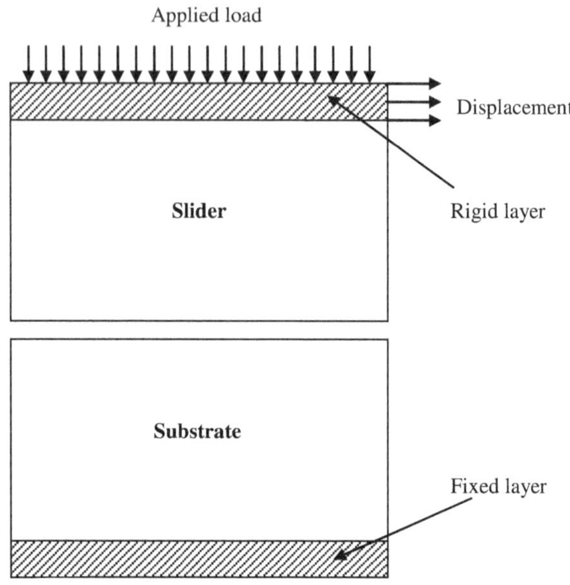

Fig. 6.3 Schematic drawing of the computational model for polymer-on-polymer sliding

Simulations were performed on bead-spring computational models of different polymer densities and molecular weights to study the effects of these two parameters. Starting with a polymer model of normalized density 0.85 and molecular weight 3,500, the density of the polymer was then changed to normalized values of 0.75 and 0.95 while keeping the molecular weight at 3,500. A normalized density of 1 corresponds to 0.95 g/cm^3. Another series of simulations were performed using polymer models of molecular weights 700 and 14,000 while keeping the density at 0.85. For each model, the simulations were conducted at different normal contact pressures by controlling the vertical displacement of the slider block (i.e., displacement in the applied load direction in Fig. 6.3).

Molecular mechanics simulation using conjugate gradient energy minimization was performed throughout the simulations, i.e., dynamics and thermal fluctuations were not included in the simulations. This greatly reduces noises from thermostat fluctuation to facilitate the detection of delicate changes in the energy profile and investigation into the mechanisms of the system. Although we have not taken into consideration heat dissipation at the surfaces during sliding, this is very minimum at low loads [70].

There are a few important data that can be extracted from the simulation. The friction or shear stress is the main force that determines the amount of resistance force needed to be overcome for continuous smooth sliding. Figure 6.4 shows the plot of the friction stresses, which are calculated as the lateral forces per unit area of contact acting on the beads of substrate in the sliding direction of the slider block, against the slider displacement. In order to study the influence of external loads on the frictional sliding of the polymeric material, three different contact pressures were simulated as indicated in Fig. 6.4. One can observe that the shear

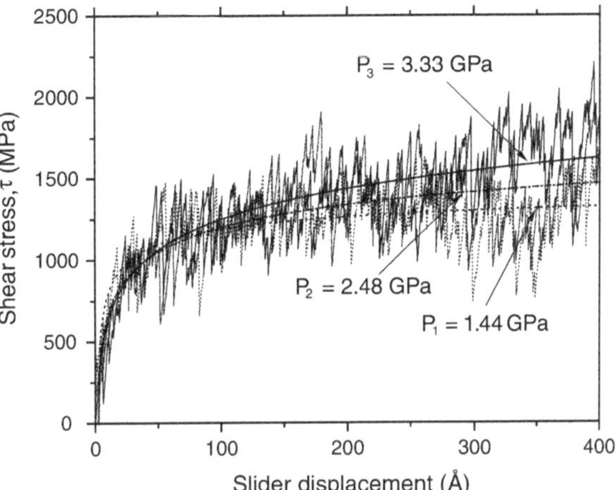

Fig. 6.4 Shear stress of substrate as a function of sliding distance at varying pressures of 1.44, 2.48 and 3.33 GPa with normalized density of 0.75 and molecular weight of 3,500

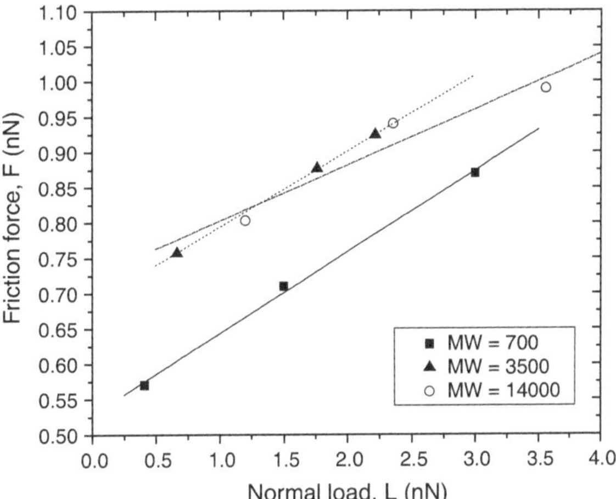

Fig. 6.5 Friction force plotted against normal load for different molecular weights

stress for each contact pressure fluctuates about a mean value after some sliding distance. The mean shear stress at that instance is taken as the friction stress for subsequent discussions.

Frictional force as a function of externally applied normal load is presented in Figs. 6.5 and 6.6 for the different polymer densities and molecular weights respectively. As observed from Figs. 6.5 and 6.6, the shear force or friction force increases linearly with applied normal load over the entire range of loading studied. There is a finite friction force at zero load, which indicates that friction is not entirely "load-controlled" but also has significant "adhesion controlled" contributions [71].

From the plots of the graphs we can express the linear dependency of friction force on externally applied load as

$$F = F_0 + \alpha L \tag{6.3}$$

where F and L is the friction force and external load respectively. Here, the gradient of the friction-load curve α is comparable to the coefficient of friction as defined by Amontons' Law [72]. F_0 is the intrinsic adhesion force between the two contacting surfaces which has to be overcome to initiate sliding [73]. This is also the force responsible for the sticking of two smooth and clean surfaces even without any externally applied pressure. This linear relationship agrees well with the commonly observed linear dependence of the interface friction force on the normal load [32, 72, 74]. For very low external loads, the adhesive force will be the dominant term in Eq. [6.3] [75]. The notion that friction should be related to adhesion seems fairly natural. The same bonding mechanism that makes it difficult to pull two contacting surfaces apart should also make it difficult to slide them over each other. The interface friction force becomes directly proportional to the normal load when the applied load becomes significantly large. Hence, Eq. [6.3] is an extension of the laws of friction of Amonton to accommodate atomically smooth and clean surfaces where adhesion is predominant.

As shown in Fig. 6.5, increasing molecular weight while keeping the polymer density constant led to an increase in F_0 but a decrease in the coefficient, α. The difference in the surface structure was believed to be the main reason for the increase of the coefficient of friction for lower molecular weight polymer. The surface interpenetration at the interface of slider and substrate can be quantified by measuring the vertical distance between the topmost substrate bead and the bottom most slider bead. From the measurements, the interpenetration depths for the models of 14,000, 3,500 and 700 molecular weight were found to be 2.8, 5.95 and 6.64 Å respectively. The inverse relationship between molecular weight and interpenetration depth has been reported previously [76].

In addition to the higher interpenetration depth, the number of dangling ends of the molecular chain for polymers of lower molecular weights was also higher when we scanned through the interfacial contacts. In concomitant, the number of chain ends that protruded across the interface was also higher. Even after sliding had progressed for some distance, more chain penetration and atom diffusion at the interfacial was observed for the model with lowest molecular weight. The depths of interpenetration at the end of the simulations were measured to be 8.27, 8.8 and 12.07 Å for molecular weights of 14,000, 3,500 and 700, respectively. Such observations suggest that slider and substrate models with lower molecular weights should experience higher level of locking.

On the other hand, the models with highest molecular weight showed the smoothest surfaces which meant contacting surfaces were in close proximity to each other over a larger area and consequently there is greater adhesion due to van der Waals interaction. Hence, for atomically smooth surfaces, such as the simulations presented, adhesion forces can dominate to give highest intrinsic adhesive force F_0, as illustrated in the plots in Fig. 6.5 for the model with highest molecular weight. However, adhesion contributes less to friction when surfaces are not as smooth or when contact pressures are increased. Friction forces then become

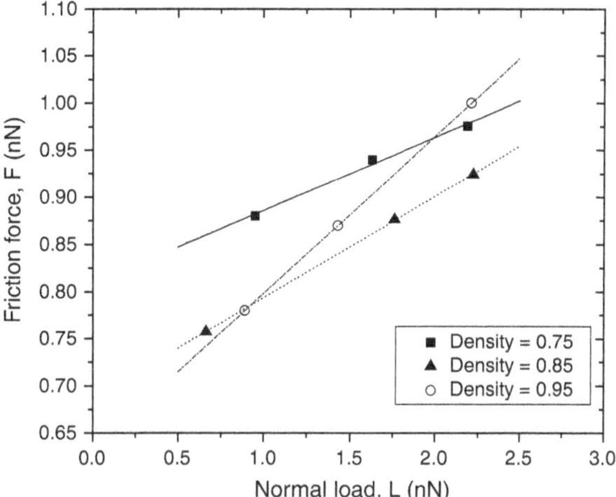

Fig. 6.6 Friction force plotted against normal load for different normalized density

dominated by normal loading which controls the gradient of the slope and effectively defines the coefficient, α. Therefore, α is the highest for the polymer sample with the lowest molecular weight. Chen et al. [77], who carried out tribological experiments on polyethylene, also found a decrease in the gradient of the friction-load curve (coefficient α) with increasing molecular weight. Therefore, it is suggested that applying high molecular weight polymer as tribological coatings can reduce friction and prolong wear life.

The effects of polymer density on friction were found to be different from the effects of molecular weight. The friction force against applied load curves for different densities of 0.75, 0.85 and 0.95 are given in Fig. 6.6. The slope of the friction-load curves showed that the lowest density polymer will have the lowest coefficient of friction, where the coefficient α was 0.078 at a density of 0.75. For densities of 0.85 and 0.95, the coefficients, α, were found to be 0.107 and 0.166, respectively.

The difference in the coefficient of friction between the linear low density and linear high density polymer can be better understood by noting the manner in which the shear strength varies with different densities of polyethylene. In order to get the two contacting surfaces to start sliding over each other, the contacting asperities must undergo elastic and plastic deformation. The localized shear stress will develop over the contacting junctions as the contacting asperities resist the deformation until they reach their strength limit. Hence, the higher the shear strength of the material, the higher the tangential force is needed to deform the asperities before sliding initiated [27]. For a constant molecular weight of 3,500, as the normalized density decreased from 0.95 to 0.85 and to 0.75, the more compliant the polymers became [78]. We can expect high density polymers to exhibit higher coefficients of friction than low density polymers because of their higher strength.

The lower compliance of surfaces for lower density polymer also encourages more asperity interpenetration and bulk flow at the contacting surface [72, 79]. These were projected by the increases in interpenetration depths from 3.05 to 5.95

to 14.21 Å with decreasing polymer density. The intrinsic adhesive force F_0 increased accordingly (Fig. 6.6) because the actual contact area becomes larger with increasing interpenetration.

Conclusion

It remains a major challenge to theorists and experimentalists to conclusively establish the origin of friction at macro-, micro and nanoscale. The physical, chemical, electrochemical, and mechanical properties of material constantly change at the interfacial contact depending on the conditions of contacting asperities and the time between successive contact events. Until these aspects are completely understood, the development of optimized materials and coatings for MEMS/NEMS applications will largely be through empirical approaches. In the advent of AFM experiments and molecular simulation, we are closer to understanding the mechanisms of friction and wear in nature.

Simulations of dry and molecularly smooth sliding of polymer-on-polymer have been performed. The interfacial friction between two sliding polymers can be attributed to two main mechanisms, i.e. adhesion and deformation. The friction coefficient of polymer-on-polymer increased with the increase of adhesive work between the substrate and slider and the shear modulus of the sliding system. We have also observed that the friction force of sliding is strongly dependent on the applied load and surface roughness. The population of chain "ends" at surfaces is the most important factor that determines the adhesion and friction between two polymer surfaces.

References

1. Hartzell, A. L., Silva, M.G.D., Shea, H.R.: MEMS reliability (MEMS reference shelf), Springer (2010)
2. Romig, A.D., Dugger, M.T., McWhorter, P.J.: Materials issues in microelectromechanical devices: science, engineering, manufacturability and reliability. Acta Mater. **51**(19), 5837–5866 (2003)
3. Komvopoulos, K.: Adhesion and friction forces in microelectromechanical systems: mechanisms, measurement, surface modification techniques, and adhesion theory. J. Adhes. Sci. Technol. **17**(4), 477–517 (2003)
4. Maboudian, R., Carraro, C.: Surface engineering for reliable operation of MEMS devices. J. Adhes. Sci. Technol. **17**(4), 583–591 (2003)
5. Timpe, S.J., Komvopoulos, K.: The effect of adhesion on the static friction properties of sidewall contact interfaces of microelectromechanical devices. J. Microelectromech. Syst. **15**(6), 1612–1621 (2006)
6. van Spengen, W.M., Puers, R., De Wolf, I.: The prediction of stiction failures in MEMS. IEEE Trans. Device Mater. Reliab. **3**(4), 167–172 (2003)
7. Luo, J.K., Fu, Y.Q., Le, H.R., Williams, J.A., Spearing, S.M., Milne, W.I.: Diamond and diamond-like carbon MEMS. J. Micromech. Microeng. **17**(7), S147–S163 (2007)

8. Bhushan, B., Liu, H.: Self-assembled monolayers for controlling adhesion, friction and wear. Nanotribology and Nanomechanics 885–928 (2005)
9. Maboudian, R., Ashurst, W.R., Carraro, C.: Self-assembled monolayers as anti-stiction coatings for MEMS: characteristics and recent developments. Sens. Actuators a-Phys. **82**(1–3), 219–223 (2000)
10. Liu, H., Bhushan, B.: Nanotribological characterization of molecularly thick lubricant films for applications to MEMS/NEMS by AFM. Ultramicroscopy **97**(1–4), 321–340 (2003)
11. Tambe, N.S., Bhushan, B.: Micro/nanotribological characterization of PDMS and PMMA used for BioMEMS/NEMS applications. Ultramicroscopy **105**(1–4), 238–247 (2005)
12. Landherr, L.J.T., Zhang, Q., Cohen, C., Archer, L.A.: Interfacial friction of thin PDMS network films. J. Polym. Sci. Part B-Polym. Phys. **46**(17), 1773–1787 (2008)
13. Satyanarayana, N., Sinha, S.K., Shen, L.: Effect of molecular structure on friction and wear of polymer thin films deposited on Si surface. Tribol. Lett. **28**(1), 71–80 (2007)
14. Feynman, R.P.: There's plenty of room at the bottom. Eng. Sci. **23**(5), 22–36 (1960)
15. Kendall, K.: Adhesion: molecules and mechanics. Science **263**(5154), 1720–1725 (1994)
16. Kim, S.H., Asay, D.B., Dugger, M.T.: Nanotribology and MEMS. Nano Today **2**(5), 22–29 (2007)
17. Schirmeisen, A., Schwarz, U.D.: Measuring the friction of nanoparticies: a new route towards a better understanding of nanoscale friction. ChemPhysChem **10**(14), 2373–2382 (2009)
18. Israelachvili, J.N., Tabor, D.: The shear properties of molecular films. Wear **24**(3), 386–390 (1973)
19. Binnig, G., Quate, C.F., Gerber, C.: Atomic force microscope. Phys. Rev. Lett. **56**(9), 930 (1986)
20. Mate, C.M., McClelland, G.M., Erlandsson, R., Chiang, S.: Atomic-scale friction of a tungsten tip on a graphite surface. Phys. Rev. Lett. **59**(17), 1942 (1987)
21. Krim, J.: QCM tribology studies of thin adsorbed films. Nano Today **2**(5), 38–43 (2007)
22. Krim, J., Solina, D.H., Chiarello, R.: Nanotribology of a Kr monolayer: a quartz-crystal microbalance study of atomic-scale friction. Phys. Rev. Lett. **66**(2), 181 (1991)
23. Carpick, R.W., Agrait, N., Ogletree, D.F., Salmeron, M.: Measurement of interfacial shear (friction) with an ultrahigh vacuum atomic force microscope. J. Vac. Sci. Technol. B **14**(2), 1289–1295 (1996)
24. Lantz, M.A., O'Shea, S.J., Welland, M.E., Johnson, K.L.: Atomic-force-microscope study of contact area and friction on NbSe2. Phys. Rev. B **55**(16), 10776 (1997)
25. Schwarz, U.D., Zworner, O., Koster, P., Wiesendanger, R.: Quantitative analysis of the frictional properties of solid materials at low loads. I. Carbon compounds. Phys. Rev. B **56**(11), 6987 (1997)
26. Schwarz, U.D., Zworner, O., Koster, P., Wiesendanger, R.: Quantitative analysis of the frictional properties of solid materials at low loads. II. Mica and germanium sulfide. Phys. Rev. B **56**(11), 6997 (1997)
27. Bowden, F.P., Tabor, D.: Friction and lubrication. Methuen, London (1967)
28. Hirano, M.: Atomistics of friction. Surf. Sci. Rep. **60**(8), 159–201 (2006)
29. Holscher, H., Schirmeisen, A., Schwarz, U.D.: Principles of atomic friction: from sticking atoms to superlubric sliding. Philos. Trans. R. Soc. Math. Phys. Eng. Sci. **366**(1869), 1383–1404 (2008)
30. Mo, Y.F., Turner, K.T., Szlufarska, I.: Friction laws at the nanoscale. Nature **457**(7233), 1116–1119 (2009)
31. Berman, A., Drummond, C., Israelachvili, J.: Amontons' law at the molecular level. Tribol. Lett. **4**(2), 95–101 (1998)
32. Israelachvili, J., Maeda, N., Rosenberg, K.J., Akbulut, M.: Effects of sub-fingstrom (pico-scale) structure of surfaces on adhesion, friction, and bulk mechanical properties. J. Mater. Res. **20**(8), 1952–1972 (2005)
33. Briscoe, B.J., Evans, D.C.B.: The shear properties of langmuir-blodgett layers. In: Proceedings of the royal society of London. Series A, mathematical and physical sciences, vol. **380**(1779), pp. 389–407 (1982)

34. Greenwood, J.A., Williamson, J.B.P : Contact of nominally flat surfaces. In: Proceedings of the royal society of London. Series A, mathematical and physical sciences, vol. **295**(1442), pp 300–319
35. Johnson, K.L., Kendall, K., Roberts A.D.: Surface energy and the contact of elastic solids. In: Proceedings of the royal society of London. A. mathematical and physical sciences, vol. **324**(1558), pp. 301–313 (1971)
36. Derjaguin, B.V., Muller, V.M., Toporov, Y.P.: Effect of contact deformations on the adhesion of particles. J. Colloid Interface Sci. **53**(2), 314–326 (1975)
37. Tabor, D.: Surface forces and surface interactions. J. Colloid Interface Sci. **58**(1), 2–13 (1977)
38. Tomlinson, G.A.: CVI. A molecular theory of friction. Philos. Mag. Ser. 7 **7**(46), 905–939
39. Braun, O.M., Naumovets, A.G.: Nanotribology: microscopic mechanisms of friction. Surf. Sci. Rep. **60**(6–7), 79–158 (2006)
40. Frenkel, Y., Kontorova T.: Phys. Z. Sowietunion **13**(1) (1938)
41. Rice, J.R., Ruina, A.L.: Stability of steady frictional slipping. J. Appli. Mech.-Trans. Asme **50**(2), 343–349 (1983)
42. Sokoloff, J.B.: Theory of dynamical friction between idealized sliding surfaces. Surf. Sci. **144**(1), 267–272 (1984)
43. Hirano, M., Shinjo, K.: Atomistic locking and friction. Phys. Rev. B **41**(17), 11837–11851 (1990)
44. Persson, B.N.J., Schumacher, D., Otto, A.: Surface resistivity and vibrational damping in adsorbed layers. Chem. Phys. Lett. **178**(2–3), 204–212 (1991)
45. Baumberger, T., Heslot, F., Perrin, B.: Crossover from creep to inertial motion in friction dynamics. Nature **367**(6463), 544–546 (1994)
46. Heslot, F., Baumberger, T., Perrin, B., Caroli, B., Caroli, C.: Creep, stick-slip, and dry-friction dynamics: experiments and a heuristic model. Phys. Rev. E **49**(6), 4973 (1994)
47. Li, Q., Kim, K.-S.: Micromechanics of friction: effects of nanometre-scale roughness. Proc. R. Soc. A: Math. Phys. Eng. Sci. **464**(2093), 1319–1343 (2008)
48. Robbins, M.O., Müser M.H.: Computer simulations of friction, lubrication and wear. Modern tribology handbook volume 1: principles of tribology. B. Bhushan, pp. 717–765 FL CRC Press, Boca Raton (2001)
49. Muser, M.H., Robbins, M.O.: Conditions for static friction between flat crystalline surfaces. Phys. Rev. B **61**(3), 2335–2342 (2000)
50. Luan, B.Q., Robbins, M.O.: The breakdown of continuum models for mechanical contacts. Nature **435**(7044), 929–932 (2005)
51. Gao, J.P., Luedtke, W.D., Gourdon, D., Ruths, M., Israelachvili, J.N., Landman, U.: Frictional forces and Amontons' law: from the molecular to the macroscopic scale. J. Phys. Chem. B **108**(11), 3410–3425 (2004)
52. Glosli, J.N., McClelland, G.M.: Molecular dynamics study of sliding friction of ordered organic monolayers. Phys. Rev. Lett. **70**(13), 1960 (1993)
53. Tupper, K.J., Brenner, D.W.: Molecular dynamics simulations of friction in self-assembled monolayers. Thin Solid Films **253**(1–2), 185–189 (1994)
54. Koike, A., Yoneya, M.: Molecular dynamics simulations of sliding friction of Langmuir-Blodgett monolayers. J. Chem. Phys. **105**(14), 6060–6067 (1996)
55. Stevens, M.J., Mondello, M., Grest, G.S., Cui, S.T., Cochran, H.D., Cummings, P.T.: Comparison of shear flow of hexadecane in a confined geometry and in bulk. J. Chem. Phys. **106**(17), 7303–7314 (1997)
56. Yew, Y.K., Minn, M., Sinha, S.K., Tan, V.B.C.: Molecular simulation of the frictional behavior of polymer-on-polymer sliding. Langmuir **27**(10), 5891–5898 (2011)
57. Satyanarayana, N., Sinha, S.K., Lim, S.C.: Highly wear resistant chemisorbed polar ultra-high-molecular-weight polyethylene thin film on Si surface for micro-system applications. J. Mater. Res. **24**(11), 3331–3337 (2009)
58. Kremer, K., Grest, G.S.: Dynamics of entangled linear polymer melts—a molecular-dynamics simulation. J. Chem. Phys. **92**(8), 5057–5086 (1990)

59. Tschop, W., Kremer, K., Batoulis, J., Burger, T., Hahn, O.: Simulation of polymer melts. I. Coarse-graining procedure for polycarbonates. Acta Polym. **49**(2–3), 61–74 (1998)
60. Tschop, W., Kremer, K., Hahn, O., Batoulis, J., Burger, T.: Simulation of polymer melts. II. From coarse-grained models back to atomistic description. Acta Polym. **49**(2–3), 75–79 (1998)
61. Baschnagel, J., Binder, K., Doruker, P., Gusev, A.A., Hahn, O., Kremer, K., Mattice, W.L., Muller-Plathe, F., Murat, M., Paul, W., Santos, S., Suter, U. W., Tries, V.: (2000). Bridging the gap between atomistic and coarse-grained models of polymers: status and perspectives. Advances in polymer science: viscoelasticity, atomistic models, statistical chemistry. Springer, Berlin **152**: 41–156
62. Faller, R., Muller-Plathe, F., Heuer, A.: Local reorientation dynamics of semiflexible polymers in the melt. Macromolecules **33**(17), 6602–6610 (2000)
63. Muller-Plathe, F.: Coarse-graining in polymer simulation: from the atomistic to the mesoscopic scale and back. ChemPhysChem **3**(9), 754–769 (2002)
64. Theodorou, D.N., Suter, U.W.: Atomistic modeling of mechanical-properties of polymeric glasses. Macromolecules **19**(1), 139–154 (1986)
65. He, G., Muser, M.H., Robbins, M.O.: Adsorbed layers and the origin of static friction. Science **284**(5420), 1650–1652 (1999)
66. Rottler, J., Robbins, M.O.: Yield conditions for deformation of amorphous polymer glasses. Phys. Rev. E **64**(5), 051801 (2001)
67. Rottler, J., Robbins, M.O.: Jamming under tension in polymer crazes. Phys. Rev. Lett. **89**(19), 195501 (2002)
68. Weeks, J.D., Chandler, D., Andersen, H.C.: Role of repulsive forces in determining equilibrium structure of simple liquids. J. Chem. Phys. **54**(12), 5237 (1971)
69. Maeda, N., Chen, N.H., Tirrell, M., Israelachvili, J.N.: Adhesion and friction mechanisms of polymer-on-polymer surfaces. Science **297**(5580), 379–382 (2002)
70. Socoliuc, A., Bennewitz, R., Gnecco, E., Meyer, E.: Transition from stick-slip to continuous sliding in atomic friction: entering a new regime of ultralow friction. Phys. Rev. Lett. **92**(13), 134301 (2004)
71. Luengo, G., Heuberger, M., Israelachvili, J.: Tribology of shearing polymer surfaces. 2. Polymer (PnBMA) sliding on mica. J. Phys. Chem. B **104**(33), 7944–7950 (2000)
72. Israelachvili, J., Giasson, S., Kuhl, T., Drummond, C., Berman, A., Luengo, G., Pan, J.M., Heuberger, M., Ducker, W., Alcantar, N.: Some fundamental differences in the adhesion and friction of rough versus smooth surfaces. Tribol. Ser. **38**, 3–12 (2000)
73. Briscoe, B.J.: Polymers, frictional properties of. encyclopedia of materials: science and technology. In: Buschow, K.H.J., Robert, W.C., Merton et al. C.F. (eds.) pp. 7675–7680. Oxford, Elsevier (2001)
74. Homola, A.M., Israelachvili, J.N., McGuiggan, P.M., Gee, M.L.: Fundamental experimental studies in tribology—the transition from interfacial friction of undamaged molecularly smooth surfaces to normal friction with wear. Wear **136**(1), 65–83 (1990)
75. Meyer, E., Colchero, J.: Friction on an atomic scale. Handbook of Micro/Nano tribology, 2nd Edn. CRC Press, USA (1998)
76. Deng, M., Tan, V.B.C., Tay, T.E.: Atomistic modeling: interfacial diffusion and adhesion of polycarbonate and silanes. Polymer **45**(18), 6399–6407 (2004)
77. Chen, N.H., Maeda, N., Tirrell, M., Israelachvili, J.: Adhesion and friction of polymer surfaces: the effect of chain ends. Macromolecules **38**(8), 3491–3503 (2005)
78. Mark, J.E.: Polymer data handbook. Oxford University Press, New York (1999)
79. Campañá, C.: Using Green's function molecular dynamics to rationalize the success of asperity models when describing the contact between self-affine surfaces. Phys. Rev E Stat. Nonlinear Soft Matter Phys. **78**(2), 026110 (2008)

Chapter 7
Atomistic Modeling of Polymeric Nanotribology

L. Dai and V. B. C. Tan

Abstract Molecular dynamics simulations, supported by experimental characterizations, show that the tribology of nano-structured polymer interfaces are largely influenced by the inter-locking of diffusion-induced polymer chains. The degree of interfacial locking was found to determine the mechanisms of tribological behavior. The instantaneous friction coefficient shows regular stick–slip behavior at low sliding speed due to the concurrency of molecular deformation. Surface melting was found at the commencement of slipping. With increasing sliding speed, the regularity of the stick–slip cycles is less obvious and the magnitude of the fluctuations in friction coefficient decreased due to increasing atomic collisions and less durable slip rebound. Stick–slip phenomenon is reduced gradually before finally converging to complete dynamic frictional sliding at high sliding speed. Based on interfacial diffusion conditions, three mechanisms of interface deformation as 'brushing', 'combing' and 'scissoring' were proposed, among which reversible 'brush' was considered as the main source of interfacial deformation and dominant mechanism of the tribological behavior. Comparatively, 'combing' and 'scissoring' only took place in case of significant interfacial diffusion, and were not reversible. The interfacial structure also imposes effects on the influences of other factors to the frictional characteristics. In models with non-diffusive interface, higher pressure from the slider flattens the substrate surface, and thereby reduces the friction coefficient. In conclusion, once the interfacial structure was known, the tribological behaviors become predictable and qualitatively consistent with experimental observations.

L. Dai (✉) · V. B. C. Tan
Department of Mechanical Engineering, National University of Singapore, Singapore, Singapore

S. K. Sinha et al. (eds.), *Nano-tribology and Materials in MEMS*,
DOI: 10.1007/978-3-642-36935-3_7,

Contents

Introduction

The development of new experimental techniques in the last two decades, such as the microbalance technique, surface force apparatus and microscopy, has brought about new opportunities in experimental characterizations in nano-tribology [1, 2]. As early as 1987, Mate et al. reported periodic stick–slip motion of a tungsten tip sliding on graphite surface [3] and measured the coefficient of friction (COF) to be 0.012. The periodicity of the stick–slip was determined by the lattice structure of graphite surface. Similarly, periodic stick–slip was also observed for Si tip sliding on the mica surface [4]. Subsequently, several tip-substrate friction studies were reported by Bhushan's group. Frictional studies involving surfaces of Si, SiO_2, Si_3N_4 and diamond were conducted and very low COF were recorded, many below 0.1 [5–8]. The COF was also found to decrease significantly for Si_3N_4 tips sliding on Si or polymer coated substrates when the size of contact was reduced from the macro to the nanometer scale [9]. Apart from tip-substrate friction, Minn and Sinha [10] investigated the sliding of Si_3N_4 ball on ultra high molecular weight polyethylene (UHMWPE) coated Si surfaces. It was found that the friction behavior was strongly influenced by the molecular orientation of the polymer and its degree of crystallinity, agreed with early works by Pooley and Tabor [11]. Double layered structures, with plasma treatment on the inner layer surface, have the lowest COF and best wear resistance because plasma treatment enhanced the inter-layer adhesion, and consequently, the structural stability and wear resistance [12]. Another notable experimental work on polymer tribology was reported by Maeda et al. [13], who found that crosslinked polymer (polystyrene and polyvinyl benzyl chlorid) has much lower COF than non-crosslinked polymer due to the less diffusion of polymer end segments when crosslinked.

At the nanometer scale, experimental works are limited by the force and displacement sensing challenges. Numerical approaches, especially molecular dynamics (MD) simulation, are establishing themselves to be effective tools to analyze nanotribological mechanisms. Robbins's group [14, 15] conducted pioneering and important nanotribology numerical investigations in early 1990s. They modeled a block of atomic film sandwiched between two rigid walls. By dragging one wall with a spring while keeping the other fixed, the motions of atomic film at the wall-film interface was observed to display stick–slip phenomenon. The wall was originally stuck to the solid film via atomic interactions. The drag force accumulated until it exceeded the critical shear stress to melt the film surface, at which point the wall surged forward for a distance. As the wall slipped, the sliding force decreased; meanwhile the wall quickly became stuck to the re-frozen film again before the next cycle of stick–slip. At higher sliding speeds, the stick–slip behavior changed to dynamic frictional sliding. Fluctuations during dynamic frictional sliding in the friction force are lower than in the case of stick–slip sliding. It was proposed that dynamic fiction was due to insufficient structural re-ordering when the sticking occurs. Later, a super kinetic friction phenomenon with continually instantaneous ultra low friction force at very high sliding rates was reported by Yoshizawa et al. [16]. More recent works of Robbins' group were extended to crystalline materials [17, 18]. They found that two clean crystalline contacting surfaces have little static friction force unless there were "third body" contaminations at the interface because mobile atoms of the third body would cause the two surfaces to be pinned together.

Studies of tip-substrate tribology were also carried out theoretically. Sorensen et al. proposed Cu–Cu model to analyze tip-crystalline surface tribology [19]. A slip is caused by nucleation and subsequent motion of dislocations. The friction force drops significantly when the tip slips until it re-sticks to the substrate. Several theoretical works were reported later focusing on various factors. Cagin et al. found that the existence of interfacial dangling bond increases the friction force [20]. Li et al. proposed that the stick–slip on crystal surfaces was mainly caused by the substrate elastic deformation and recovery in their Ni/Al model [21]. Yang and Komvopoulos analyzed tip geometries and claimed that COF is sensitive to the anisotropy of the tip shape [22]. Both experimental and theoretical works on the tip-substrate tribology have been recently reviewed by Kim and Kim [23].

Recently, the tribology of polymers has seen many investigations on self-assembled mononlayer (SAM) deposited on rigid substrates. Glosli and McClelland proposed two mechanisms for the stick–slip friction between two SAM surfaces [24]- a plucking mechanism associated with sudden release of shear strain via molecular deformation, and a more continuous viscous mechanism arising from continuous collisions with atoms from the opposite film. The effects of molecule chain length and sliding velocity was studied by Chandross et al. [25], who reported that the friction force was found to be independent of chain length at 2 GPa contact pressure. The friction force was weakly correlated to sliding velocity only when the pressure dropped to 0.2 GPa. The tribology of SAMs was recently reviewed by Harrison's group [26]. Besides SAMs, Tanaka and Kato reported that COF drops as the normal load increases in their perfluoropolyether lubricated interface model [27]. Heo et al. set up models of crystalline polyethylene pipes and found that the friction behavior is strongly affected by sliding direction relative to polymer crystalline orientation [28].

Theoretical models on the tribology problems in literature are limited to orderly structured materials like simple atomic layers, SAM-like materials or crystalline structures. However, materials of current interest are those with non-crystalline structures, such as amorphous polymers. Tribological behaviors and, more importantly, the underlying mechanisms remain an area of active research for amorphous polymers. In this chapter, MD simulations, supported with experimental characterizations, are presented to study the tribology phenomena of amorphous polyethylene interfaces.

Numerical Modeling

Two atomistic polyethylene film models, a 50 Å thick substrate film and a 20 Å thick slider film, in a periodic cell, are constructed for this investigation. The amorphous films have the same density as experimental samples, and are constructed with 50-ethylene-monomer linear polyethylene chains. Three slider-substrate models, with the same substrate film but different conditions imposed on the slider, were built as shown in Fig. 7.1. A layer of molecules at the bottom of the

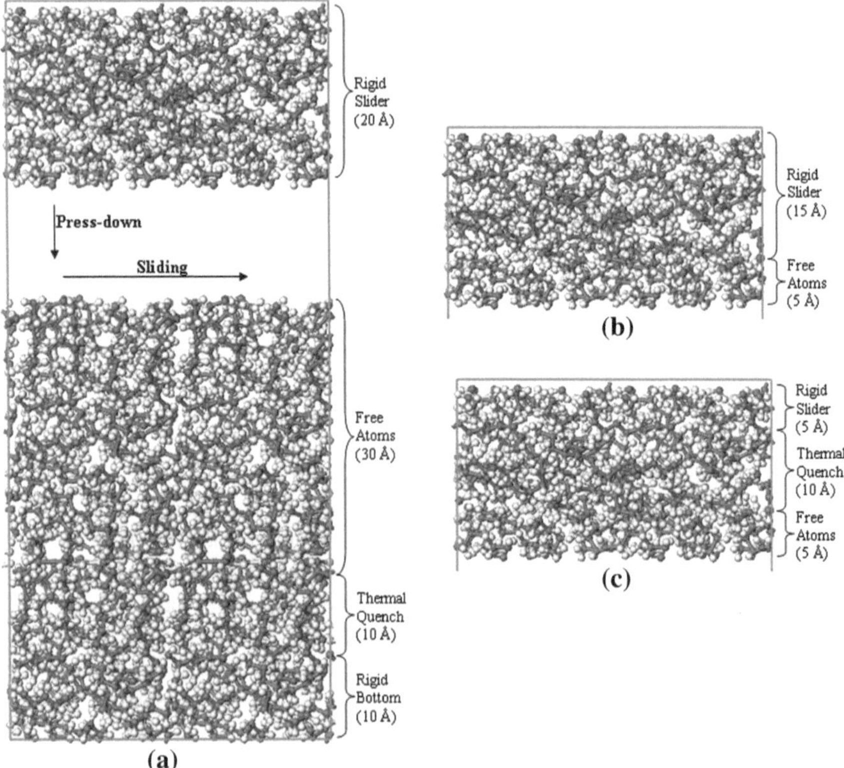

Fig. 7.1 Front views of polymer models. **a** Model I, **b** Slider for model II and, **c** Slider for model III. The substrates for all three models are the same. The width of the cell is 32 Å, and the thickness of each layer is marked in the charts. Grey dots stand for C atoms, white dots for H atoms

substrate were spatially fixed. Another layer of thermally quenched atoms lay above them and are topped by a free layer as the substrate. Model I has a completely rigid slider whereas in model II a thin layer of atoms at the bottom of the slider are free while the rest of the slider is still rigid. The slider in model III has the same layered structure as the substrate.

MD simulations were carried out using the LAMMPS code [29] with the COMPASS interatomic forcefield at a 10 Å cut-off distance [30]. Atoms in the rigid region of the substrate were fixed while those in the rigid region of the slider were set to move as a rigid block. Periodic scaling of the atomic velocities was performed every 0.1 ps to maintain a temperature of 300 K for the thermal quench layer. The remaining atoms were not constrained.

The slider started from a position sufficiently far above the substrate so that it did not interact with the substrate atoms at the start of the simulations. The slider was then pressed downwards and a constant contact pressure of 1 GPa was applied on the top of the slider throughout the simulation. As the slider approached the substrate, slider and substrate atoms started to interact each, and a new interface was created. The interface would take some time initially to oscillate to locate the ground energy morphology. Once a stable interfacial interface was reached, sliding of the slider could be carried out to study the tribology phenomena.

The equilibrated interface configurations of the three models give rise to different types of interface conditions as shown in Fig. 7.2a and have some common features in agreement with interface conditions proposed by Maeda et al. [13]. Type-A has a distinct interface with atoms from the two film atoms located cleanly on opposite side of the interface. Friction for type-A interface is often expected to be weak. In

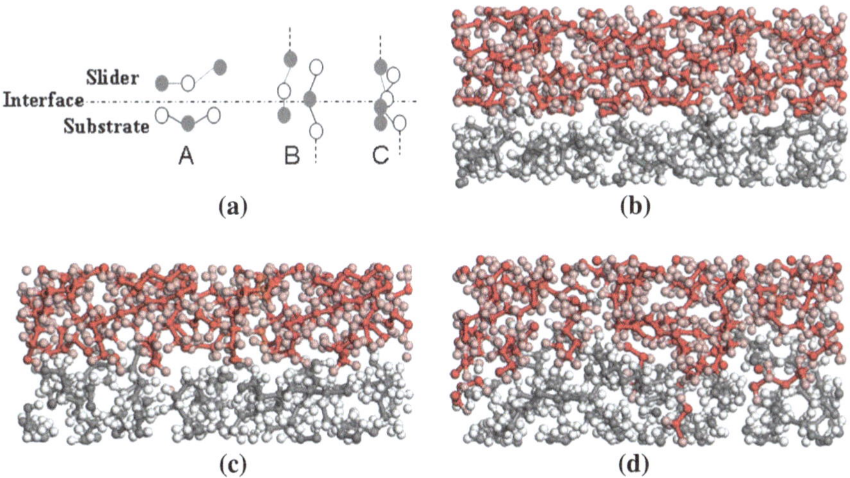

Fig. 7.2 Views of interfaces. **a** Schematic of interface types-A: non-diffused; B: slightly diffused but non-entangled; C: diffused and entangled. Equilibrated interfacial views of three models as, **b** model I, **c** model II, and, **d** model III. The sliders atoms were marked differently as *red* (C atoms) and *pink* (H atoms) for a clear view of interfacial diffusion conditions

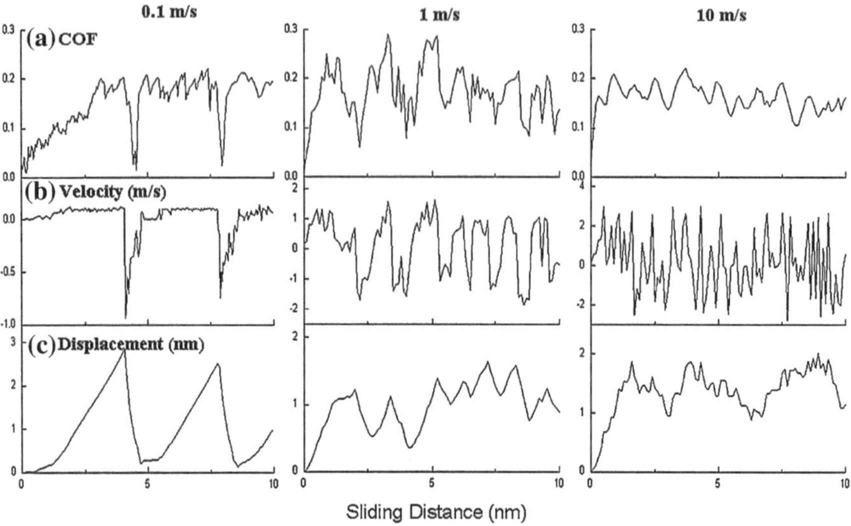

Fig. 7.3 Time histories of the data as **a** COF between slider and substrate (*top*), **b** instantaneous average velocities of atoms on the surface of the substrate closest to the slider (*middle*), and, **c** average horizontal displacements of substrate surface atoms from their original positions (*bottom*), for model I for sliding velocities of 0.1 m/s (*left*), 1 m/s (*centre*) and 10 m/s (*right*)

type-B, the end segments of the polymer chains from both sides diffuse slightly across the interface. The tribological interaction is moderately stronger than type-A interface due to stearic barriers posed by the diffused polymer chain ends during sliding. Type-C interface represents interfaces with strong inter-diffusion accompanied by chain entanglement of molecules from both slider and substrate. The interfacial interactions would be the strongest of the three. Figure 7.2b–d are snapshots of the equilibrated interfacial structures of three models. Obvious different interfacial diffusion states induced by different slider conditions are seen. The interfacial structures almost entirely resemble type-A in model I whereas mainly type-B with some type-A and a small fraction type-C interactions are found in model II. In model III, a combination of type-C and type-B interactions are present.

After the stabilization of the interface, a constant horizontal sliding velocity was applied to the rigid atoms in the slider. Figure 7.3 presents simulation results for model I at sliding speeds of 0.1, 1 and 10 m/s for a total sliding distance of 10 nm. The charts show the time histories of the COF between slider and substrate (Fig. 7.3a), the instantaneous average velocities of atoms on the surface of the substrate (Fig. 7.3b) and the average horizontal displacements of these surface atoms from their original positions (Fig. 7.3c). Surface atoms are defined as substrate atoms which lie within the interatomic forcefield cut-off distance from slider atoms.

For sliding at 0.1 m/s, an obvious periodic stick–slip phenomenon is observed. The cause of stick–slip is interpreted in Fig. 7.4. Figure 7.4a shows the initial equilibrated interfacial configuration prior to sliding. The polymer chains within

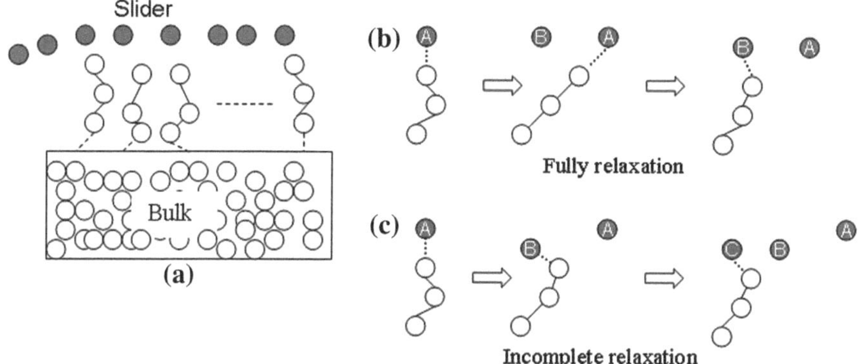

Fig. 7.4 Initial interface configuration **a**, and **b** interpretation of full relaxation, and **c** incomplete relaxation during slip. *Unfilled circles* denote substrate atoms and *filled circles* denote atoms in the slider

the substrate are strongly entangled in the bulk with a few end segments protruding loosely out at the interface. Figure 7.4b is a schematic of the relative interfacial deformation. It shows atom A from the surface of the slider interacting with the end atom of a polymer chain as the slider is dragged across the substrate. The chain is straightened due to the attraction from atom A. During the period of sticking, the friction force increased until it attains a threshold value corresponding to the peak interatomic attractive force from atom A. The peak friction force is attained at intervals of 4.1 nm (Fig. 7.3) sliding distance, the point at which the substrate polymer chain detached abruptly from atom A and retracted on being released. The low sliding speed provides ample time for the molecules to relax almost completely before they are stretched again. Because of the high degree of entanglement of the polymer chains in the bulk and the low sliding speed, the release of a single molecule triggered the release of a series of other molecules within a stick–slip cycle. This concurrent deformation and full relaxation of a significant number of atoms led to the abrupt drops in COF (Fig. 7.3) between 'stick' and 'slip'.

Since the slip of surface atoms on the substrate was in the opposite direction to the sliding, the average of their instantaneous velocity was negative at the commencement of slip (Fig. 7.3). From the thermal dynamics of individual surface atoms ($E_k = \frac{3}{2}K_BT$, where E_k: kinetic energy; K_B: Boltzmann constant; T: temperature), the instantaneous temperature of the surface atoms at the onset of slip was determined to have reached the regime of melting (417–425 K), indicating a melting slip, an observation in congruence with Debye–Waller factor analysis from Robinson's model [14]. With sufficient time to relax, the surface atoms returned close to their original positions before they are stuck to and dragged along by the next slider atom (atom B in Fig. 7.4b) to commence the next cycle of stick–slip. It can be seen from Fig. 7.3 that all the results for sliding at 0.1 m/s show a regular periodicity of 4.1 nm sliding distance due to simultaneous relaxation of polymer chains under such slow sliding.

At a higher sliding velocity of 1 m/s, the spatial frequency of the stick–slip is higher. The reduced relaxation time meant that the polymer chains of the substrate have insufficient time to relax substantially between periods of sticking. During slip, no clear concurrency of polymer chain deformations was observed because of the insufficient time to trigger slippage of many molecular chains within a stick–slip cycle. Therefore, the periodicity of stick–slip cycles became irregular, and fluctuations in the COF (Fig. 7.3) are of smaller magnitude. The higher sliding speed also brought more frequent and stronger atomic collisions at the interface, leading to higher instantaneous speed of surface atoms (SAs). As illustrated in Fig. 7.4c, the surface atoms of the substrate collided repeatedly with the slider atoms, providing no opportunity for complete relaxation. This is similar to the 'viscous mechanism' reported in Glosli's work [24]. Compared to the simulations of 0.1 m/s sliding velocity, the surface atoms were dragged along further, i.e. rebounded less during slip, for 1 m/s sliding velocity. These atoms did not return to their original positions even when the slider was removed, indicating the existence of permanent plastic deformation.

When the sliding velocity increased to 10 m/s, the tribological characteristics conformed to pure dynamic frictional sliding (Fig. 7.3), i.e. zero stiction. It is noted that the maximum velocity recorded for the SAs is much lower than the slider velocity, indicating that the surface no longer adheres to the slider. This also suggests that there is a limiting sliding velocity beyond which sticking is not observed, and the stick–slip phenomenon will switch to dynamic frictional sliding. We have tried even faster sliding speed, but the frictional characteristic was little changed, indicating that it has reached a sliding speed dependent convergent dynamical tribological behavior at 10 m/s sliding speed.

The simulations using model I suggest that tribological behavior is dependent on sliding velocity. Regular periodic stick–slip is characteristic of slow sliding, but with increasing sliding speed, the polymer substrate no longer showed evidence of periodic relaxation due to the absence of sufficient molecular reconfiguration. Instead, the frequency of the stick–slip cycles becomes higher and fluctuations in COF are smaller in magnitude as shown in Fig. 7.3. Beyond a critical sliding speed, characteristics of convergent dynamic frictional sliding were observed.

Due to the significant interfacial diffusion, the sliding tribology in model II and III was dominated by the interactions of the chains that have diffused across the interface. As shown in Fig. 7.5, three mechanisms of deformations—'brushing', 'combing' and 'chain scission' —were identified for the substrate polymer chains. Brushing occurs when the substrate chains only interfered with a small segment of the slider chains (no more than the length of three PE monomers). The substrate chains are likely to be bent and pressed down by the collisions from the sliding units (Fig. 7.5a) as they brush across the opposite surface. At low sliding velocities (e.g. 0.1 m/s) the substrate chains were observed to repeatedly bend and straighten and diffuse back into the slider region again, giving rise to periodic tribological characteristics. However, in the case of fast sliding, for example at 10 m/s, the chains were pressed down throughout the sliding process because the continuous atomic collisions did not allow appreciable straightening of substrate chains.

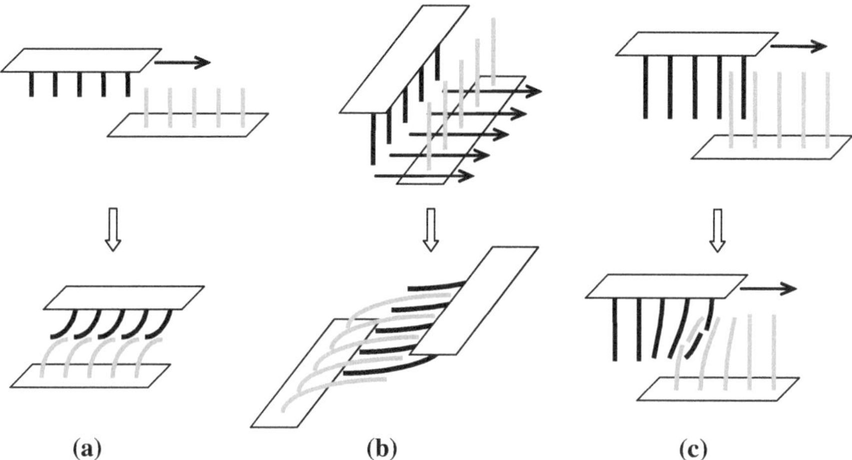

Fig. 7.5 Schematic description of interfacial polymer chain deformation mechanisms for diffused interface. The *dark* and *light grey lines* represent polymer chains from the slider and substrate, respectively, **a** Brushing, **b** Combing, **c** Chain scission

When the diffused substrate segments did not directly obstruct the sliding chains, the substrate chains will normally follow zigzag motions as they weave and squeeze in between neighboring slider chains (Fig. 7.5b), i.e. the molecular chains from opposite surfaces combed through one another. 'Combing' requires little interface deformation and therefore does not contribute much to the frictional forces. Besides the widely observed 'brushing', 'combing' was more noticeable in model III than model II, because of the deep surface diffusion in model III.

Very rarely did scission of molecular chains occur. 'Chain scission' (Fig. 7.5c), only happened when the substrate chains were deeply diffused into the slider, and more importantly, almost completely blocked the path of the sliding units. Some chains were even entangled with one other. During sliding, the diffused substrate chains were held tightly and dragged along in the sliding direction. The diffused chains would then fully straighten before finally sheared into two parts. 'Chain scission' only appeared in model III and only for fast sliding speeds of 1–10 m/s. After breakage, the sheared off segment further diffused into the slider and joined the sliding units of the slider. The cropped substrate chains were not long enough to diffuse deep into the slider again and experienced 'brushing' with continued sliding. There was significant reconfiguration of the interfacial structure following the scission of molecular chains resulting in reduced fiction.

Figure 7.6 presents the time histories of COF, surface atom velocities and surface atom displacement for models II. Under 0.1 m/s sliding, the COF still shows periodicity of stick–slip. However, comparing to model I, it has higher vibration during stick, and retardant retreat when slip, as a result of the diffused interface. During the stick event, the deformation of SAs in model I was totally enforced by the interaction from the slider located at the other side of the interface.

Meanwhile, in model II, besides the interactions across the interface, the diffused slider and substrate atoms imposed each other significant collision effect to accelerate the movement and vibration of SAs. Therefore, strong vibrations took place during sticking event. This also led to the retard and less magnitude of atomic rebound at the stage of slip. As mentioned earlier, regular periodic stick–slip is the manifestation of the deformation and relaxation of molecular chains 'brushing' over one another—a dominant feature of slow sliding for model I. Whilst 'brushing' continued to be widespread for models II, the additional phenomena of 'combing' masked the stick–slip cycles. This is because 'brushing' is largely reversible deformation and relaxation of molecular chains whereas there is no recovery from 'combing' during sliding. Furthermore, it also caused the SAs in model II to displaced further and further away from their original positions even after each slip compared to model I.

With increasing the sliding speed, the COF curves became irregular and gradually transformed to dynamic frictional sliding at 10 m/s sliding. With interface diffusion, SAs in model II displaced significantly in the sliding direction at the initial stage and then oscillated at equilibrium configurations with quite small pace of moving further as shown by the plots of surface atom displacements in Fig. 7.6. This shall be attributed to the high frequency of atomic collision at the beginning, and then slowing down after some time of sliding. 'Brushing' continued to be the most prevalent interface deformation and the main cause of the fluctuations, meanwhile the lower amplitude of the fluctuations for models II compared to model I (Fig. 7.3) is again due to the presence of 'combing'.

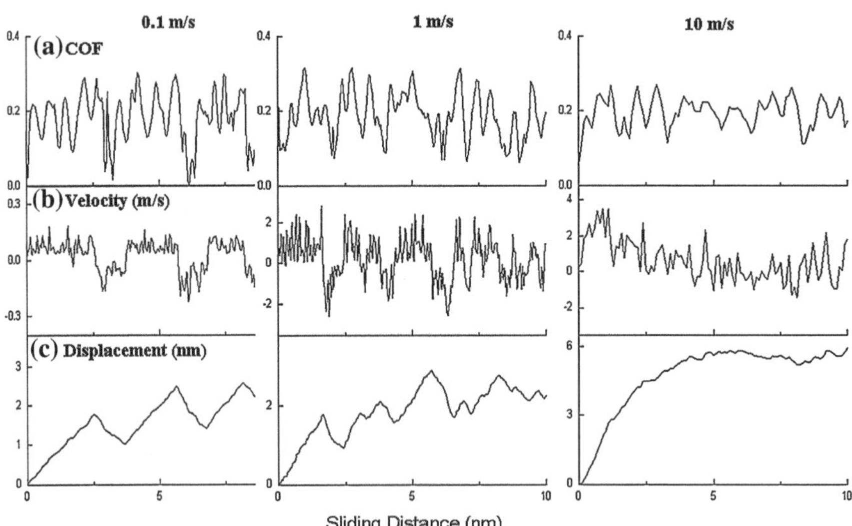

Fig. 7.6 Time histories of, **a** COF between slider and substrate, **b** instantaneous average velocities of atoms on the surface of the substrate closest to the slider, and, **c** average horizontal displacements of substrate surface atoms from their original positions for model II, from top to bottom rows, respectively

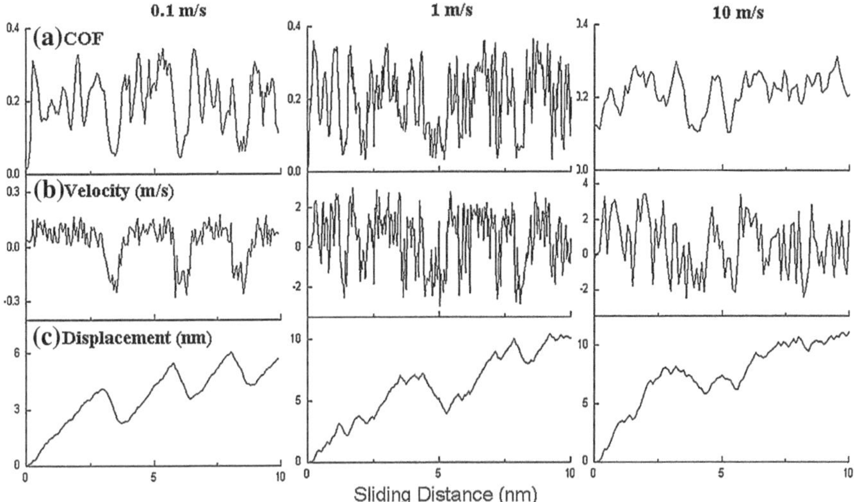

Fig. 7.7 Time histories of **a** COF between slider and substrate, **b** instantaneous average velocities of atoms on the surface of the substrate closest to the slider, and **c** average horizontal displacements of substrate surface atoms from their original positions for model III, from top to bottom rows, respectively

The same data plots for model III were shown in Fig. 7.7 accordingly. Due to the strongest interfacial diffusion among the three models, the periodic stick–slip behavior at 0.1 m/s sliding was rarely observable, with high magnitude of data fluctuations. The other diffusion induced factors, such as retardant slip, initial displacement of SAs, etc., were most strongly reflected in the data curves. Higher sliding rates (1 m/s and above) brought about the unrecoverable chain scissions. At 10 m/s sliding in model III, the scission of chains exhausted after a certain sliding distance (5.2 nm) as reflected by a reduction in the fluctuations of the COF plot (Fig. 7.7).

Figure 7.8 shows the time-averaged COF of the three models as a function of sliding speed. As expected, models with stronger interfacial diffusion had higher COF, due to the higher frequency of atomic collisions from the diffused segments. Denser and stronger interfacial locking also downgraded the effect from the 'brushing' mechanism, which dominates the tribology behavior transformation by means of sliding speed. Thus, stronger diffusions require higher sliding speeds to induce the transformation from stick–slip to dynamic frictional sliding. It is also noticed that the COF increased during the transformation from periodic stick–slip to irregular stick–slip, but dropped when pure dynamic frictional sliding occurred. This was due to the less percentage of increase in friction forces to normal forces during the transformations although both frictional and normal forces increase. For all three models, the COF became independent of sliding velocity in the pure dynamic frictional sliding regime, in agreement with Glolsi's model [24].

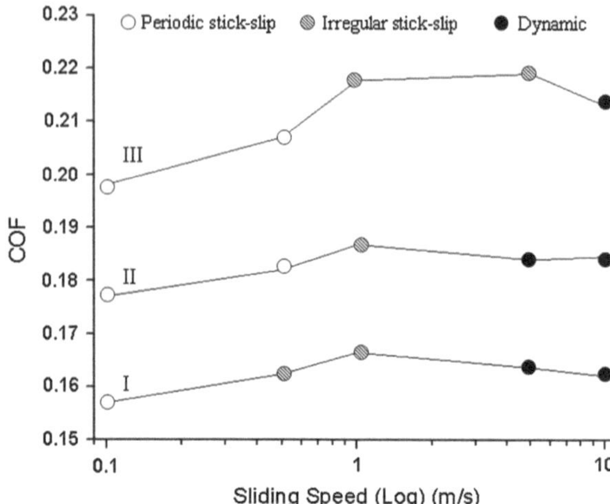

Fig. 7.8 Time averaged COF of three models as a function of sliding speed

Experiment

Numerical works were carried out at very small size and time scale. To verify whether our results are consistent at macro scale, experimental tests were conducted using a ball-on-disc tribometer where normal and friction forces (converted from normal and lateral laser displacements, respectively) were measured with laser displacement sensors (Fig. 7.9). With amorphous polyethylene, a rotating disc and a sliding ball with 3.2 mm diameter as a counterface were built. Due to the much stronger compaction for the sliding ball, the interface contact was actually a hard ball sliding on the soft substrate. No obvious interfacial diffusion was observed during the slidings. For different tests, the linear sliding speeds from the rotational motions of the disc were set as 0.42, 4.2 and 42 mm/s. The normal loads were applied on the top of the ball with 100, 125 and 150 mN, respectively. The Hertzian contact pressure was accordingly calculated as 23, 24.9 and 26.4 MPa, assuming a perfect fully contact between the sliding ball and disc. The sampling rate used in recording data was 50 Hz. Three repeated sliding processes were conducted for the same test, and results were averaged as the final values.

Under constant loading pressure as 23 MPa, the experimental sliding speed dependence of frictional mechanics was presented in Fig. 7.10. Due to the macro time and size scale, the atomic stick–slip phenomenon was not observable, measurements was mainly to record the force vectors, and for reporting the COF.

At 0.42 m/s sliding, the time history of COF was recorded as a slightly vibrating curve, with an average value of 0.094. With the sliding increased to 4.2 mm/s, two stages of COF values were recorded: COF was initially measured to be around 0.162 up to a sliding distance of 1 mm, where a sudden drop occurred to a value of around 0.15. We suspect that an atomic re-location happened in order to smooth the contact sliding. At the high speed 42 mm/s sliding, data fluctuation was significantly enhanced, probably due to the instability from continuous

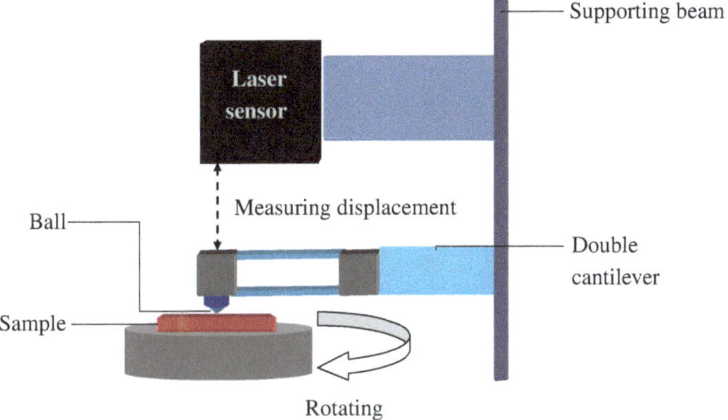

Fig. 7.9 Schematic of the ball-on-disc tribometer

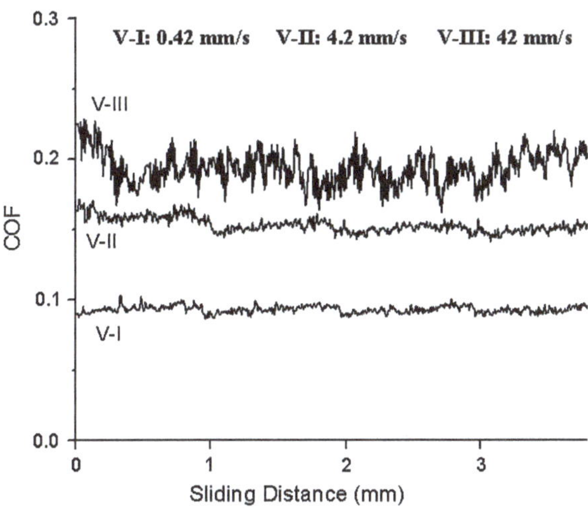

Fig. 7.10 Experimental COF curves as a function of sliding speed. The loading pressure was kept constant as 23 MPa

colliding effect, which also promotes average COF to be around 0.2, twice as that of 0.42 mm/s sliding. The recorded values of COF qualitatively agree with simulation values in Fig. 7.3.

The loading pressure is also considered as a factor for the tribology behavior, because it affects interfacial structures, and thereby the atomic interactions at the interface. Figure 7.11 shows the friction data curves with 1 m/s sliding, but under various loading pressures for model I. It can be seen that higher loads increased interfacial forces in both normal and horizontal directions. However, we notice a greater flattening-compressed substrate top surface under higher pressure, shown as the surface roughness curve in Fig. 7.11. Flat contact surface will cause the interfacial atomic interaction force vectors to rotate toward the normal direction, thus the normal force was much more enhanced than the friction force. Such re-orientation of interfacial forces leads to a drop of COF with pressure increase.

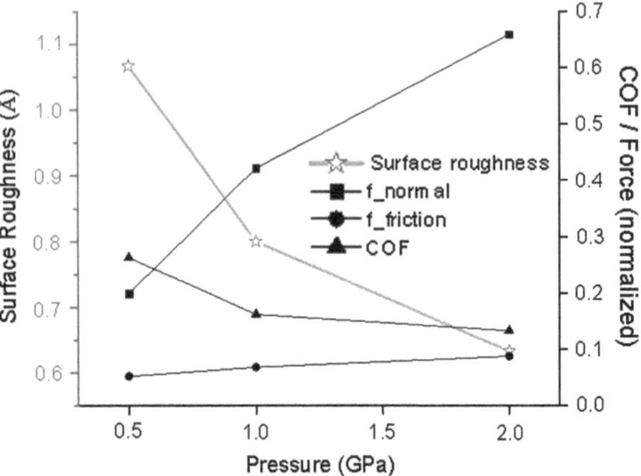

Fig. 7.11 Average substrate surface roughness, normal force, friction force and COF plots against normal loading pressures. The normal force refers to the total load on the substrate atoms in the normal direction. Force values were normalized to between 0 and 1. All the data were obtained from simulations of 1 m/s sliding using model I

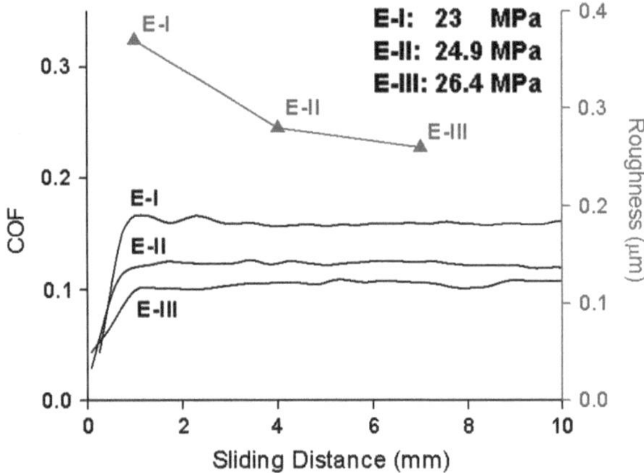

Fig. 7.12 Experimental results of COF and average surface roughness under different loading pressure. Values of COF were read from the left axis (black), and roughness from the right axis (grey)

However, in model II and III, high pressure enhanced the interfacial diffusion as well as the flattening of the substrate surface, leading to a case by case effect.

Due to different hardness between the sliding ball and the substrate, interfacial diffusion was almost curbed. Therefore, the experimental friction condition was most close to model I. The pressure effect analysis from experimental characterizations was shown in Fig. 7.12 with a fixed sliding speed of 4.2 mm/s. It clearly shows that COF drops with increasing the loading pressure. Due to the incommensurate hardness of the contacting surface, lower roughness of substrate surfaces and decreasing COF were measured when higher loading pressure was applied. Similar pressure based COF trend has been observed in experiments [6, 31] and numerical works [17, 27].

Conclusion

In conclusion, the tribology of polymer–polymer interface has been studied by MD simulations and several friction mechanisms were observed. Friction is dependent on interfacial lock status, material property, sliding speed, external pressure, etc. The interfacial structure was considered as the determinant factor, which determined speed dependent tribological behaviors, from periodic stick–slip, to irregular stick–slip and to the final dynamic friction with the function of sliding acceleration. Three mechanisms, namely, 'brushing', 'combing' and 'chain scission', govern interfacial deformations during frictional sliding. Higher external pressure reduced the COF due to a flattened surface in case of non-diffusive interface. However, at diffusive interface, high pressure enhanced the interfacial diffusion as well as the flattening of the substrate surface, leading to a case by case effect. These factors influenced frictional mechanics that was consistent at macro scale, supported by our experimental characterizations.

Acknowledgments The authors wish to acknowledge the financial support given to this work by the National Research Foundation (NRF), Singapore (Award no. NRF-CRP 2-2007-04).

References

1. Kaneko, R., et al.: Recent progress in microtribology. Wear **200**, 296 (1996)
2. Persson, B.N.J.: Sliding friction. Surf. Sci. Rep. **33**, 83 (1999)
3. Mate, C.M., et al.: Atomic-scale friction of a tungsten tip on a graphite surface. Phys. Rev. Lett. **59**, 1942 (1987)
4. Akamine, S., Barrett, R.C., Quate, C.F.: Improved atomic force microscope images using microcantilevers with sharp tips. Appl. Phys. Lett. **57**, 316 (1990)
5. Ruan, J.A., Bhushan, B.: Frictional behavior of highly oriented pyrolytic graphite. J. Appl. Phys. **76**, 8117 (1994)
6. Bhushan, B., Kulkarni, A.V.: Effect of normal load on microscale friction measurements. Thin Solid Films **278**, 49 (1996)
7. Bhushan, B., Li, X.D.: Micromechanical and tribological characterization of doped single-crystal silicon and polysilicon films for microelectromechanical systems devices. J. Mater. Res. **12**, 54 (1997)
8. Bhushan, B., Sundararajan, S.: Micro/Nanoscale friction and wear mechanisms of thin films using atomic force and friction force microscopy. Acta Mater. **46**, 3793 (1998)
9. Tambe, N.S., Bhushan, B.: Scale dependence of micro/nano-friction and adhesion of MEMS/NEMS materials, coatings and lubricants. Nanotechnology **15**, 1561 (2004)
10. Minn, M., Sinha, S.K.: Molecular orientation, crystallinity, and topographical changes in sliding and their frictional effects for UHMWPE film. Tribol. Lett. **34**, 133 (2009)
11. Pooley, C.M., Tabor, D.: Transfer of ptfe and related polymers in a sliding experiment nature-physical. Science **237**, 88 (1972)
12. Satyanarayana, N., Lau, K.H., Sinha, S.K.: Nanolubrication of poly(methyl methacrylate) films on Si for microelectromechanical systems applications. Appl. Phys. Lett. **93**, 261906 (2008)
13. Maeda, N., et al.: Adhesion and friction mechanisms of polymer-on-polymer. Science **297**, 379 (2002)

14. Thompson, P.A., Robbins, M.O.: Origin of stick-slip motion in boundary lubrication. Science **250**(4982), 792 (1990)
15. Robbins, M.O., Thompson, P.A.: Critical velocity of stick-slip motion. Science **253**(5022), 916 (1991)
16. Yoshizawa, H., McGuiggan, P., Israelachvili, J.: Identification of a second dynamic state during stick-slip motion. Science **259**(5099), 1305 (1993)
17. He, G., Muser, M.H., Robbins, M.O.: Adsorbed layers and the origin of static friction. Science **284**, 1650 (1999)
18. Muser, M.H., Robbins, M.O.: Conditions for static friction between flat crystalline surfaces. Phys. Rev. B **61**, 2335 (2000)
19. Sorensen, M.R., Jacobsen, K.W., Stoltze, P.: Simulations of atomic-scale sliding friction. Phys. Rev. B **53**, 2101 (1996)
20. Cagin, T., et al.: Simulation and experiments on friction and wear of diamond: a material for MEMS and NEMS application. Nanotechnology **10**, 278 (1999)
21. Li, B., et al.: Molecular dynamics simulation of stick-slip. J. Appl. Phys. **90**, 3090 (2001)
22. Yang, J., Komvopoulos, K.: A molecular dynamics analysis of surface interference and tip shape and size effects on atomic-scale friction. J. Tribol. **127**, 513 (2005)
23. Kim, H.J., Kim, D.E.: Nano-scale friction: a review. Int. J. Precis. Eng. Manufact. **10**, 141 (2009)
24. Glosli, J.N., McClelland, G.M.: Molecualr dynamics study of sliding friciton of ordered organic monolayers. Phys. Rev. Lett. **70**(13), 1960 (1993)
25. Chandross, M., Grest, G.S., Stevens, M.J.: Friction between Alkylsilane monolayers: molecular simulation of ordered monolayers. Langmuir **18**, 8392 (2002)
26. Harrison, J.A., et al.: Friction between solids. Phil. Trans. R. Soc. A **366**, 1469 (2008)
27. Tanaka, K., Kato, T., Matsumoto, Y.: Molecular dynamics simulation of vibrational friction force due to molecular deformation in confined lubricant film. J. Tribol. **125**, 587 (2003)
28. Heo, S.J., et al.: Effect of the sliding orientation on the tribological properties of polyethylene in molecular dynamics simulations. J. Appl. Phys. **103**, 083502 (2008)
29. Plimpton, S.J.: Fast parallel algorithms for short-range molecular dynamics. J. Comput. Phys. **117**, 1 (1995)
30. Sun, H.: Compass: an ab initio force-field optimized for condensed-phase applicationssoverview with details on alkane and benzene compounds. J. Phys. Chem. B **102**, 7338 (1998)
31. Kustandi, T.S., et al.: Texturing of UHMWPE surface via NIL for low friction and wear properties. J. Phys. D Appl. Phys. **43**, 015301 (2010)

Chapter 8
Probing the Complexities of Friction in Submicron Contacts Between Two Pristine Surfaces

Wun Chet Davy Cheong and Anna Marie Yong

Contents

In the 1970s, scientist discovered that friction on the nanometre scale is very different from friction on the micrometre or larger scale. Amonton/Coulomb's law of friction does not apply at that scale. The control of friction on all length scales is important. On the macro scale, wear and tear in industries amounts to billions of dollars. For state of the art micro and nanodevices however, we are not even talking about monetary losses. These devices which involve sliding interfaces (e.g., actuators or nanogears) simply cannot work because wear and friction damage these tiny devices beyond repair. The much greater surface-to-volume ratio characteristic of these devices leads to serious adhesion and wear problems. In addition, conventional methods to reduce friction such as the use of lubricants cannot work on the nanometre scale because the traditional liquid lubricants become too viscous when confined to the layers of molecular thickness.

Amonton's law of friction describes the sum effect of many mechanisms of friction working simultaneously on many different length scales. The understanding of the mechanisms governing friction is still far from complete. In this theoretical study, atomistic simulation is used to show the different mechanisms of non-wear friction in effect as the contact area between two pristine surfaces changes from a few atomic radii to a few orders of nanometres. The various

W. C. D. Cheong (✉) · A. M. Yong
Institute of Materials Research and Engineering, A-STAR (Agency for Science, Technology and Research), 3 Research Link, Singapore 117602, Singapore
e-mail: davy-cheong@imre.a-star.edu.sg

S. K. Sinha et al. (eds.), *Nano-tribology and Materials in MEMS*,
DOI: 10.1007/978-3-642-36935-3_8,

mechanisms of friction results in different friction stresses generated between the surfaces.

The mechanisms of friction and wear on the nanometre scale can be very different from their micrometre or larger scales counterpart. Recent developments in nanotechnology have revealed great potential in modifying surfaces on the nanometre scale. Researchers have proven that clusters of nanometre scale contacts can provide strong adhesion just like the spatula at the ends of the setae on lizard's feet that allow them to walk on walls and ceilings [1]. On the other hand, lotus leaves inspired nano-structured surfaces with the right surface energy and nanoscopic bumps can ensure that dust particles do not stick to the surface easily, giving it self-cleansing capabilities [2].

A micro-mechanical dislocation model of frictional slip between two asperities was presented by Hurtado and Kim [3], which predicts that the frictional stress is constant and of the order of the theoretical shear strength, when the contact size is small. However, at a critical contact size there is a transition beyond which the frictional stress decreases with increasing contact size, until it reaches a second transition where the frictional stress gradually becomes independent of the contact size. Hence, the mechanisms of slip are size-dependent, or in other words, there exists a scale effect. Before the first transition, the constant friction is associated with concurrent slip of the atoms without the aid of dislocation motion. The first transition corresponds to the minimum contact size at which a single dislocation loop is nucleated and sweeps through the whole contact interface, resulting in a single-dislocation-assisted slip. This mechanism is predicted to prevail for a wide range of contact sizes, from 10 nm to 10 μm in radius for typical dry adhesive contacts; however, there are no available experimental data in this size range. The second transition occurs for contact sizes larger than 10 μm, beyond which friction stress is once again constant due to cooperative glide of dislocations within dislocation pileups. The above dislocation model excludes wear or plastic deformation of either surface.

There are still many findings to be uncovered on the mechanisms of friction. In this chapter, molecular dynamics method (MD) is employed to discuss the variation of friction force with respect to contact length and its associated mechanisms.

Two aspects of friction are studied in this chapter. Firstly, the effect of coating thickness on frictional wear and secondly contact size effect on non-wear friction.

Frictional Wear on Coated Substrate

The simulations of sliding on diamond coated silicon are discussed in this section. The diamond asperity used is hemispherical and has a radius of 2.14 nm, consisting of 1,818 atoms with 0.357 nm bond length. The silicon substrate consists of 46,993 atoms, each separated away by 0.543 nm, and the dimension is ($120 \times 68 \times 44$) atoms, or ($16.29 \times 9.23 \times 5.97$) nm^3. Four variations of coating thickness are used.

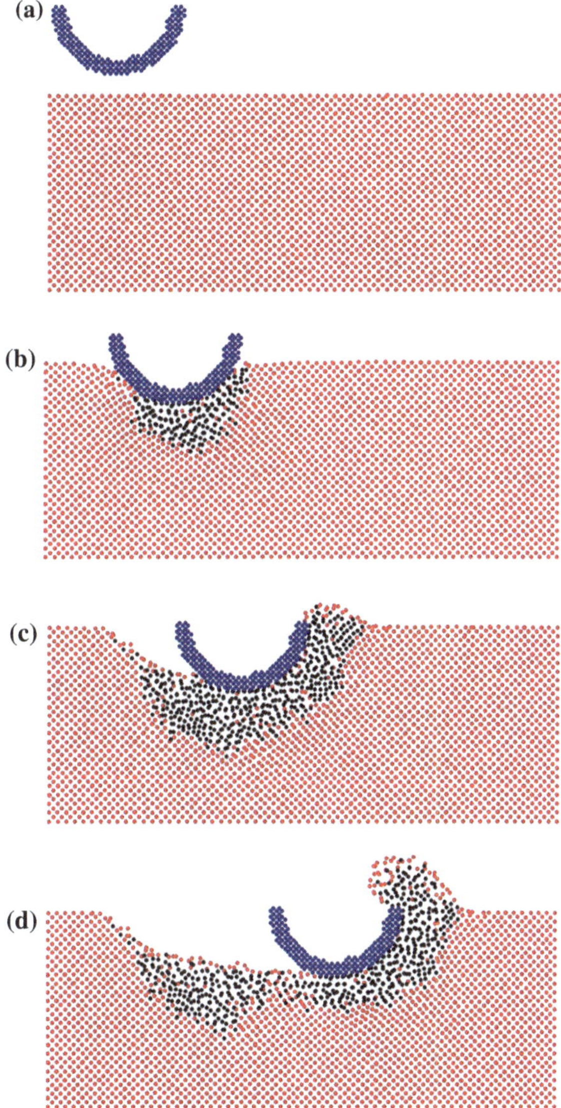

Fig. 8.1 Cross-sectional views at different stages for sliding *without coating* ($t^* = 0$). **a**, **b**, **c**, **d**

The Figs. 8.1, 8.2, 8.3, 8.4, 8.5 are cross sectional snap shots of four simulations of a carbon asperity sliding and cutting into diamond coated substrates (sliding speed of 100 m/s and maximum cut depth of 2 Nm) of four different coating thicknesses. For the same examples, Figs. 8.6a–d show the variation of sliding force versus time and Figs. 8.7a–d show the variation of normal force versus time.

In all four cases of different coating thicknesses, there were no signs of dislocations within the substrate. The inelastic deformation observed under the tool tip and along the sliding path is caused by viscous flow in the amorphous region

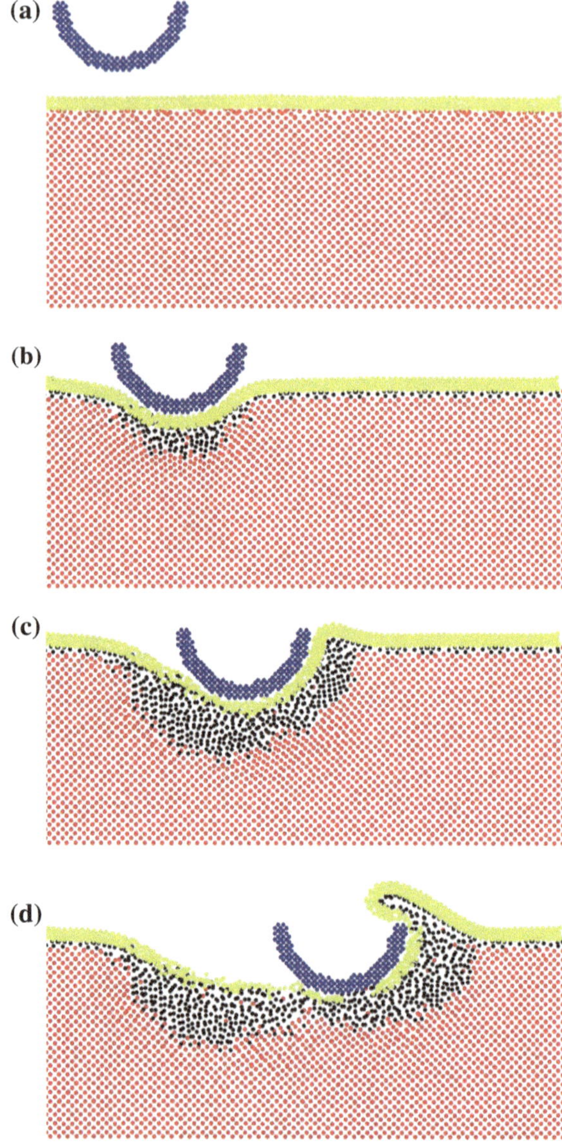

Fig. 8.2 Cross-sectional views at different stages for sliding with *very thin coating* ($t^* = 0.17$). **a**, **b**, **c**, **d**

rather than other failure mechanism such as through dislocations. Such transformation mechanism is similar to that of nano-indentation [4]. Silicon having diamond cubic structure first transforms into a metallic body centered β-silicon, and upon removal of stresses, β-silicon further transforms into amorphous silicon.

The following examples demonstrate the variations in deformation mechanism in nano-sliding with different coating thickness to asperity radius ratios.

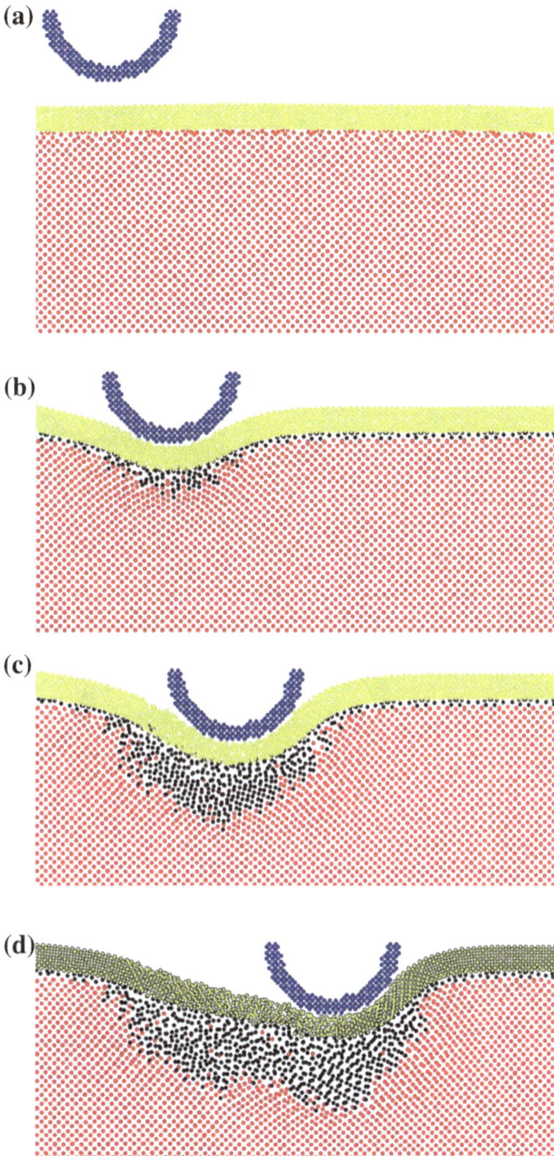

Fig. 8.3 Cross-sectional views at different stages for sliding with *thin coating* ($t^* = 0.35$). **a**, **b**, **c**, **d**

(a) No Coating ($t^* = 0$)

The non-coated system shows similar general wear mechanisms as those described by Zhang and Tanaka [5]. As the depth of cut of the asperity increases, the mechanism of wear progresses from no-wear to adhering, ploughing, and finally cutting. Formation of chip is observed when the asperity is cutting the substrate. (See Fig. 8.1)

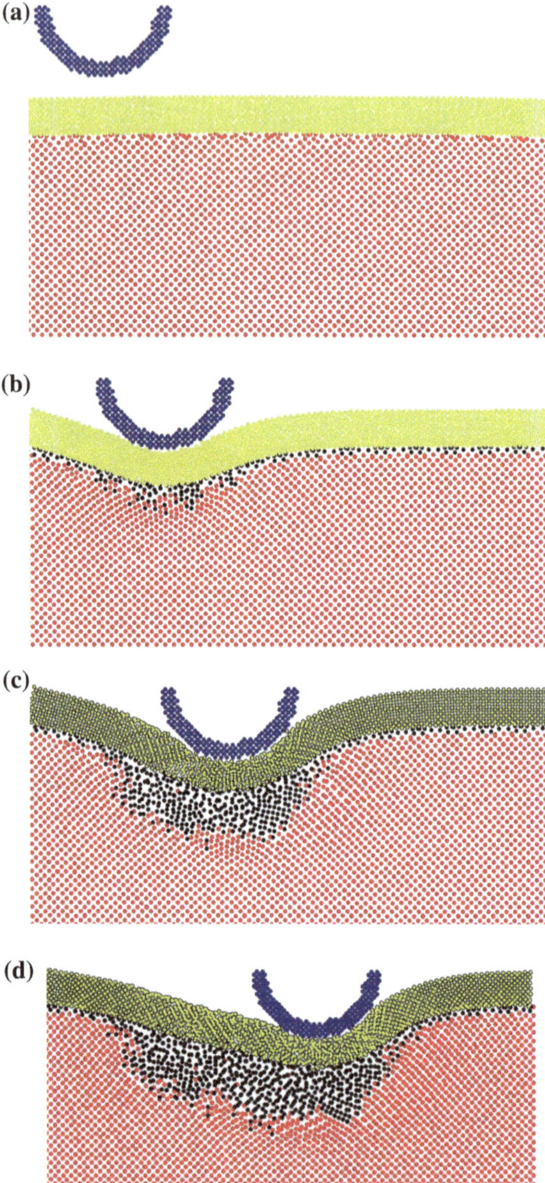

Fig. 8.4 Cross-sectional views at different stages for sliding with *medium thickness coating* ($t^* = 0.70$). **a**, **b**, **c**, **d**

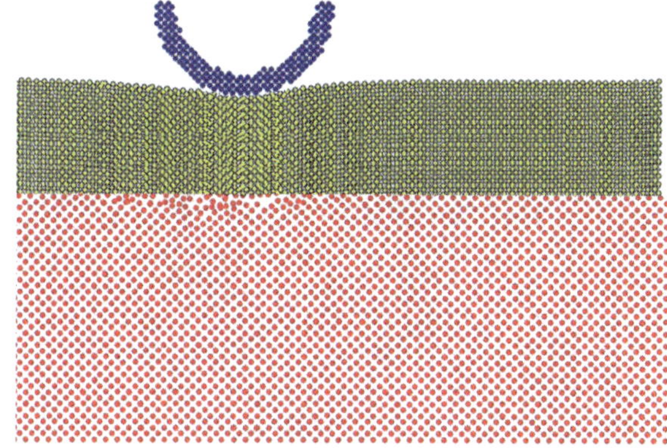

Fig. 8.5 Cross-sectional view for sliding with *thick coating* ($t^* = 1.4$)

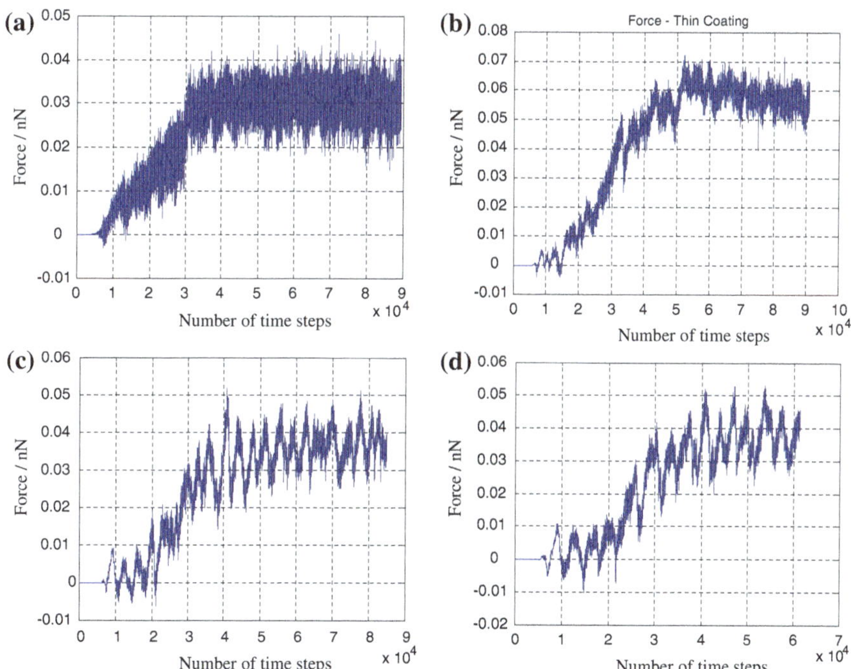

Fig. 8.6 Sliding force versus number of time steps for coatings of various thicknesses: **a** no coating; **b** very thin coating ($t^* = 0.17$); **c** thin coating ($t^* = 0.35$); **d** moderate thickness coating ($t^* = 0.7$)

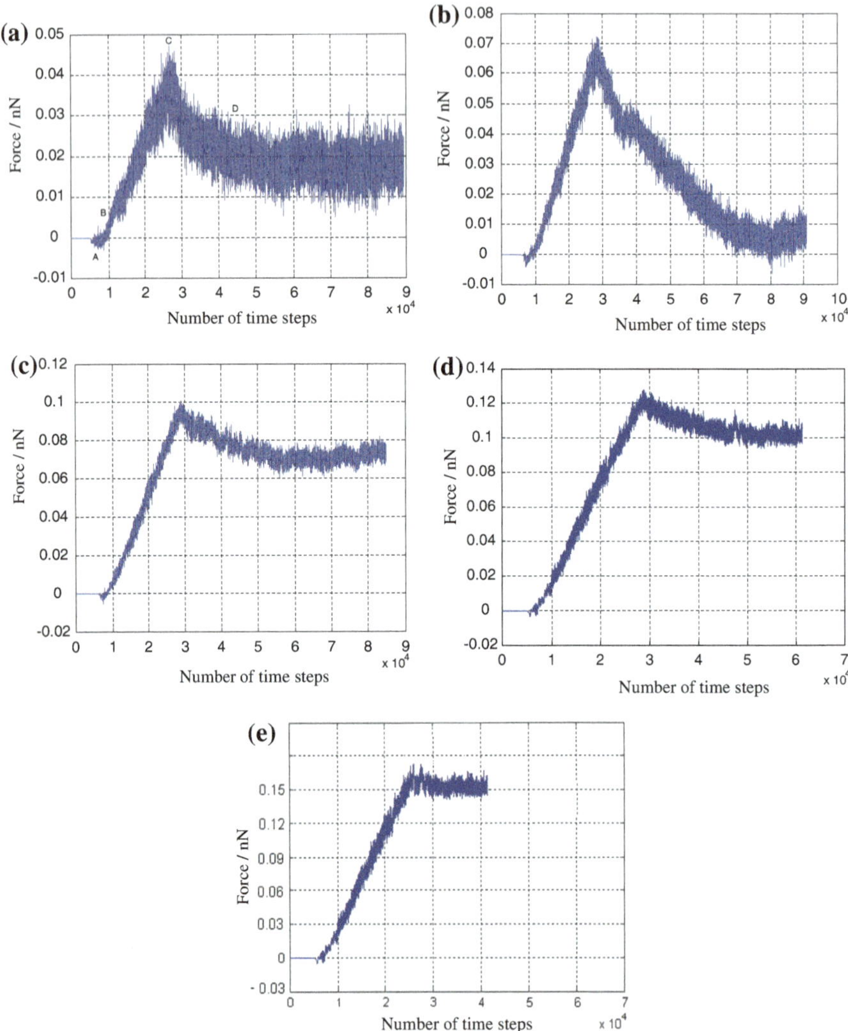

Fig. 8.7 Normal force versus number of time steps for coatings of various thicknesses: **a** no coating; **b** very thin coating ($t^* = 0.17$); **c** thin coating ($t^* = 0.35$); **d** moderate thickness coating ($t^* = 0.7$); **e** thick coating ($t^* = 1.4$)

The sliding force increases linearly from 0 to 0.03 nN as the depth of cut of the asperity increases to and remains constant at that value as the asperity ploughs through it. (See Fig. 8.6a)

Observing the normal force on the asperity (see Fig. 8.7a), it is found that the asperity experience and initial attractive force when the asperity is first in contact with the silicon surface. The normal force increases linearly to 0.04 nN as the depth of cut of the asperity increases. As the asperity ploughs through the silicon causing the formation of a chip, the normal force stabilizes at 0.02 nN.

(b) Very thin coating (t* = 0.17)

The plastic deformation mechanism in this case is very similar to that of sliding without coating. The presence of the very thin coating does not change the depth of phase transformation in the silicon substrate. The coating experiences plastic deformation through plastic flow and the silicon substrate experience plastic deformation as in the case of sliding without coating. As the asperity slides over the carbon coated silicon substrate, it begins to plough into the substrate forming a chip (see Fig. 8.2). The main difference between sliding with coating of this thickness and sliding without coating is that the thinner chip had formed only in the case of sliding with coating. The coating contains the chips formed and hence the chip formed is thinner than that of sliding without coating. The coating fails as the asperity slides over it ploughing into the substrate. No delamination is observed.

Initially, the sliding force remains constant because the contact area is small (Hurtado and Kim). The sliding force increases as the depth of cut increases to 0.06 nN. Stick-slip phenomenon is observed as the asperity slides over the coating. However, the slip-stick pattern is irregular because of the damage to the carbon coating as the asperity slides over it. The ploughing force of 0.06 nN is much higher in this case than in sliding without coating because the coated chips provide more resistant to the asperity. Figure 8.6b shows the asperity sliding force versus time steps plot.

The maximum normal force is much higher at 0.07 nN due to the carbon coating on top of the silicon substrate. There is a large drop in the normal force experienced by the asperity due to failure and tearing of the carbon coating indicated by the point Q on the figure. The normal stress stabilizes at 0.01 nN because the chips formed are still contained by the carbon coating and hence do not exert that much a normal force on the asperity as in the case of sliding with coating. Figure 8.7b shows the asperity normal force versus time steps plot.

(c) Thin coating (t* = 0.35)

As the thickness is increased, no ploughing is observed at the same depth of cut. Contrary to what is expected with a thicker coating, the size of the amorphous region in the substrate is surprisingly larger when t* = 0.35 where there is no ploughing than in the case when t* = 0.17, when ploughing occurs. This could be due to the absence of chip formation. In the case of t* = 0.17, the formation of chips allows the amorphous silicon to flow out of the substrate thereby relieving the stresses under the asperity. However in the case of t* = 0.35, the thicker coating prevents the formation of chips and hence the higher stresses lead to the formation of a deeper amorphous region under the asperity. Though the coating undergoes plastic deformation, it adheres to the amorphous silicon without tearing or debonding. Figure 8.3 shows snap shots of the plastic deformation of diamond coated silicon due to sliding with t* = 0.35 at various time steps.

The initial sliding force is once again constant because of the initial small contact area in non-wear sliding. The sliding force increases from 0 to 0.03 nN as the depth of cut of the asperity increases. The final sliding force is much lower than the case with diamond coated substrate due to the absence of ploughing and chipping. The stick-slip pattern is regular because the coating remains intact throughout the entire no-wear sliding process. Figure 8.6c shows the asperity sliding force versus time steps plot.

The maximum normal force reaches 0.1 nN which is higher than in the case of thin coating. This implies that the maximum normal force on the asperity increases with coating thickness. There is little drop in normal force as no chipping occurs and the drop in normal force can be explained by the plastic deformation experienced by the silicon substrate beneath the coating. Figure 8.7c shows the asperity normal force versus time steps plot.

(d) Medium thickness coating ($t^* = 0.7$)

As the thickness is increased further, the thicker coating takes more of the load of the sliding asperity. This results in a smaller amorphous zone in the silicon substrate. There is no plastic deformation in the coating as the coating recovers elastically as the asperity passes. In both these and the previous cases, the silicon substrate fails before coating. Figure 8.4 shows snap shots of the plastic deformation of diamond coated silicon due to sliding with $t^* = 0.7$ at various time steps.

It can be seen from the figure that a further increase in the coating thickness does not change the sliding force plot. Figure 8.6d shows the asperity sliding force versus time steps plot.

The maximum normal force reaches 0.12 nN which is even higher than the case of thick coating. This is consistent with the observation that the maximum normal force on the asperity increases with coating thickness. This might be explained by the increase in the rigidity of the coating with increasing thickness. There is little drop in normal force as no chipping occurs and the normal force stabilises at 0.1 nN as the asperity slides over the carbon coating. The drop in the normal force can be explained by the plastic deformation experienced by the silicon substrate beneath the coating. Figure 8.7d shows the asperity normal force versus time steps plot.

(e) Thick coating ($t^* = 1.4$)

When $t^* = 1.4$, there is no observable plastic deformation in the silicon substrate except for the coating substrate interface (see Fig. 8.5). Plastic deformation of the carbon coating occurs in a way similar to that of sliding on a diamond substrate. The sliding force versus time plot is no different from those of $t^* = 0.7$ and $t^* = 0.35$, respectively, as no chips were formed during sliding.

The normal force stabilizes at the maximum value after a linear increase due to the increasing depth of cut. There is no drop in the normal force because there is no observable plastic deformation in the substrate (see Fig. 8.7e).

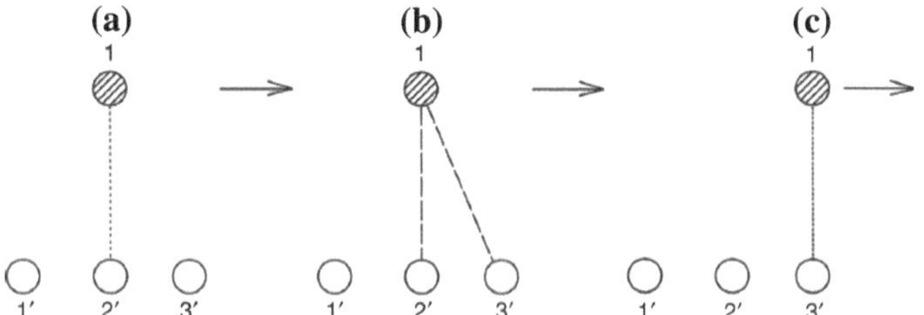

Fig. 8.8 The friction system consisting of four atoms numbered by *1*, and *1′*, *2′* and *3′*. All atoms are assumed to interact with each other. Atom 1 non-adiabatically (abruptly) changes its position during sliding. The non-adiabaticity leads to transforming the elastic energy into the vibrational or kinetic energy of atom 1. The vibrational energy of atom 1 may be considered to dissipate into the vibrational energies of other atoms, i.e., into the thermal energy. This picture involves the irreversible physical process, i.e., the energy dissipation in its natural form

Non-wear Friction on a Non-coated Contact

Tomlinson has described a mechanism of the energy dissipation to explain non-wear dynamic frictional force, based on the adhesion model [6]. In the Tomlinson model, it is assumed that the non-adiabatic (or abrupt) change of the positions of atoms during sliding subsequently transforms the elastic energy into vibrational energy. It assumes an irreversible physical process i.e., the energy dissipation in its natural form.

Tomlinson proposed an atomistic picture for the origin of the frictional forces. Suppose the friction system consist of four atoms numbered by 1, and 1′, 2′, and 3′ as seen in Fig. 8.8. All atoms are assumed to interact with each other. Atom 1 forms a part of the upper body, which interacts with the other atoms of the upper body (not illustrated in Fig. 8.8), and the atoms 1′, 2′, and 3′ form the lower body.

We shall concentrate on the behaviour of atom 1 when the upper body slowly slides against the lower.

When atom 1 is over atom 2′, atom 1 experiences the attraction from atom 2′, as seen in Fig. 8.8a. During sliding, atom 1 moves towards the right direction. When the sliding displacement is small, this is a process of storing the elastic energy, as seen in Fig. 8.8b. When atom 1 goes beyond a certain distance, the attraction from atom 3′ exceeds that from atom 2′. Atom 1 equilibrates in the position over atom 3′.

Then, the model assumes that atom 1 non-adiabatically (abruptly) changes its position. The non-adiabaticity leads to transforming the elastic energy into the vibrational or kinetic energy of atom 1, as seen in Fig. 8.8c. The vibrational energy of atom 1 may be considered to dissipate into the vibrational energies of other atoms, i.e., into the thermal energy. This picture involves the irreversible physical process, the energy dissipation in its natural form. If atom 1 is assumed only to change its position slowly, atom 1 may not take an excess kinetic energy, which is concluded from the adiabatic theorem [7].

Although the Tomlinson model adequately explains that the stick-slip phenomena in friction arises due to the lateral surface potential and elastic potential in relation to one another, it is not able to capture the effect of bond breaking and formation mechanisms at the interface.

Hence we employ a molecular dynamics method to study the effect of the contact length for non-wear sliding.

An atomically flat diamond tip is sliding over copper (111) surface in [110] direction. The diamond atomic lattice is considered a rigid body in comparison to copper in this study.

The interaction between atoms i and j is described by the modified [5]:

$$\phi(r_{ij}) = \lambda_i D_e\left[exp\{-2\lambda_2\alpha(r_{ij} - r_0)\} - 2exp\{\lambda_2\alpha(r_{ij} - r_0)\}\right]$$

where r_{ij} is the separation, r_0 is the equilibrium distance when ϕ is minimum, and D_e is the dissociation energy between the atom pair. The modified Morse potential consists of a short-range repulsion interaction and a long-range attractive tail. At separation $r_{ij} > r_0$, the interaction is considered an attractive interaction and at $r_{ij} < r_0$ a repulsive one. The parameter α is a material constant defined by: $\alpha = \sqrt{k_e/2D_e}$, where k_e is the force constant at well minimum. The difference between the standard Morse potential and the modified one is that the latter includes non-dimensional parameters λ_1 and λ_2 to take into consideration the cohesive strength of the diatomic bond for greater accuracy. Contacts between the diamond-copper atomic pair are contaminated or lubricated and hence reduce the adhesive strength between the diatomic pair. Therefore, to take into account of this effect, Liangchi et al. [8] assigned λ_1 to take the value of $0 \leq \lambda_1 < 1$ and $\lambda_2 \geq 1$. It is essential to note that, when λ_1 ranges from $0 \leq \lambda_1 < 1$, the interaction potential curve will shift in the negative direction of separation r. Hence, for simplicity in our present analysis, we assume a chemically clean environment where, $\lambda_1 = \lambda_2 = 1$.

Figure 8.9 shows a complex variation in the friction force between 2 flat surfaces as they slide past each other. Variations in the friction force can be as large

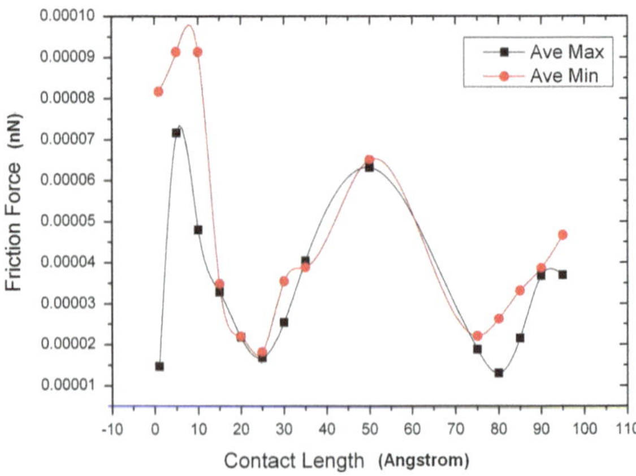

Fig. 8.9 Variation of average friction force experienced by 2 flat sliding surfaces as a function of contact length

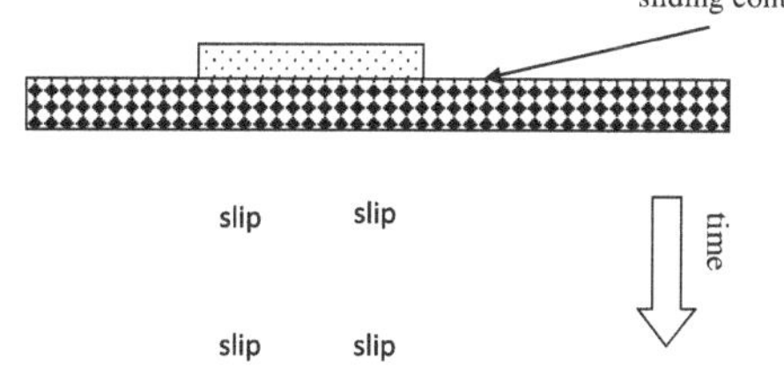

Fig. 8.10 Concurrent slip in contact surface

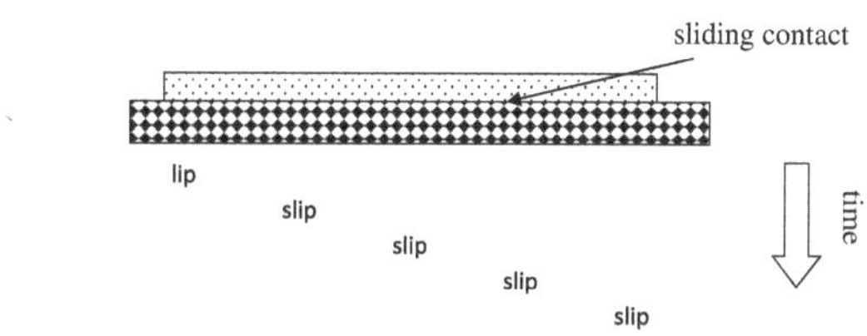

Fig. 8.11 Single dislocation assisted slip

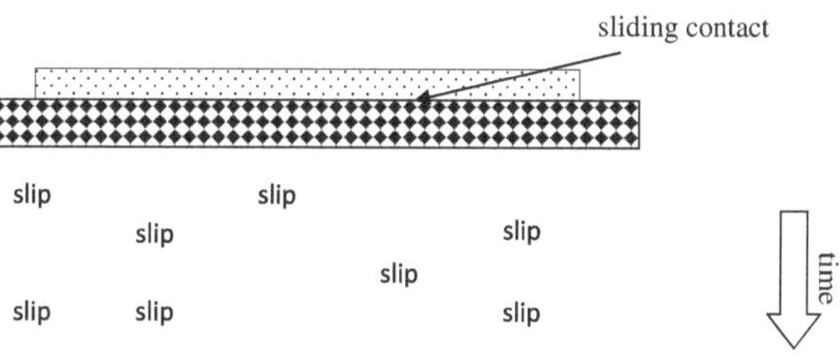

Fig. 8.12 Dislocation pile ups occur in contact surfaces

as one order of magnitude as the contact length increases from 0 to 100 angstrom. The details of the different mechanisms of friction are as follows:

Within the contact length range of 0–10 angstrom, the atoms on the two sliding surfaces sticks and slips coherently. Hence, the longer the contact the more the atoms interact and stick. That is why the average friction force between the two surfaces increases as contact length increase from 0 to 10 angstrom. Hence when the contact length is small, the difference in the atomic spacing between the atoms of the sliding surfaces do not affect the mechanism of friction (Fig. 8.10).

When the contact length further increase to the range of 10–30 angstrom, the difference in atomic spacing between the two sliding surfaces results in a dislocation forming between the two contacting surface. The dislocation nucleates from one end of the contact and propagates to the other end as the sliding occurs leading to a reduction in the friction force (Fig. 8.11).

When the contact length is between the range of 30–50 angstrom, the dislocations begin to pile up. There is insufficient driving force to move the dislocations within the contact and the dislocation jams in the contact surface causing it to slip discontinuously (Fig. 8.12).

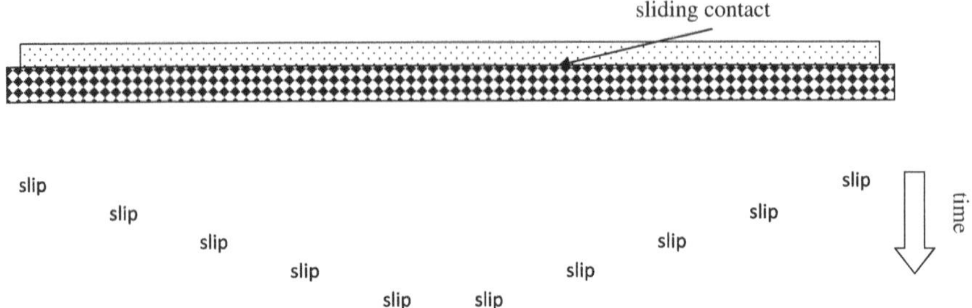

Fig. 8.13 Dislocations nucleates from both end of the contact and meets near the centre of the contact

When the contact length is in the range of 50–80 angstrom, dislocation now initiates from both ends of the contact as asymmetry of the crystal structures of the top and bottom surfaces becomes insignificant with a large contact surface. The dislocations propagates towards the centre of the contact surface and annihilates. This results again in low friction between the sliding surfaces.

When the contact length is more than 80 angstroms, dislocations again start to pile up as the driving force to propagate the dislocations through the entire contact is insufficient because of an increase in the contact surface (Fig. 8.13).

Hence the friction mechanism of the contact is governed by 2 competing forces namely the driving force and the number of dislocations in the sliding contact.

Conclusions

In this chapter, it is shown that molecular dynamics simulation can be used to study the mechanism of friction on the nanometre scale.

In the case of sliding the occurrence of plastic deformation in carbon coating and silicon substrate is due to amorphous phase transformation and plastic flow and the size of the residual amorphous zone in the silicon doesn't necessarily decrease with increasing coating thickness. As discussed in the previous section, a thicker coating restricts the formation of chips and hence inhibits stress relieving due to chip formation in the deformation zone leading to a larger residual amorphous zone.

Hence, using the molecular dynamics simulation, it is possible to observe in real-time the effects of carbon coating on silicon substrate deformation due to sliding.

Molecular dynamics was also used to elucidate a complex contact size dependent transition in the mechanism of friction.

References

1. Freller, H., Hempel, A., Lilge, J., Lorenz, H P.: Influence of intermediate layers and base materials on adhesion of amorphous carbon and metal-carbon coatings. Diamond Relat. Mater. **1**, 563 (1992)
2. Zehnder, T., Balmer, J., Luthy, W., Weber, H P.: Determination of limits in deposition of adhering a-C films on silicon produced by pulsed laser deposition. Thin Solid Films **263**, 198 (1995)
3. Hurtado, J.A., Kim, K.S.: Scale effects in friction of single-asperity contacts. I. From concurrent slip to single-dislocation-assisted slip. Proc. R. Soc. London. A. **455**, 3363–3400 (1999)
4. Cheong, W.C.D., Zhang, L.C.: "Molecular dynamics simulation of phase transformations in silicon mono-crystals due to nano-indentation. Nanotechnology **11**, 173 (2000)
5. Zhang, L.C., Tanaka, H.: Atomic scale deformation in silicon monocrystals induced by two-body and three-body contact sliding. Tribol. Int. **31**, 425 (1998)
6. Tomlinson G.A.: A molecular theory of friction. Phil. Mag. **7,** 905 (1929)
7. Goldstein, H.: Classical Mechanics, 2nd edn. Addison-Wesley, Massachusetts (1980). (Reading)
8. Zhang, L.C., Tanaka, H.: Towards a deeper understanding of wear and friction on the atomic scale—a molecular dynamics analysis. Wear, **44**, 211 (1997)

Chapter 9
Nano/Micro-Tribological Properties of MEMS/NEMS Materials

R. Arvind Singh, N. Satyanarayana and Sujeet K. Sinha

Abstract Micro/Nano-Electro-Mechanical Systems (MEMS/NEMS) are miniaturized devices built at micro/nano-scales. At these scales, surface forces such as adhesion and friction undermine the smooth operation and operating lifetimes of the MEMS/NEMS components. MEMS/NEMS devices are traditionally made from silicon, whose tribological properties are poor. In order to enhance the tribological performance of silicon, researchers have investigated various thin films/coatings that include diamond-like carbon (DLC) coatings, self-assembled monolayers (SAMs), polymers (PMMA, PDMS and SU-8) and perfluoropolyether films (Z-15, Z-DOL and PFPE). The nano/micro-tribological properties of these materials are governed by several parameters such as their physical structure, chemical composition, surface properties that include surface energy/wettability and interfacial shear strength, and mechanical properties such as elastic modulus that influences the contact area. In this chapter, a brief overview on the nano/micro-tribological properties of adhesion, friction and wear durability of these MEMS/NEMS materials has been presented. Further, recent advancements in MEMS/NEMS structural materials and novel surface modification methods have also been mentioned.

R. Arvind Singh (✉)
Energy Research Institute @ NTU (ERIAN), Nanyang Technological University (NTU), Singapore, # 06-09, CleanTech One, 1 CleanTech Loop 637141, Singapore
e-mail: arvindsingh@ntu.edu.sg; r.arvindsingh@gmail.com

N. Satyanarayana
Department of Mechanical Engineering, National University of Singapore, 9 Engineering Drive 1, Singapore 117576, Singapore

S. K. Sinha
Department of Mechanical Engineering, Indian Institute of Technology, Kanpur 208016, India
e-mail: sujeet@iitk.ac.in

S. K. Sinha et al. (eds.), *Nano-tribology and Materials in MEMS*,
DOI: 10.1007/978-3-642-36935-3_9,

Contents

Introduction

Miniaturization of devices has led to the development of micro/nano-electro-mechanical systems (MEMS/NEMS). These systems can be largely classified into two groups, namely sensors-based and actuators-based systems. In the sensors-based systems, the devices have sensing elements only, whereas in the actuators-based systems, the devices have tiny elements that undergo relative mechanical motion such as micro-motors, micro-gears and micro-shutters [1]. When surfaces are subjected to actuating motion, tribological issues such as adhesion, friction and wear/removal of material arise. These tribological issues strongly manifest in the actuators-based devices and they confound the smooth operation and operating lifetimes of the devices [1]. Therefore, there arises a great need to solve the tribological issues in order to realize the commercialization of actuators-based MEMS/NEMS devices.

When the size of devices shrink to micro/nano-scales, the surface area to volume ratio increases, due to which body forces such as inertia and gravity become insignificant when compared to those of surface forces such as capillary, van der Waals, electrostatic, and chemical bonding [1, 2]. Hence, in the actuators-based systems that are built at micro/nano-scales, surface forces namely adhesion and friction become comparable with the forces driving the device motion, rendering the devices completely inoperable [1, 2]. Further, most of the actuators-based systems are often made from silicon, mainly due to the large amount of process knowledge developed for the material in semiconductor industries [1, 2]. However, silicon does not have good tribological properties, owing to its hydrophilic nature/high surface energy and inherent brittleness [1, 2]. In order to enhance the tribological properties of silicon, researchers have investigated various special films/coatings and protective layers with thicknesses ranging from a few monolayers to nano/micrometers. It is necessary that the thickness of the films/coatings must be less than a few micrometers, as the gap/clearance between the elements of actuators-based devices is usually about few microns [2, 3]. Therefore, conventional solid or liquid lubricants cannot be used in actuators-based devices as their sizes are of the same order as those of the device elements [4].

Examples of thin films/coatings investigated by various researchers to protect silicon and improve its tribological performance include diamond-like carbon

(DLC) coatings, self-assembled monolayers (SAMs), polymers and perfluoropolyether films [4]. The nano/micro-tribological properties of these films/coatings are influenced by parameters such as their physical structure, chemical composition, surface nature i.e. hydrophilic/hydrophobic, surface mechanical properties like elastic modulus and hardness, and interfacial shear strength when in contact with other surfaces. In this chapter, a brief overview on the nano/micro-tribological properties of adhesion, friction and wear durability of these MEMS/NEMS materials has been presented, so as to bring forth an overall understanding of the materials and their tribological performance for their application in actuators-based MEMS/NEMS devices. Most of these investigations have been conducted at nano- and micro-scales using commercial atomic force microscopes (AFM) and various custom-built or commercially available micro-tribo testers. Conventional tribological testers, such as pin/ball-on-disk equipments are not suitable to evaluate tribological properties of materials used in MEMS/NEMS applications as the contact areas involved in these devices are only a few hundreds of nm^2 and the contact loads are in μN or mN range [5].

Nano-Tribological Properties

Table 9.1 shows the water contact angle (WCA), elastic modulus, and tribological properties of adhesion force and coefficient of friction at nano-scale (μ_n) of silicon and various thin films/coatings, taken from the literature [6–9]. Nano-scale adhesion values shown in the table have been obtained by different methods using atomic force microscopes (AFM), namely: (1) the value of the negative intercept of the friction force versus normal load was taken as the adhesion force, when square

Table 9.1 Water contact angle (WCA), elastic modulus, and tribological properties of adhesion force and coefficient of friction at nano-scale (μ_n) of silicon and various thin films/coatings, taken from the literature [6–9, 12]. The adhesion force values have been obtained by considering the negative intercept of the friction force versus normal load as the adhesion force [6], and by direct measurements using force distance curves [8, 9]

Material	WCA (deg)	Elastic modulus (GPa)	Nano-scale adhesion force (nN)	Nano-scale friction coefficient (μ_n)	Ref
Silicon	30	130	50	0.05	[6, 7]
DLC	60	280	40	0.03	[6]
Z-15	46	–	84	0.01	[6, 7]
Z-DOL	72	–	35	0.03	[6, 7]
PMMA	74	7.7	26	0.06	[6]
PDMS	104	360–870 kPa	190	0.1	[6]
Silicon	<5	–	150	–	[8]
PMMA	72	–	40	–	[8]
SU-8	75	5–6	50	–	[8, 12]
OTS	100	–	5.6	0.04	[9]

pyramidal Si_3N_4 tips with nominal radii of 30–50 nm were slid against the test materials [6, 7] and (2) the adhesion force was obtained from force distance curves using square pyramidal Si_3N_4 tips with nominal radii of 20–60 nm [8] and using Si_3N_4 tips with nominal radii of 15–50 nm [9]. It could be seen from the table that the adhesion force at the nano-scale is dependent on the water contact angle values that are indicative of surface energy [10]. At the nano-scale, adhesion force arises due to the contribution of various attractive forces such as capillary, electrostatic, van der Waals, and chemical forces under different circumstances [1, 2]. Amongst these forces, the capillary force that arises due to the condensation of water from the environment is the strongest [2]. Silicon is hydrophilic in nature due to the presence of its native oxide, which has high surface energy owing to surface hydroxyl groups [1], and therefore it shows low WCA values [6–8]. Due to this reason it supports strong capillary force that causes high adhesion through the formation of meniscus/water-bridges at the interface, rendering the tiny elements of actuators-based MEMS/NEMS devices inoperable [1, 2, 6–8].

When compared to silicon that is hydrophilic, thin films/coatings have higher WCA values/lower surface energies and hence, they exhibit lower values of adhesion force (except for Z-15 and PDMS polymer) (Table 9.1) [6–9]. Diamond-like carbon (DLC) films have higher WCA value due to their Sp^3 (diamond-like) content [11]. Higher WCA value/lower surface energy of DLC film suppresses capillary force and lowers its adhesion force [6]. The octadecyltrichlorosilane (OTS) SAM is hydrophobic in nature and thus the capillary force gets suppressed to a large extent resulting in low adhesion force value [9]. Amongst the perfluoropolyethers, Z-DOL that is thermally bonded to the substrate shows lower adhesion force value due to its higher WCA value [6, 7]. However, Z-15 shows higher adhesion force value even though its WCA is high, which is attributed to the formation of meniscus bridges by its mobile lubricant molecules at the interface [6, 7]. In the case of polymeric films, poly(methyl methacrylate) (PMMA) shows lower adhesion force owing to its higher WCA value [6]. Interestingly, although polydimethylsiloxane (PDMS) and SU-8 polymers have higher WCA values when compared to PMMA, their adhesion force values are higher than that of PMMA [6, 8]. The reason for such an occurrence is that these two polymers have lower elastic modulus when compared to PMMA, thereby they readily conform to counterface surfaces (AFM tips) resulting in higher real area of contact that increases adhesion force [6, 8, 12]. Amongst PDMS and SU-8 polymers, the adhesion force value of SU-8 is lower than that of silicon [8], whereas PDMS shows adhesion force value higher than that of silicon [6]. There are two reasons for the PDMS polymer to show higher adhesion force values: (1) it has a very low elastic modulus value, which gives rise to large contact areas and (2) it exhibits long range attraction due to electrostatic surface charge [6]. Thus, from the above mentioned works [6–9], it is clearly seen that the adhesion force of the surfaces at nano-scale is dependent on the nature of their surfaces (hydrophilic/hydrophobic) and the real contact area at the interface.

Table 9.1 shows the coefficient of friction at nano-scale (μ_n) of silicon and various thin films/coatings, taken from the literature [6, 7, 9]. The slope of the friction force versus normal load was taken as μ_n [6, 7]. It is well-known that adhesion strongly influences friction at small-scales. At micro/nano-scales, friction is in a regime where the contribution from adhesion can outweigh that from the asperity deformation [2, 13, 14]. Silicon shows higher μ_n value owing to its higher adhesion and higher contact area due to its higher interfacial energy [2, 6, 13]. When compared to silicon, thin films/coatings have lower adhesion values, and hence they exhibit lower values of μ_n (except for PMMA and PDMS polymers) (Table 9.1) [6]. DLC film has lower μ_n value when compared to that of silicon due to its lower adhesion and lower contact area [13]. Further, DLC films are known to have high resistance to wear. Scratch and wear resistance tests conducted on ultra-thin hard amorphous DLC films of thicknesses 3.5–20 nm using AFM [15] have shown that thicker films support the entire load, whereas thin films share the load with their substrate causing the substrate to deform, and very thin coatings undergo delamination.

When compared to silicon, the OTS SAM shows lower μ_n value because of its lower adhesion, and most importantly due to its molecular chains that act as molecular springs [16]. Under an applied normal load, these molecular springs undergo orientation along the sliding direction and reduce the shearing force at the interface, which in turn reduces the friction property [16]. The nano-scale friction behavior of SAMs is strongly dependent on their physical/chemical properties. Investigations have revealed that SAMs having phenyl terminal groups showed higher friction property than those with methyl terminal groups, due to the high stiffness induced by the benzene rings [16, 17]. Further, it has been found that fluorinated SAMs have higher friction property because of the effect of larger van der Waals radii of the fluorine atom [17]. The friction property of SAMs is also influenced by their chain length, as has been observed earlier that SAMs with longer chain lengths showed lower friction values [18]. With the increase in chain length, the packing density of the chains increases which enables them to retain molecular scale order during shear, resulting in lower friction [18]. Interestingly, it has been identified that mixtures of SAMs with different chain lengths exhibit improved nano-tribological properties over their pure counterparts [18]. The smaller number of chains in the outer region of the mixtures plays a crucial role in giving rise to their improved tribological properties [18].

Amongst the perfluoropolyethers, Z-DOL shows lower μ_n value than that of Z-15 [6, 7]. The molecules of Z-DOL that are thermally bonded to the substrate generally have some orientation which facilitates sliding and reduces the μ_n value [6, 7]. Z-15 shows μ_n value higher than that of silicon [6, 7]. Z-15 film is soft when compared to silicon, and hence as the AFM tip comes in contact with the film, it penetrates into the film resulting in large area of contact for the formation of meniscus bridges by its mobile lubricant molecules. This increases the adhesion force between the tip and the film, which in turn increases the friction property [6, 7]. In addition, Z-15 has higher viscosity than water and therefore the film provides higher resistance to motion and thus it shows higher μ_n value [7]. Considering the polymers, although PMMA has lower adhesion when compared to silicon [6, 8], its μ_n value is similar to that of silicon [6]. PMMA has lower elastic modulus than silicon, due to which the contact area increases during sliding that results in higher μ_n value [6]. PDMS has higher μ_n value when compared to silicon as it has high adhesion, and a very low elastic modulus value that gives rise to large contact areas [6].

Micro-Tribological Properties

Although investigations using scanning probe techniques (e.g. AFM) have brought forth fundamental understanding of the tribological behaviour of MEMS/NEMS materials at nano-scale, it is important to note that these techniques do not simulate the real conditions as found in the miniaturized devices. Usually in these techniques, sharp tips are used to measure surface forces. Because of the sharpness of the tips, contact pressures of the order of a few gigapascals are generated, which is several orders of magnitude higher than the contact pressures expected in the real devices [14]. Further, the typical contact in these devices is in the micrometer range and does not always meet the single asperity contact model that is applicable in the case of the scanning probe techniques [14]. Also, the sliding velocities in these techniques are about 2–4 orders of magnitude lower than those in the devices [1]. Therefore, several investigations of the MEMS/NEMS materials have been conducted at micro-scale, in order to evaluate their tribological performance at conditions almost similar to those that prevail in real-time devices.

Table 9.2 shows the WCA and coefficient of friction at micro-scale (μ_m) of silicon and various thin films/coatings, taken from the literature [13, 17–20]. Silicon being hydrophilic, it shows high μ_m value owing to high adhesion and wear [13, 14]. Similar to that at the nano-scale, silicon supports large capillary force at micro-scale, which increases its friction property [14]. Also, at this scale, it undergoes severe wear due to its brittle nature and generates large amount of

Table 9.2 Water contact angle (WCA) and coefficient of friction at micro-scale (μ_m) of silicon and various thin films/coatings, taken from the literature [13, 17–20]. The SAMs with varying chain length, namely OTS, DTS and HTS were tested for their micro-friction property at the applied normal load of 4 mN [18], whereas the rest of the test materials were tested at the load of 3 mN [13, 17, 19, 20]. The sliding speed and the scan length were kept constant at 1 mm/s and 3 mm, respectively for all the test materials [13, 17–20]

Material	WCA (deg)	Micro-scale friction coefficient (μ_m)	Ref
Silicon	22	0.5	[13]
DLC	66	0.17	[13]
OTS	105	0.1	[18]
DTS	105	0.15	[18]
HTS	105	0.22	[18]
DMDC	103	0.21	[17]
DPDC	84	0.27	[17]
PFOTS	106	0.35	[17]
PFDA	95	0.6	[17]
Z-DOL	74	0.13	[19]
PMMA	69	0.65	[20]

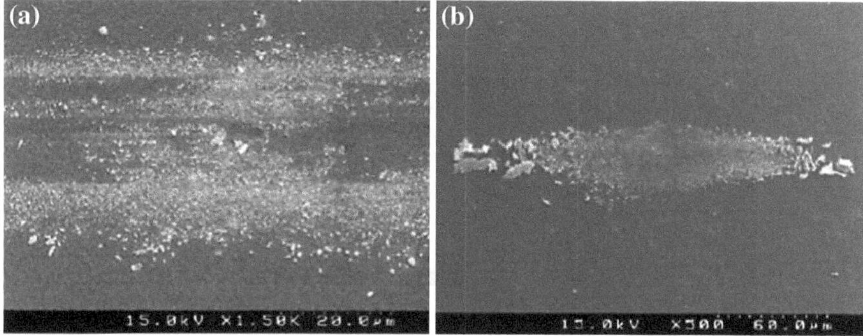

Fig. 9.1 SEM images of: **a** worn surface of silicon wafer after the test, and **b** tip of the soda lime counterface ball slid against the silicon surface [13]. Large amounts of wear debris can be seen on both the worn silicon surface and the tip of the soda lime ball [13]

debris [13]. Figure 9.1a and b show representative scanning electron microscope (SEM) images of the wear track on silicon surface and the tip of the counter face ball slid against the silicon surface, respectively [13]. Large amounts of wear debris can be seen on both the worn surface and the tip of the counter face ball surface [13]. Compared to silicon, DLC film shows lower friction property owing to its higher WCA/lower surface energy that reduces adhesion force, which in turn reduces the μ_m value [13]. In addition, DLC films exhibit transfer film/layer formation on the counterface surfaces that lower their friction property [21, 22]. Further, it has been found that the friction mechanism in DLC films depends on their mechanical properties [23]. In this work [21], DLC films with varying thicknesses (100 nm to 1 μm), with the elastic modulus increasing with the thickness, were investigated for their micro-friction behavior using soda lime balls with different radii. Results showed that the friction force increased with the applied normal load, whereas with respect to the ball size, two different trends were observed. In the case of 100 nm thick sample, friction showed an increase with the ball size at any given normal load, while for thicker films, friction showed an inverse relation with the ball size at all applied normal loads. The friction behaviour in the case of the 100 nm thick film was adhesive in nature, whereas for the thicker films plowing was dominant. Figure 9.2a and b show SEM images of worn surfaces of 100 nm and 1 μm thick DLC films, respectively [23]. In the case of the thicker film, the morphology of the wear track shows ridges (material flow) on both the sides of the wear track, which are due to the plowing effect [23]. This difference in the operating mechanisms in the DLC films has been attributed to the variation in the contact areas influenced by their mechanical property, namely, the elastic modulus [23]. Furthermore, DLC coatings are hard coatings [24]. However, if the coatings are too hard it may be deleterious in terms of tribological performance. Figure 9.2c and d show the surface of the tip of a soda lime ball tested against a DLC coating of hardness ~50 GPa deposited on silicon wafer using the filter vacuum arc method, at lower and higher magnifications, respectively. It could be seen from these figures that the soda lime ball has been abraded by the

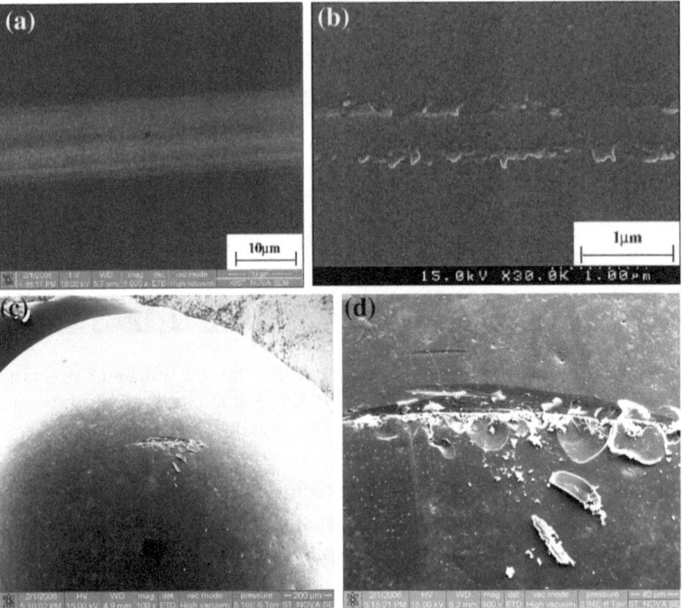

Fig. 9.2 (**a**) and (**b**) SEM images of worn surfaces of 100 nm and 1 μm thick DLC films, respectively [23]. In the case of the thicker film, the ridges (material flow) on both the sides of the wear track are due to the plowing effect [23]. (**c**) and (**d**) show the surface of the tip of a soda lime ball tested against a DLC coating of hardness ~50 GPa, at *lower* and *higher* magnifications, respectively. The fractured particles of the ball are seen on the tip

hard DLC coating, and the fractured particles of the ball are seen on the tip. Hence, selection of DLC coatings in terms of their mechanical properties of elastic modulus and hardness are important for critical applications.

Compared to silicon, OTS SAM showed lower μ_m value due to its higher WCA/ lower surface energy and its molecular chains that act as molecular springs and reduce the friction (Table 9.2) [18]. A comparative study of the micro-tribological properties of silicon and silicon coated with DLC and OTS [14] showed that: (1) the friction of silicon rapidly increases with the rise in humidity as it is hydrophilic, whereas in DLC and OTS friction shows negligible dependence on humidity due to their higher WCA values, and (2) the friction of silicon continually decreases with increase in the sliding speed, owing to the change in the capillary action with sliding speed. Higher sliding speeds prevent the formation of stable capillary neck between silicon and its counter face, thereby decreasing the friction force with the increase in speed. Compared to silicon, DLC and OTS show no variation in their friction as a function of speed.

At the micro-scale, the friction behavior of SAMs is strongly dependent on their physical/chemical properties, similar to that at the nano-scale. Table 9.2 shows the μ_m values of SAMs with varying chain lengths, namely hexyltrichlorosilane (HTS), decyltrichlorosilane (DTS) and octadecyltrichlorosilane (OTS) having carbon atoms of 6, 10 and 18 in the molecular chains, respectively [18]. The μ_m value decreases with the increase in the chain length due to increase in the packing density. Longer chain SAMs stabilized by van der Waals attraction retain molecular scale order during shear and act as better lubricants [18]. The micro-friction behaviour of SAMs also depends on the nature of their molecular chain. For example, diphenyldichlorosilane (DPDC) shows higher μ_m value than dimethyldichlorosilane (DMDC) (Table 9.2) [17], as DPDC has benzene rings in its molecular chain which are stiffer in nature when compared to that of the linear chain in DMDC. Further, the type of chemical groups present in the molecular chain of SAMs also affects their μ_m value, as was observed in the case of fluorinated SAM, namely perfluorooctyltrichlorosilane (PFOTS) when compared to its non-fluorinated counterpart OTS [17]. PFOTS showed higher value of μ_m than OTS which indicates the effect of fluorine addition (Table 9.2) [17]. Fluorine atoms have larger van der Waals radii and consequently, they interact more strongly with the neighbouring chains, giving rise to long-range multimolecular interactions. To overcome the energetic barrier due to the long-range multi-molecular interactions, more energy gets imparted to the molecular film during sliding that would result in higher frictional property [17]. Some SAMs are affected by the presence of moisture/humidity. As an example, perfluorodecanoicacid (PFDA) shows high μ_m value (Table 9.2), followed by wear (Fig. 9.3a) at the relative humidity of 45 ± 5 % (Table 9.2) [17]. When PFDA is exposed to moisture, the moisture breaks the surface and solvates it thereby removing it from the surface. When tests were conducted at the lower humidity of ~ 8 %, the μ_m value reduces by almost 5 times [17]. PFDA is used as a lubricant in digital micromirror devices (DMD) [25]. However, to ensure that PFDA acts as an effective lubricant, a hermetic chip package is used, to avoid moisture [23]. Z-DOL has low μ_m value, which indicates that it can act as a good lubricant [19]. Further, PMMA polymer shows high μ_m value and undergoes severe wear (Fig. 9.3b), owing to its higher contact area and lower hardness [20].

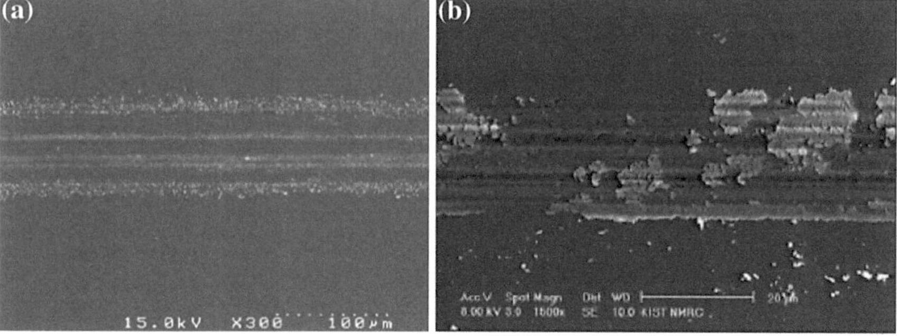

Fig. 9.3 SEM images of the worn surfaces of: **a** PFDA tested at the relative humidity of 45 ± 5 % [17], and **b** PMMA polymer [20]. Both these materials show severe wear at microscale

Recent Advancements

Silicon, as mentioned earlier does not have good wear resistance. It undergoes wear at very low loads within a short duration of time due to its inherent brittleness (Fig. 9.1a) [13]. Although thin films/coatings such as DLC and SAMs reduce adhesion and friction at nano/micro-scale, they too do not have good wear durability (Table 9.3) [26, 27]. In real MEMS/NEMS devices, it is necessary that the protective films/coatings should perform for hundreds of thousands of cycles without any wear/material removal. A reliable route to achieve high durability is to chemically bind the lubricant molecules to silicon surfaces. As an example, perfluoropolyether (PFPE) lubricant bound to silicon through intermediary layers of aminopropyltrimethoxysilane (APTMS) and glycidoxypropyltrimethoxysilane (GPTMS) has shown considerable enhancement in wear durability [27, 28]. On the similar line, durability of PMMA polymer can be enhanced significantly by bonding PFPE on chemically activated polymeric surface [29].

Although a lot of research has been conducted on various thin films/coatings to improve the tribological properties of adhesion, friction and wear durability of silicon, over time silicon has become less attractive for MEMS/NEMS applications owing to its inherent disadvantages. Silicon is hydrophilic/has high surface energy, brittle and non-biocompatible. In search for a better candidate for MEM/NEMS and Bio-MEMS devices, the SU-8 polymer has been identified as the best replacement to silicon, as SU-8 is hydrophobic, bio-compatible and can be made into complex shapes with low/high aspect ratios at micro/nano-scales [30–32]. However, SU-8 polymer does not have good tribological properties [33, 34].

Recently, a simple yet robust surface modification method has been applied to SU-8 polymer to enhance its tribological properties for MEMS/NEMS applications

Table 9.3 Water contact angle (WCA), steady-state coefficient of friction (μ_s) and wear durability in terms of number of sliding cycles (n) of the SU-8 thick films, modified SU-8 thick films, silicon and modified silicon surfaces [35]. Experiments were stopped when the surfaces of the test materials showed evidences of surface failure (coefficient of friction >0.3 and/or traces of wear at the surfaces with fluctuating friction values). The WCA, μ_s and wear durability values of DLC and OTS are taken from the Refs. [26] and [27], respectively. Note that n ~100 cycles corresponds to the time duration of 30 s and 100,000 cycles corresponds to 8 h

Material	WCA (deg)	Steady state friction coefficient (μ_s)	Durability (n)
Si	12	0.7	<100
DLC	–	0.2	400
OTS	108	0.18	1,600
SU-8 ThF	94	0.64	<100
SU-8 ThF/PFPE	100	0.29	22,000
SU-8 ThF/O_2-25 W/PFPE	59	0.17	>100,000
Si/PFPE	66	0.17	5,000
Si/SU-8 TF	75	0.73	<100
Si/SU-8 TF/PFPE	87	0.24	1,000
Si/SU-8 TF/O_2-100 W/PFPE	92	0.19	>100,000

[35]. Results from the tribological investigation revealed that the modified SU-8 films showed significant improvement in their tribological properties when compared to those of the unmodified films by several orders: (1) the modified films showed steady-state coefficient of friction values that were ~2.5 times lower than those of the unmodified films, (2) the modified films exhibited wear durability >1,000 times higher than those of the unmodified films, and (3) the modified films effectively performed for the duration of >8 h under the applied normal load that was 5 times higher than the load at which the unmodified films underwent severe wear in <30 s [35]. In this work [35], SU-8 thick films (50 μm thick) were spin coated on silicon wafers, and were treated with the same processes such as those used in the fabrication of SU-8 MEMS components that include pre-baking, UV exposure, post baking and hard baking. As the first step towards the surface modification, the thick films were exposed to oxygen plasma (25 W) for 1 min and were subsequently coated with perfluoropolyether (PFPE) nanolubricant for 1 min. Table 9.3 shows the WCA values, steady-state coefficients of friction (μ_s) and wear durability in terms of number of sliding cycles (n) of the unmodified and modified films [35]. Experiments were stopped when the surfaces of the test materials showed evidences of surface failure (coefficient of friction >0.3 and/or traces of wear at the surfaces with fluctuating friction values). From the table, it could be seen that the unmodified SU-8 thick film (SU-8 ThF) showed very high μ_s value and very low durability (n <100 cycles) at the applied normal load of 0.3 N and sliding speed of 42 mm/s. Figure 9.4a shows the worn surface of this material. Coating of SU-8 ThF surface with PFPE (SU-8 ThF/PFPE), showed significant reduction in μ_s value and high wear durability (n >100,000 cycles), which is >1,000 times higher than that of the SU-8 ThF at the same load and speed. SU-8 ThF exposed to O_2 plasma and coated with PFPE (SU-8 ThF/O_2-25 W/PFPE) also showed considerable reduction in μ_s value and high wear durability (n >100,000 cycles). With the increase in load, SU-8 ThF/PFPE showed decrease in its wear durability. The surface fails at ~87,700 cycles at 0.8 N load, and at ~22,000 cycles at 1 N. Figure 9.4b shows an optical image of the worn surface of SU-8 ThF/PFPE after the test conducted at 1 N. In comparison, SU-8 ThF/O_2-25 W/PFPE showed high wear durability of >100,000 cycles, even at the high load of 1.5 N. Figure 9.4c shows an optical image of the surface of SU-8 ThF/O_2-25 W/PFPE after the test conducted at 1.5 N. From this image, it could be seen that the surface shows almost 'no wear'. The main reason for such an excellent wear durability of SU-8 ThF/O_2-25 W/PFPE is attributed to the strong anchoring of PFPE molecules to SU-8 film surface through chemical bonding. In SU-8 ThF/PFPE, the nanolubricant was applied over the regular SU-8 film, whereas in SU-8 ThF/O_2-25 W/PFPE, the nanolubricant was coated over chemically active SU-8 film. The molecules of PFPE nanolubricant have hydroxyl (-OH) groups at both their terminal ends, and the O_2 plasma treated SU-8 film surfaces have carboxyl (-COOH) functional groups [35]. As both the PFPE molecules and O_2 plasma treated SU-8 film surfaces have polar reactive chemical groups, strong chemical bonds form between them. Due to this reason, PFPE molecules get anchored strongly onto the SU-8 film surfaces which increase their wear durability.

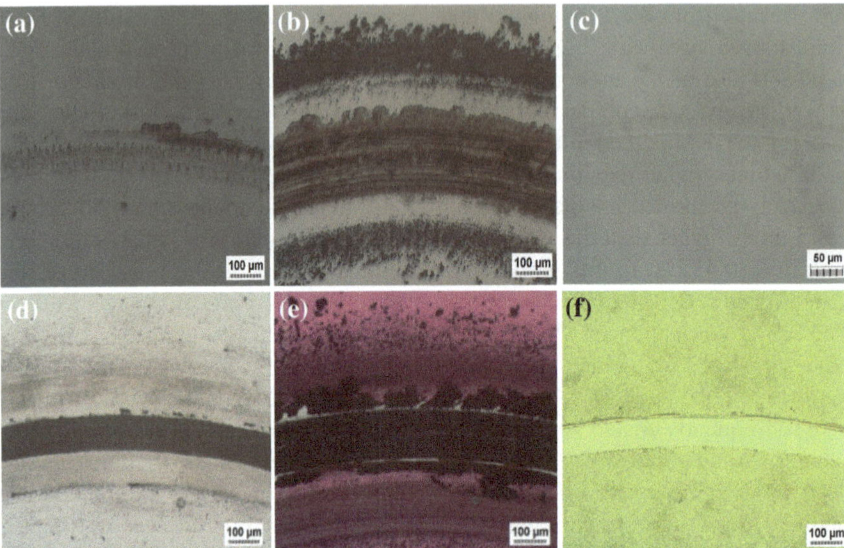

Fig. 9.4 Optical images of the surfaces of the materials taken after the tests: (**a**) SU-8 ThF (at ~3,000 cycles) (**b**) SU-8 ThF/PFPE (at ~54,000 cycles) (**c**) SU-8 ThF/O_2-25 W/PFPE (at ~100,000 cycles) (**d**) Si (at ~400 cycles) (**e**) Si/SU-8 TF (at ~1,400 cycles), and (**f**) Si/SU-8 TF/O_2-100 W/PFPE (at ~100,000 cycles) [35]. SU-8 ThF/PFPE and SU-8 ThF/O_2-25 W/PFPE were tested at 1 N and 1.5 N, respectively. The rest of the materials were tested at 0.3 N. The sliding speed was kept constant at 42 mm/s for all the tests

The surface modification method mentioned above has been extended for enhancing the tribological properties of silicon [35]. Instead of a thick SU-8 film, thin films (500 nm thick) were spin coated on silicon wafers. Upon spin coating, the thin films were exposed to oxygen plasma (100 W) for 1 min and were subsequently coated with PFPE nanolubricant for 1 min. Table 9.3 shows the WCA value, steady-state coefficient of friction (μ_s) and wear durability in terms of number of sliding cycles (n) of the unmodified silicon and modified silicon surfaces, at the applied normal load of 0.3 N and sliding speed of 42 mm/s [35]. It could be seen from the table that unmodified silicon (Si) showed very high μ_s value and very low wear durability (n <100 cycles). Figure 9.4d shows the worn surface of silicon. SU-8 thin film on silicon (Si/SU-8 TF) also shows high μ_s value and very low wear durability (n <100 cycles). Figure 9.4e shows the worn surface of Si/SU-8 TF. Silicon coated with PFPE (Si/PFPE) showed an increase in its durability, but is not sufficient for the real time application in miniaturized devices. Silicon coated with SU-8 thin film and subsequently coated with PFPE (Si/SU-8 TF/PFPE) does not have good durability. Compared to these surfaces, silicon coated with SU-8 thin film that was exposed to O_2 plasma and subsequently coated with PFPE (Si/SU-8 TF/O_2-100 W/PFPE) showed low μ_s value and very high wear durability of >100,000 cycles. Figure 9.4f shows an optical image of the Si/SU-8 TF/O_2-100 W/PFPE surface, taken after the test. The surface showed only a sliding mark and no recognizable wear in the form of wear debris/particles was seen. The surface modification method explained above can also work for any kind of structural material from which MEMS/NEMS are/can be made. Similar to the present process of surface modification to silicon, for any material, a thin SU-8 film can be coated and treated by the method, which would enhance the tribological performance of the material. The method is cost effective and time effective, and is a commercially viable tribological solution for MEMS/NEMS devices.

Summary

In this chapter, we have presented a brief overview on the nano/micro-tribological properties of adhesion, friction and wear durability of MEMS/NEMS materials that includes various films/coatings such as diamond-like carbon (DLC) coatings, self-assembled monolayers (SAMs), polymers (PMMA, PDMS and SU-8) and perfluoropolyether films (Z-15, Z-DOL and PFPE). It is seen that the nano/micro-tribological properties of these materials are governed by several parameters such as their physical structure, chemical composition, surface properties that include surface energy/wettability and interfacial shear strength, and mechanical properties such as elastic modulus that influences the contact area. Although most of these materials show enhanced tribological performance in terms of reducing surface forces and can protect silicon in MEMS/NEMS devices, it is important that they exhibit long term wear durability. The wear durability of a tribological surface is a critical factor, arguably even more than low friction, as it defines the useful operating lifetime and hence, modifications of surfaces for tribological applications in MEMS/NEMS devices should not only result in the reduction of surface forces, but should also effectively increase their wear durability. It is also seen that polymers like SU-8 are replacing silicon as MEMS/NEMS structural material and robust solutions towards enhancing the tribological performance of the polymers would support commercialization of MEMS/NEMS actuator-based devices.

References

1. Kim, S.H., Asay, D.B., Dugger, M.T.: Nanotribology and MEMS. Nanotoday **2**, 22–29 (2007)
2. Maboudian, R., Howe, R.T.: Critical review: Adhesion in surface micromechanical structures. J. Vac. Sci. Technol. B **15**, 1–20 (1997)
3. Tanner, D.M., Walraven, J.A., Irwin, L.W., Dugger, M.T., Smith, N.F., Eaton, W.P., Miller, W.M., Miller, S.L.: The effect of humidity on the reliability of a surface micromachined microengine. IEEE international reliability physics symposium, March 21–25, San Diego, 189–197 (1999)
4. Bhushan, B.: Modern tribology handbook, vol. 2. CRC Press, Boca Raton (2001)

5. Bhushan, B.: Applications of micro/nano tribology to magnetic storage devices and MEMS. Tribol. Int. **28**, 85–96 (1995)
6. Tambe, N.S., Bhushan, B.: Scale dependence of micro/nano-friction and adhesion of MEMS/NEMS materials, coatings and lubricants. Nanotechnology **15**, 1561–1570 (2004)
7. Liu, H., Bhushan, B.: Nanotribological characterization of molecularly thick lubricant films for applications of MEMS/NEMS by AFM. Ultramicroscopy **97**, 321–340 (2003)
8. Zhuang, Y.X., Menon, A.: On the stiction of MEMS materials. Tribol. Lett. **19**, 111–117 (2005)
9. Yoon, E.S., Yang, S.H., Han, H.G., Kong, H.: An experimental study on the adhesion at a nano-contact. Wear **254**, 974–980 (2003)
10. Bain, C.D., Evall, J., Whitesides, G.M.: Formations of monolayers by coadsorption of thiols on gold: variation in the head group, tail group, and solvent. J. Am. Chem. Soc. **111**, 7155–7164 (1989)
11. Paul, R., Das, S.N., Gayen, R.N., Roy, R.K., Bhar, R., Pal, A.K.: Synthesis of DLC films with different sp2/sp3 ratios and their hydrophobic behavior. J. Phys. D Appl. Phys. **41**, 055309 (2008)
12. Al-Halhouli, A.T., Kampen, I., Krah, T., Buttgenbach, S.: Nanoindentation testing of SU-8 phototresist mechanical properties. Microelectron. Eng. **85**, 942–944 (2008)
13. Yoon, E.S., Arvind Singh, R., Oh, H.J., Kong, H.: The effect of contact area on nano/micro-scale friction. Wear **259**, 1424–1431 (2005)
14. Liu, H., Ahmed, I.U., Scherge, M.: Microtribological properties of silicon and silicon coated with diamond like carbon, octadecyltrichlorosilane and stearic acid cadmium salt films: A comparative study. Thin Solid Films **381**, 135–142 (2001)
15. Sundararajan, S., Bhushan, B.: Micro/nanotribology of ultra-thin hard amorphous carbon coatings using atomic force/friction force microscopy. Wear **225–229**, 678–689 (1999)
16. Bhushan, B., Liu, H.: Nanotribological properties and mechanisms of alkylthiol and biphenyl thiol self-assembled monolayers studied by AFM. Phys. Rev. B **63**, 245412–245423 (2001)
17. Arvind Singh, R., Yoon, E.S., Han, H.G., Kong, H.: Friction behavior of chemical vapor deposited self-assembled monolayers on silicon wafer. Wear **262**, 130–137 (2007)
18. Arvind Singh, R., Kim, J., Yang, S.W., Oh, J.E., Yoon, E.S.: Tribological properties of trichlorosilane-based one- and two-component self-assembled monolayers. Wear **265**, 42–48 (2008)
19. Arvind Singh, R., Pham, D.C., Kim, J., Yang, S., Yoon, E.S.: Bio-inspired dual surface modification to improve tribological properties at small-scale. Appl. Surf. Sci. **255**, 4821–4828 (2009)
20. Arvind Singh, R., Yoon, E.S., Kim, H.J., Kim, J., Jeong, H.E., Suh, K.Y.: Replication of surfaces of natural leaves for enhanced micro-scale tribological property. Mater. Sci. Eng. C **27**, 875–879 (2007)
21. Erdemir, A., Bindal, C., Fenske, G.R., Zuiker, C., Wilbur, P.: Characterization of transfer layers forming on surfaces against diamond-like carbon. Surf. Coat. Technol. **86–87**, 692–697 (1996)
22. Moolsradoo, N., Watanabe, S., Deposition and tribological properties of sulphur-doped DLC films deposited by PBII method, Advances in materials science and engineering, 201 (2010)
23. Arvind Singh, R., Yoon, E.S., Kim, H.J., Kong, H., Park, S.J., Lee, K.R.: Friction behavior of diamond-like carbon films with varying mechanical properties. Surf. Coat. Technol. **201**, 4348–4351 (2006)
24. Robertson, J.: Diamond-like amorphous carbon. Mater. Sci. Eng. R **37**, 129–281 (2002)
25. Henck, S.A.: Lubrication of digital micromirror devices. Tribol. Lett. **3**, 239–247 (1997)
26. Minn, Myo., Sinha, Sujeet.K.: DLC and UHMWPE as hard/soft composite film on Si for improved tribological performance. Surf. Coat. Technol. **202**, 3698–3708 (2008)
27. Satyanarayana, N., Sinha, S.K.: Tribology of PFPE over coated self-assembled monolayers deposited on Si surface. J Phys D: Appl Phys **38**, 3512–3522 (2005)

28. Satyanarayana, N., Gosvami, N.N., Sinha, S.K., Srinivasan, M.P.: Friction, adhesion and wear durability of an ultra-thin perfluoropolyether-coated 3-glycidoxypropyltrimethoxy silane self-assembled monolayer on a Si surface. Philos. Mag. **87**, 3209–3227 (2007)
29. Satyanarayana, N., Lau, H.K., Sinha, S.K.: Nanolubrication of poly(methyl methacrylate) films on Si for micromechanical systems applications. Appl. Phys. Lett. **93**, 261906–261908 (2008)
30. Abgrall, P., Conedera, V., Camon, H., Gue, A.M., Nguyen, N.T.: SU-8 as a structural material for labs-on-chips and micro electromechanical systems: Review. Electrophoresis **28**, 4539–4551 (2007)
31. Seidemann, V., Rabe, J., Feldmann, M., Buttgenbach, S.: SU8-micromechanical structures with in situ fabricated movable parts. Microsyst. Technol. **8**, 348–350 (2002)
32. Foulds, I.G., Parameswaran, M.: A planar self-sacrificial multilayer SU-8-based MEMS process utilizing a UV-blocking layer for the creation of freely moving parts. J. Micromech. Microeng. **16**, 2109–2115 (2006)
33. Jiguet, S., Judelewicz, M., Mischler, S., Hofmann, H., Bertsch, A., Renaud, P.: SU-8 nano composite coatings with improved tribological performance for MEMS. Surf. Coat. Technol. **201**, 2289–2295 (2006)
34. Jiguet, S., Judelewicz, M., Mischler, S., Bertsch, A., Renaud, P.: Effect of filler behavior on nano composite SU8 photoresist for moving micro-parts. Microelectron. Eng. **83**, 1273–1276 (2006)
35. Arvind Singh, R., Satyanarayana, N., Kustandi, T.S., Sinha, S.K.: Tribo-functionalizing Si and SU8 materials by surface modification for application in MEMS/NEMS actuator-based devices. J. Phys. D Appl. Phys. **44**, 015301–015310 (2011)

Chapter 10
Friction and Wear Studies of Ultra-Thin Functionalized Polyethylene Film Chemisorbed on Si with an Intermediate Benzophenone Layer

Myo Minn and Sujeet K. Sinha

Abstract A successful attachment of an ultra-thin (thickness approximately 20 nm) functionalized polyethylene (fPE) film onto Si substrate via a reactive benzophenone (Ph_2CO) layer is presented in this chapter. The wear durability results obtained under 40 mN applied load and 0.052 m/s sliding speed show an improvement in the wear life of Si/Ph_2CO/fPE by one order of magnitude over Si/Ph_2CO. When perfluoropolyether (PFPE) is applied as a top mobile lubricant layer coated onto Si/Ph_2CO and Si/Ph_2CO/fPE, a significant improvement in the wear durability is observed as Si/Ph_2CO/fPE/PFPE does not fail until one million cycles whereas Si/Ph_2CO/PFPE has shown a wear life of 250,000 cycles. Si/Ph_2CO/fPE/PFPE can withstand a minimum applied load of 150 mN at a sliding speed of 0.052 ms^{-1} without failure, providing a pressure $\times$ velocity (PV) limit of greater than 106.6 MPa ms^{-1}.

Contents

M. Minn
Department of Mechanical Engineering, National University of Singapore,
9 Engineering Drive 1, Singapore 117576, Singapore

S. K. Sinha (✉)
Department of Mechanical Engineering, Indian Institute of Technology,
Kanpur 208016, India
e-mail: sujeet@iitk.ac.in

S. K. Sinha et al. (eds.), *Nano-tribology and Materials in MEMS*,
DOI: 10.1007/978-3-642-36935-3_10,

Introduction

Silicon is widely used as a structural material in micro-electro-mechanical system (MEMS) and other micro devices. However, there are some wear durability issues in the application of Si as it has poor tribological properties which can lead to early failure of devices. In an attempt to overcome the poor wear durability, many researchers have been trying to modify the Si surface with a protective thin solid or liquid film [1–6]. Nonetheless, a small increase in the surface tension or the surface energy from the protective film could lead to an increase in adhesion, stiction, friction and wear as a result of pull-off force and menisci formation [7, 8]. One of the effective ways to reduce these unwanted properties is to use an organic thin film. Major constraint of these organic thin films is that though they can reduce stiction and friction, their long-term wear performance does not improve significantly [9–11]. Hence, the exploration of new films that can provide reliability of devices in terms of low friction and long wear durability is still necessary. In recent studies, it is observed that the application of polymer films on Si substrate is a promising way to achieve low friction and high wear performance [12–15]. In our previous studies, we have applied different surface modification techniques to ultra high molecular weight polyethylene (UHMWPE) film and extended the wear durability by several orders [16–18], showing that polymer films are very promising coating materials in lowering friction and improving wear durability of Si substrate. However, for the application of such coatings in micro/nano devices such as micro-electro-mechanical systems (MEMS), the film thickness has to be brought down to a few nanometers range.

Therefore, in this chapter, we present a coating of an ultra-thin polymer film on Si substrate and investigate its friction and wear properties in combination with other layers. We chose the functionalized polyethylene (fPE), polyethylene-graft-maleic anhydride, as a polymer film because of its excellent tribological performance in bulk form and another advantage is its terminal groups that can chemically attach to a substrate. In order to attach fPE onto Si substrate which has no functional group, the Si substrate needs to be modified with a benzophenone (abbreviated as Ph_2CO) layer serving as an intermediate between Si and fPE. Benzophenone was chosen because it can easily attach to C–H bonds of fPE in different chemical environments [19–25].

The chemisorption of fPE film onto Si/Ph_2CO can promote wear resistance to some extent. However, as the thickness of the fPE film used in this study is very thin, it is difficult to obtain long-term lubrication effects. If there is no replenishment layer, wear can easily initiate from the contact area despite excellent bond strength. The importance of using a mobile layer of perfluoropolyether (PFPE) in replenishing process for longer wear durability has been well studied [26]. As far as the wear durability is concerned, PFPE coated self-assemble monolayers (SAMs) have shown higher durability than either SAMs or PFPE layer alone [27, 28]. Hence, in this study, we have applied PFPE as the top mobile layer for effective lubrication.

Experimental Procedures

Materials

Polished n-type silicon (100) wafer was used as the substrate. 4-hydroxybenzophenone, allyl bromide and potassium carbonate were obtained from Alfa Aesar and used as received. Polyethylene-graft-maleic anhydride (PE-g-MAH) was obtained from Aldrich Inc. and perfluoropolyether (PFPE, Z-dol 4000) was obtained from Solvay Solexis.

Sample Preparation

Synthesis of Benzophenone Layer on Si Substrate

Si substrates (1×1 cm) were rinsed for 1 min and sonicated for 1 h in acetone. The substrates were blow-dried with pure nitrogen gas and given plasma treatment under an air environment in order to remove any contaminants from the substrates. For the air-plasma treatment, a Harrick Plasma cleaner/steriliser was used with an exposure time of 15 min and an RF power of 30 W under vacuum.

4-hydroxybenzophenone (19.8 g) and allyl bromide liquid (13.3 g) were dissolved in acetone (60 mL). After that, 14 g of potassium carbonate was added into the solution. The extraction of 4-allyloxybenzophenone crystals from the solution was performed as described in Ref. [29]. The extracted crystals (2 g) were added to dimethyl chlorosilane (20 mL) in the presence of Pt–C [10 mg, (10 %)] catalyst and the mixture was then refluxed at about 75 °C for 4 h. After removing the catalyst by filtration of the mixture in toluene, 4-(3′-chlorodimethylsilyl) propyloxybenzophenone solution was obtained.

The cleaned Si substrates were immersed in the obtained solution with triethylamine (Et_3N) catalyst for 24 h. The samples coated with benzophenone layer were obtained after rinsing the Si samples with chloroform and then drying with pure nitrogen gas.

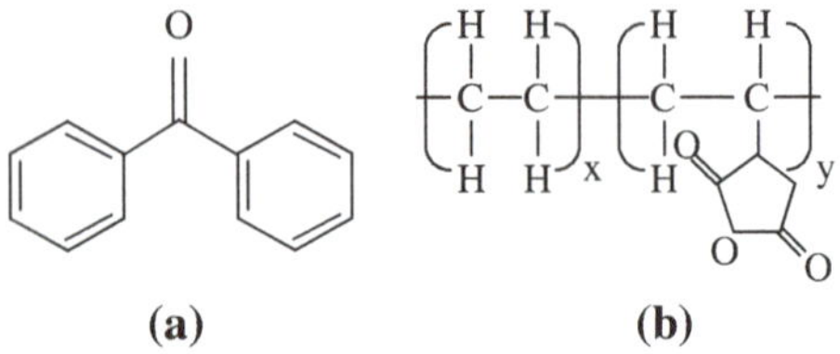

Fig. 10.1 The chemical structures of **a** benzophenone and **b** polyethylene-graft-maleic anhydride (fPE)

(a) **(b)**

Preparation of Functionalized Polyethylene on Benzophenone Coated Si Samples

The fPE solution was prepared by dissolving polyethylene-graft-maleic anhydride, PE-g-MAH (1 wt %) in toluene at a temperature of 75 °C for 15 min. A magnetic stirrer was used to increase the dissolution rate. The chemical structures of benzophenone and PE-g-MAH (fPE) are provided in Fig. 10.1. The benzophenone coated Si samples (Si/Ph_2CO) were immersed into the fPE solution for 5 h. The samples were then ultrasonically cleaned with toluene and methanol for 10 min each in order to remove fPE not chemically attached to the samples. After blow-drying with pure nitrogen at room temperature, the chemically attached fPE layer was formed on Si/Ph_2CO as Si/Ph_2CO/fPE.

Preparation of PFPE Mobile Layer on Si/Ph_2CO and Si/Ph_2CO/fPE Samples

Perfluoropolyether (PFPE, Zdol 4000) was prepared from a solution of 0.2 wt % PFPE in H-Galden ZV60 solvent. Si/Ph_2CO and Si/Ph_2CO/fPE samples were dipped in the PFPE solution at dipping and withdrawal speeds of 2.4 mm/s with a fixed dipping duration of 30 s. The chemical structures of PFPE and H-Galden ZV60 are $HOCH_2CF_2O\text{-}(CF_2CF_2O)_p\text{-}(CF_2O)_q\text{-}CF_2CH_2OH$ and $HCF_2O\text{-}(CF_2O)_p\text{-}(CF_2CF_2O)_q\text{-}CF_2H$, respectively, where the ratio p/q is quoted as 2/3. A simple schematic diagram of Si substrate with different modified layers is provided in Fig. 10.2.

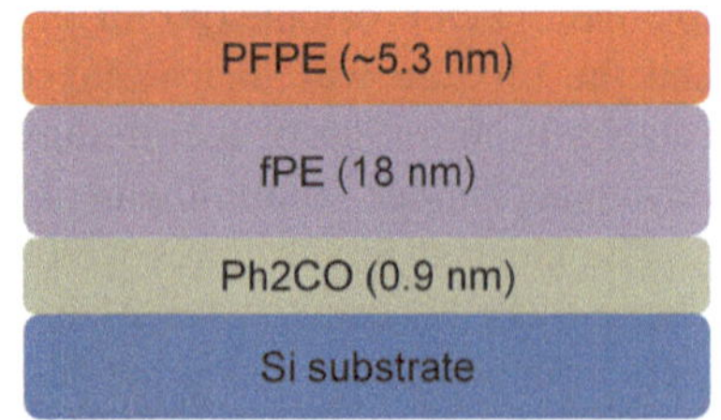

Fig. 10.2 A schematic diagram of Si substrate with different modified layers (not to scale)

Surface Characterizations

Contact angle and Surface Energy Measurements

The presence of different surface layers on Si substrate was examined by contact angle measurement. VCA Optima Contact Angle System (AST product, Inc., USA) was used for the measurement of contact angles with de-ionized water, ethylene glycol and hexadecane droplets. A droplet of 0.5 μl was used for each contact angle measurement. A total of five independent measurements were performed randomly on the samples and an average value was taken for every sample. The measurement error was within ±3°. After measuring the contact angle with different liquids, the surface energies were calculated using the equipment software. The relationship between the contact angle and the surface free energy was in the form,

$$\gamma_{SV} = \gamma_{SL} + \gamma_{LV} \cos\theta$$

where θ is the contact angle, γ is the surface energy and the subscripts SV, SL and LV represent surface–vapor, surface–liquid and liquid–vapor interfaces, respectively. We used the acid–base approach in which at least three liquids are required; two have known acid and base fractions larger than zero and at least one must have basic and polar parts.

Roughness and Thickness Measurements

The average roughness (R_a) of the samples was measured using dynamic optical profiler (Veeco Wyko NT1100) (a non-contact profiling device). The scanning area for the measurement was 124 × 93 μm in vertical scan interferometry (VSI) mode. As the total film thickness of overall coating was less than 25 nm, the reflectivity of the Si substrate was not affected by the coated polymer films as the film appeared transparent to naked eyes. However, the roughness measurement of the coated films was carried out without any additional gold sputtering for optical reflectivity, as any additional process could damage the film itself because of heating effect.

The thickness of each coated layer was determined using a variable-angle-spectroscopic ellipsometry (VASE, J. A. Woollam. Co., USA) at wavelengths from 400 to 1,000 nm at 10 nm intervals and at incident angles of 65°, 70°, and 75°. The refractive indices were assumed as 1.46 for SiO_2 [30], 1.5 for benzophenone [29], 1.54 for fPE [31] and 1.296 for PEPE [32].

XPS Analysis

The chemical state of the layers was observed by x-ray photoelectron microscopy (XPS) (Kratos Analytical AXIS His). XPS (Al Kα source) tests were performed with an X-ray source (1486.6 eV photons) at a constant dwell time of 100 ms and a pass-energy of 40 eV. The core level signals were obtained at a photoelectron take-off angle of 90° (with respect to the sample surface). All binding energies (BE) were referenced to the C1 s hydrocarbon peak at 284.6 eV. In peak combination, the line width (full width at half-maximum or FWHM) for the Gaussian peaks was maintained constant for all components in a particular spectrum. The curve de-convolution of the XP spectra was performed by using XPS Peak Fitting Program XPSPEAK41 (Freeware for XPS Community, Chemistry, CUHK).

Friction and Wear Tests

Friction and wear tests were conducted on a custom built ball-on-disc tribometer which can measure the normal and frictional displacements of the cantilever with laser sensors. The displacements were converted to the respective forces with the calibration chart. The sensitivity of the laser sensor was 0.5 μm which was equivalent to 0.125 mN force on the cantilever. A silicon nitride (Si_3N_4) ball of 4 mm diameter with a surface roughness of 5 nm was used as the counterface. The normal load used was 40 mN. The initial coefficient of friction was measured at a sliding track radius of 1 mm with a fixed disc rotational speed of 2 rpm (relative linear speed 0.21 mm/s). The sampling rate for the data recording was set at 10 Hz. The wear durability was determined at a fixed disc rotational speed of 500 rpm (relative linear speed was 0.052 m/s) and recorded at a sampling rate of 5 Hz. The wear durability was defined as the number of cycles completed when either the coefficient of friction exceeded 0.3 or large fluctuations in the coefficient of friction (indicative of film failure) or a clear wear track occurred. The ambient temperature and the relative humidity were fixed at 25 °C and 65 %, respectively.

Results

Surface Energy and Roughness

The water contact angle, the surface energy, the surface roughness and the thickness data for bare Si and modified Si are provided in Table 10.1. The bare Si surface shows the most hydrophilic nature with highest surface energy among all the surfaces. After modifying with benzophenone layer (Ph_2CO) on Si, the surface of Si/Ph_2CO becomes nearly hydrophobic with the water contact angle of 86°. This

Table 10.1 Water contact angle, surface energy, surface average roughness and thickness data for bare Si and coated Si

Sample	Water contact angle (°)	Surface energy (mJ/m^2)	Surface average roughness, R_a (nm)	Thickness of top layer (nm)
Si	6	42.9	0.4	–
Si/Ph_2CO	86	25.6	0.86	0.7 ± 0.2
Si/Ph_2CO/fPE	80	28.9	6.7	15 ± 3
Si/Ph_2CO/PFPE	82	31.8	3.59	4.3 ± 1
Si/Ph_2CO/fPE/PFPE	71	24.5	4.09	4.3 ± 1

value correlates with the measurement of Shen et al. [33]. fPE layer on Si/Ph_2CO sample has a contact angle of 80° which is slightly less than the reported values for the bulk polyethylene (91°–97°) [34, 35]. This difference appears to be because of the effect of underlying benzophenone layer. After coating PFPE film on Si/Ph_2CO and Si/Ph_2CO/fPE samples, the water contact angles further reduce to 82° and 71°, respectively. These noticeable changes in the water contact angle after modifying with different layers confirm the successful formation of the layers on the Si substrate.

Roughness and Thickness

The surface roughness (R_a) of the bare Si is the lowest at about 0.4 nm whereas roughness values become much higher for modified layers in the range of 0.86–6.7 nm. Most layers have shown very similar roughness values except that fPE makes the surface slightly rougher because of large molecules whereas the surface becomes smoother with the fillings of PFPE molecules in the original valleys.

The thickness of the modified benzophenone layer on Si is 0.7 ± 0.2 nm which is slightly lower than the value reported in Ref. [29]. The fPE and PFPE films have thickness values of 15 ± 3 nm and 4.3 ± 1 nm respectively.

XPS Results

The formation of different layers on Si substrate was examined by XPS chemical analysis and their C1 s spectra are shown in Fig. 10.3. The two C1 s peaks are observed in the spectrum of Si/Ph_2CO (Fig. 10.3a) in which the peak at 285 eV formed due to the presence of C–C/C–H bonds from the benzophenone film [36]. This peak is not observed in the C1 s spectrum of bare Si. The C1 s peak (O–C = O) in Fig. 10.3b represents the functional group of anhydride from the fPE film [37]. The existence of the O–C = O peak confirms the existence of fPE on Si/

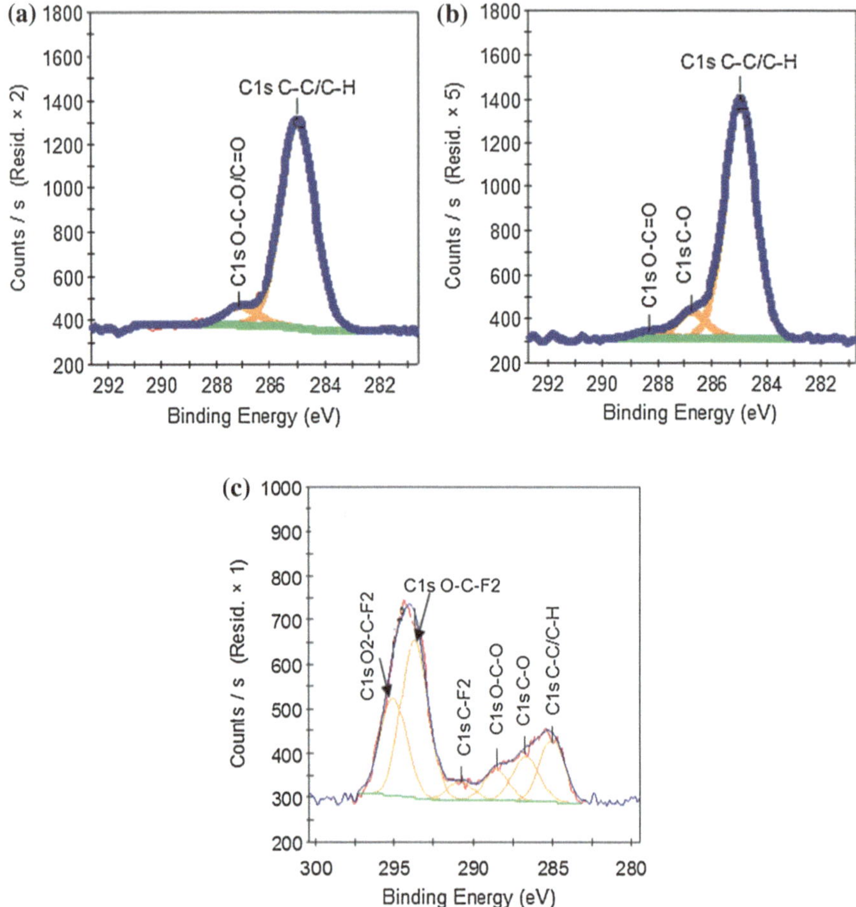

Fig. 10.3 High-resolution C1 s X-ray photoelectron spectrum (XPS) of **a** Si/Ph_2CO, **b** Si/Ph_2CO/fPE, and **c** Si/Ph_2CO/fPE/PFPE

Ph_2CO. After the PFPE film is coated onto Si/Ph_2CO/fPE, three extra C1 s peaks (O–C–F2, F–C–F and O–C–F) are observed in Fig. 10.3c which verifies the formation of the PFPE film.

Micro-Tribological Properties

Initial Coefficient of Friction

The initial coefficients of friction of bare Si and modified Si are studied and the results are summarized in Fig. 10.4. The highest surface energy of bare Si (provided in Table 10.1) leads to much higher initial coefficient of friction than

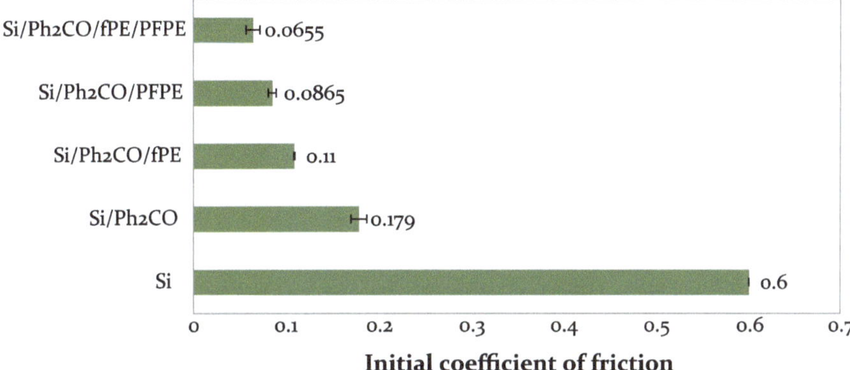

Fig. 10.4 The initial coefficient of friction of Si with different modified layers tested at a normal load of 40 mN and a rotational speed of 2 rpm

any modified film. One of the reasons for the higher friction is the strong adhesion between bare Si and the silicon nitride ball due to high surface energy of bare Si. The lower coefficients of friction are observed for other modified layers as their surface energies are also low.

The coefficient of friction of Si/Ph_2CO is higher than that of Si/Ph_2CO/fPE and Si/Ph_2CO/PFPE though its surface energy is lower. A possible reason is that the benzene structure in Si/Ph_2CO could hinder the sliding and increase the friction. Si/Ph_2CO/fPE shows a lower coefficient of friction of 0.11 and a further reduction in friction is observed for Si/Ph_2CO/PFPE and Si/Ph_2CO/fPE/PFPE, respectively. The additional reduction in friction is due to the presence of the mobile fraction of PFPE molecules which provides a low shear stress.

Wear Durability

The wear durability of different modified Si samples is shown in Fig. 10.5. Bare Si fails immediately as expected because of its high surface energy, high adhesion and brittle nature. The wear durability of Si/Ph_2CO increases slightly to 100 cycles whereas that of Si/Ph_2CO/fPE extends to 1,000 cycles. For the comparison purpose, the wear life of Si/fPE is also studied and has shown a wear life of 150 cycles. Though the wear life of either Si/Ph_2CO or Si/fPE film is around 100 cycles, the application of a dual Si/Ph_2CO/fPE film extends the wear durability to 1,000 cycles. When PFPE is applied as an ultra-thin (2–4 nm) lubricant layer, the wear life of Si/Ph_2CO/fPE increases significantly. The optical images of Si/Ph_2CO, Si/Ph_2CO/fPE and Si/Ph_2CO/PFPE on failure at 100 cycles, 1,000 cycles and 250,000 cycles, respectively, are shown in Fig. 10.6. Some debris observed on the surface of Si/Ph_2CO after 100 cycles of sliding were examined with energy dispersive spectroscopy (EDS) and confirmed as Si. A clear wear track is also observed on the image of Si/Ph_2CO/fPE after 1,000 cycles of sliding. Satyanarayana et al. [38] have studied the wear

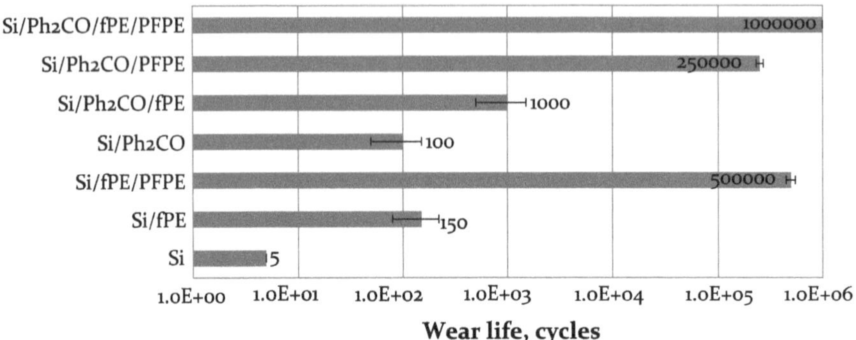

Fig. 10.5 The wear durability of Si with different modifications under the application of 40 mN applied normal load and 500 rpm rotational speed

resistance of Si/APTMS/fPE film (APTMS stands for 3-aminopropyltrimethoxysilane SAM) and observed a wear life of 4,000 cycles under similar contact conditions. The results show that though Si/Ph_2CO and Si/Ph_2CO/fPE can extend the wear resistance of Si to a certain limit, however, neither Ph_2CO nor Ph_2CO/fPE alone is sufficient to protect Si substrate for long term applications.

A significant improvement in wear durability is observed for Si/Ph_2CO/PFPE film which fails after around 250,000 cycles of sliding (Fig. 10.6c). Figure 10.6e shows the transferred materials on the ball surface but there is no sign of damage or scratching on the ball surface after cleaning with acetone (Fig. 10.6f). A great improvement is observed for Si/Ph_2CO/fPE/PFPE film as it does not fail until one million cycles, when the experiment was stopped. The optical image of the ball slid against the film is shown in Fig. 10.6g. The ball image had a small patch which was transferred lubricant to the counterface surface but no wear debris was observed (the ball surface after cleaning with acetone is shown in Fig. 10.6h). Thus, it is confirmed that for Si/Ph_2CO/fPE/PFPE film neither the ball nor the film was damaged for the entire 1 million cycles of sliding. To clearly understand the role of substrate modifications on the wear performance of PFPE, the wear durability of PFPE alone on bare Si (without any intermediate layer) was also measured and it failed in less than 1,000 cycles, showing that Ph_2CO and fPE films provide better chemical attachment of PFPE to Si/Ph_2CO and Si/Ph_2CO/fPE and also the polymer layer protects the substrate from any damage and wear debris generation. Thus, a strongly attached fPE layer with the intermediate benzophenone layer is necessary for extremely high wear durability of Si substrate.

The Role of Applied Loads on the Tribological Behaviors of Si/Ph_2CO/fPE/PFPE

The results of the effect of applied load on the coefficient of friction of Si/Ph_2CO/fPE/PFPE film are shown in Fig. 10.7. The coefficient of friction does not change much

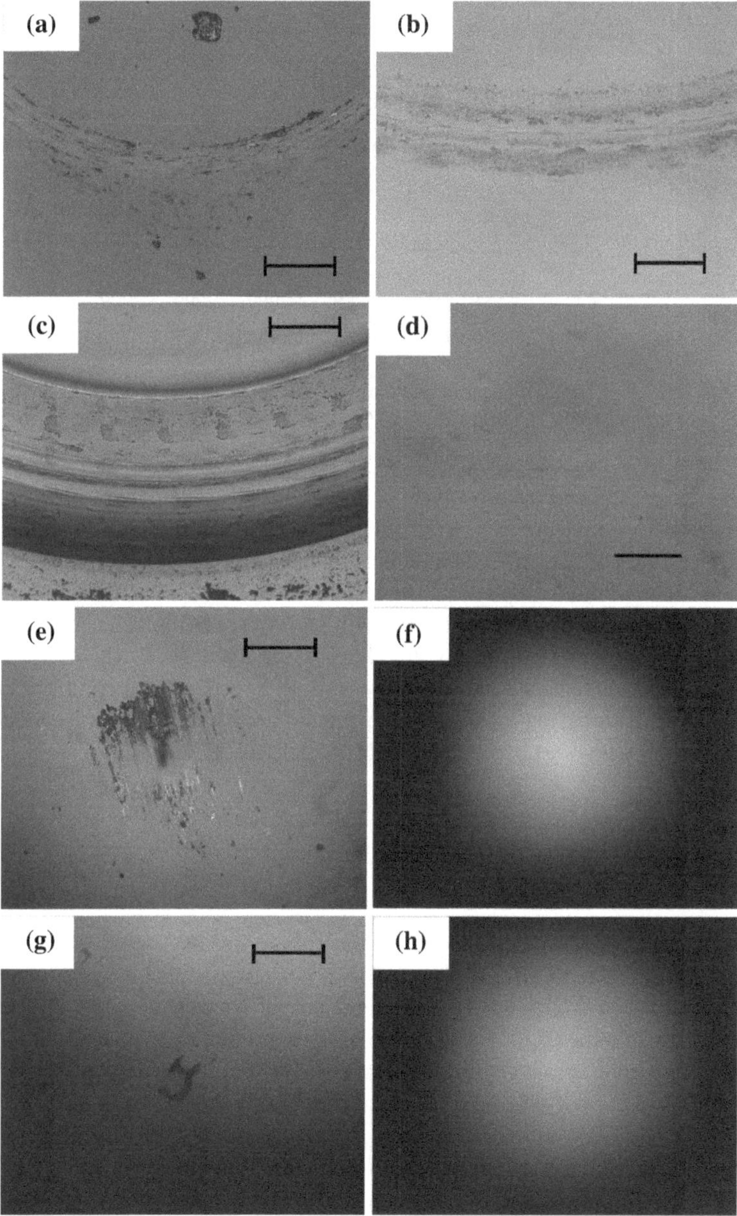

Fig. 10.6 The optical images of **a** Si/Ph_2CO, **b** Si/Ph_2CO/fPE, **c** Si/Ph_2CO/PFPE when they had failed, and **d** Si/Ph_2CO/fPE/PFPE after the experiment was stopped. The optical images of Si_3N_4 ball **e** slid against (**c**) and (**f**) after cleaning with acetone, **g** slid against (**d**) and (**h**) after cleaning with acetone. The applied load was 40 mN and the scale bars shown are 100 μm

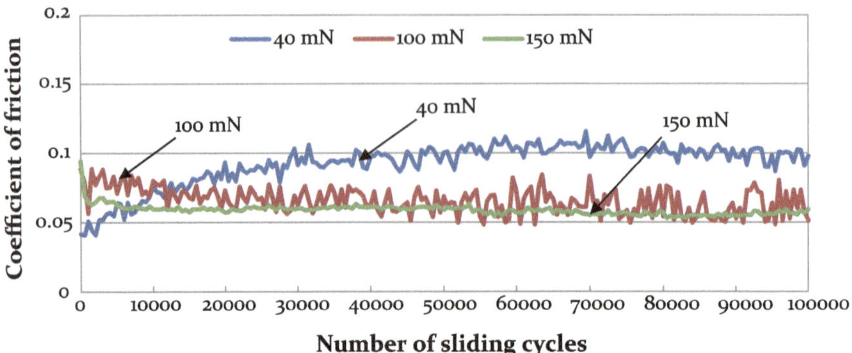

Fig. 10.7 Variation of the coefficient of friction with respect to the number of sliding cycles at three different normal loads for Si/Ph_2CO/fPE/PFPE

throughout the tests until 100,000 cycle of sliding for all applied loads. However, the coefficient of friction decreases (from 0.1 to 0.06) with an increase in applied load from (40–100 mN or 150 mN). The wear durability shows that Si/Ph_2CO/fPE/PFPE withstands a maximum applied normal load of 150 mN at a sliding speed of 0.104 ms^{-1} for one million cycles or more with no sign of failure. Hence, a minimum pressure × velocity (PV) limit for the film is obtained as 106.6 MPa ms^{-1} by multiplying the contact pressure and the sliding speed. The theoretical contact pressure was calculated using the Hertzian contact theory with the elastic modulii of 190 and 310 GPa for Si substrate and Si_3N_4 ball, respectively.

Discussion

In this chapter, we presented results on the tribological advantages of ultra-thin layers of fPE and fPE/PFPE in the presence of benzophenone as intermediate layer on Si. It is observed that the initial coefficient of friction depends on both the surface energy and the chemical structure of the modified layers. Generally, higher surface energy gives stronger adhesion force and consequently higher friction [39]. In a comparison between the lower surface energy of Si/Ph_2CO and Si/Ph_2CO/fPE, the presence of bulky and rigid benzene structures restrict the sliding movement and hence increase the friction, whereas the linear and flexible backbone structure of fPE provides the "smooth molecular profile" and low friction. One reason for this is also the alignment of the linear molecules in the direction of sliding. Our earlier study has also shown that a polymer with benzene side groups such as polystyrene would give high coefficient of friction despite high water contact angle [38]. Further reduction in friction was observed as PFPE was applied as the top layer due to its low surface energy and superior lubricating properties.

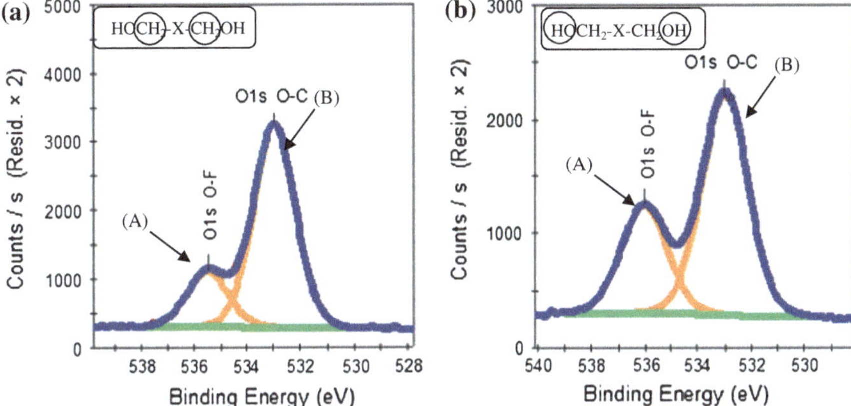

Fig. 10.8 High-resolution O1 s X-ray photoelectron spectrum (XPS) of **a** Si/Ph_2CO/PFPE, and **b** Si/Ph_2CO/fPE/PFPE. $HOCH_2$-X-CH_2OH is the chemical structure of PFPE where X is CF_2O-$(CF_2CF_2O)_p$-$(CF_2O)_q$-CF_2. *Circles* show the elements that mostly take place in reactions

On the wear durability, although the presence of Si/Ph_2CO or Si/Ph_2CO/fPE could slightly extend the wear performance, the application of PFPE film on both Si/Ph_2CO and Si/Ph_2CO/fPE films increases the wear resistance by more than three orders. One reason for the enhancement in the wear durability is that PFPE molecules can chemically attach to either Si/Ph_2CO or Si/Ph_2CO/fPE. The bonded and mobile PFPE combination provides efficient lubrication on the Si/Ph_2CO/fPE surface. The chemical interaction between PFPE and Si/Ph_2CO or Si/Ph_2CO/fPE studied with XPS and the O1 s peaks are provided in Fig. 10.8. Peak A represents the oxygen of O–F groups and Peak B reveals that of O–C groups. The number of O1 s counts for Peak A is invariant for both films, showing that the O–F groups do not involve any chemical reaction. However, a great difference in the number of O1 s counts for Peak B is observed. As the number of O1 s counts of Peak B for Si/Ph_2CO/PFPE film is much higher than that for Si/Ph_2CO/fPE/PFPE film, we can state that the reaction probability of OH groups of PFPE in the former film is higher than that of the latter film (see boxes in Fig. 10.8). The extra OH groups of PFPE provide higher adhesion to the film which in turn reduces the wear resistance [40]. In the case of Si/Ph_2CO/fPE/PFPE, the anhydride groups in fPE react extensively with hydroxyl (OH) groups in PFPE [41] and thus the intensity of O1 s drops. The reaction enhanced the wear resistance due to greater bonding and it could be the main reason for lower static friction and higher wear durability of Si/Ph_2CO/fPE/PFPE film in comparison with Si/Ph_2CO/PFPE film. Hence, the presence of fPE bonded to Si with the help of benzophenone has helped to increase the wear performance. This would not have been possible with a polymer layer such as polystyrene because PE has linear molecular chains which provide low friction and excellent wear resistance in the coated form [38]. Benzophenone acts as a very effective bonding agent for the Si substrate and the fPE layer.

Conclusions

The functionalized polyethylene (fPE) was coated onto Si substrate by reacting with benzophenone as the intermediate bonding layer. Perfluoropolyether (PFPE) lubricant was applied as the top layer onto Si/Ph_2CO/fPE. The formation of each layer was confirmed by water contact angle measurement and XPS analysis. The friction and wear durability of different modified layers were measured with a ball-on-disc tribometer and the final conclusions are as follow:

1. Though there was a small deviation in contact angle between thin fPE film and bulk PE, the occurrence of C1 s O–C = O confirmed that fPE attached to the silicon substrate via a benzophenone reactive group.
2. The initial coefficient of friction depended on both the surface energy and the molecular structure of the films. Generally, the higher surface energy of the Si substrate provided higher static friction than any other modified films. In addition, though Si/Ph_2CO has lower surface energy than Si/Ph_2CO/fPE, due to its bulky benzene structure, Si/Ph_2CO shows higher coefficient of friction.
3. The wear durability of Si/Ph_2CO/fPE shows ten times increase over that of Si/Ph_2CO. The existence of PFPE film on Si/Ph_2CO, Si/fPE and Si/Ph_2CO/fPE increased the wear life to 200,000 cycles, 500,000 cycles and more than one million cycles, respectively, where the applied load was 40 mN at a sliding speed of 0.052 ms^{-1}.
4. The coefficient of friction showed a minor decrease with increasing applied load (40–150 mN) for Si/Ph_2CO/fPE/PFPE film. The PV limit that the film could safely withstand was 106.6 MPa ms^{-1}, which is probably a minimum as the film did not fail in our tests.

Acknowledgments This chapter is based on research work supported by the Singapore National Research Foundation under CRP Award no. NRF-CRP 2-2007-04. We would like to thank Dr. N. Satyanarayana for his valuable suggestions.

References

1. Bhushan, B., Israelachvili, J.N., Landman, U.: Nanotribology: friction, wear and lubrication at the atomic scale. Nature **374**, 607–616 (1995)
2. Bhushan, B.: Tribology and Mechanics of Magnetic Storage Devices, 2nd edn. Springer, New York (1996)
3. Man, K.F., Stark, B.H., Ramesham, R.: A Resource Handbook for MEMS Reliability. California Institute of Technology, Pasadena (1998)
4. Kayali, R.L., Stark, B.H.: MEMS reliability assurance activities at JPL. EEE Links **5**, 10–13 (1999)
5. Tanner, D.M., Smith, N.F., Irwin, L.W., et al.: MEMS Reliability: Infrastructure, Test Structure, Experiments, and Failure Modes. Sandia National Laboratories, Albuquerque (2000)
6. Arney, S.: Designing for MEMS reliability. MRS Bull. **26**, 296–299 (2001)

7. Bhushan, B.: Principles and Applications of Tribology. Wiley, New York (1999)
8. Bhushan, B.: Introduction to Tribology. Wiley, New York (2002)
9. Tsukruk, V.V.: Nanocomposite polymer layers for molecular tribology. Tribol. Lett. **10**(1–2), 127–132 (2001)
10. Tsukruk, V.V.: Molecular lubricants and glues for micro- and nanodevices. Adv. Mater. **13**, 95–108 (2001)
11. Satyanarayana, N., Sinha, S.K.: Tribology of self-assembled monolayers on silicon surfaces. J. Phys. D Appl. Phys. **38**, 3512–3522 (2005)
12. Sidorenko, A., Ahn, H.S., Kim, D.I., Yang, H., Tsukruk, V.V.: Wear stability of polymer nanocomposite coatings with trilayer architecture. Wear **252**(11–12), 946–955 (2002)
13. Satyanarayana, N., Sinha, S.K., Ong, B.H.: Tribology of a novel UHMWPE/PFPE dual-film on Si surface. Sens. Actua. A-Phys **128**(1), 98–108 (2006)
14. Morgan, S.E., Misra, R., Jones, P.: Nanomechanical and surface frictional characteristics of a copolymer based on benzoyl-1,4-phenylene and 1,3-phenylene. Polym. **47**, 2865–2873 (2006)
15. Sun, C., Zhou, F., Shi, L., Yu, B., Gao, P., Zhang, J., Liu, W.: Tribological properties of chemically bonded polyimide films on silicon with polyglycidyl methacrylate brush as adhesive layer. Appl. Surf. Sci. **253**, 1729–1735 (2006)
16. Minn, M., Sinha, S.K.: DLC and UHMWPE as hard/soft composite film on Si for improved tribological performance. Surf. Coat. Tech. **202**, 3698–3708 (2008)
17. Minn, M., Leong, Y.H., Sinha, S.K.: Effects of interfacial energy modification on the tribology of UHMWPE coated Si. J. Phys. D Appl. Phys. **41**, 055307 (2008)
18. Minn, M., Sinha, S.K.: Tribology of ultra high molecular weight polyethylene film on Si substrate with chromium nitride, titanium nitride and diamond like carbon as intermediate layers. Thin Solid Films **518**, 3830–3836 (2010)
19. Horie, K., Ando, H., Mita, I.: Photochemistry in polymer solids. 8. Mechanism of photoreaction of benzophenone in poly(vinyl alcohol). Macromolecules **20**(1), 54–58 (1987)
20. Smets, G.J., Hamouly, S., Oh, T.J.: Photochemical reactions on polymers. Pure Appl. Chem. **56**(3), 439–446 (1984)
21. Horie, K., Morishita, K., Mita, I.: Photochemistry in polymer solids. 3. Kinetics for nonexponential decay of benzophenone phosphorescence in acrylic and methacrylic polymers. Macromolecules **17**(9), 1746–1750 (1984)
22. Horie, K., Mita, I.: Photochemistry in polymer solids-4. On the intensity dependence of non-exponential decay of benzophenone phosphorescence in poly(methyl methacrylate). Eur. Polym. J. **20**, 1037–1039 (1984)
23. Bräuchle, C., Burland, D.M., Bjorklund, G.C.: Hydrogen abstraction by benzophenone studied by holographic photochemistry. J Phys Chem **85**, 123–127 (1981)
24. Dorman, G., Prestwich, G.D.: Benzophenone photophores in biochemistry. Biochemistry **33**(19), 5661–5673 (1994)
25. Turro, N.J.: Modern Molecular Photochemistry. University Science Books, Mill Valley (1991)
26. Ahn, H.-S., Julthongpiput, D., Kim, D.-I., Tsukruk, V.V.: Dramatic enhancement of wear stability in oil-enriched polymer gel nanolayers. Wear **255**(7–12), 801–807 (2003)
27. Eapen, K.C., Patton, S.T., Zabinski, J.S.: Lubrication of microelectromechanical systems (MEMS) using bound and mobile phases of Fomblin Zdol®. Tribol. Lett. **12**(1), 35–41 (2002)
28. Choi, J., Morishita, H., Kato, T.: Frictional properties of bilayered mixed lubricant films on an amorphous carbon surface: effect of alkyl chain length and SAM/PFPE portion. Appl. Surf. Sci. **228**(1–4), 191–200 (2004)
29. Prucker, O., Naumann, C.A., Ruhe, J., Knoll, W., Frank, C.W.: Photochemical attachment of polymer films to solid surfaces via monolayers of benzophenone derivatives. J. Am. Chem. Soc. **121**, 8766–8770 (1999)
30. Julthongpiput, D., LeMieux, M., Tsukruk, V.V.: Micromechanical properties of glassy and rubbery polymer brush layers as probed by atomic force microscopy. Polymer **44**(16), 4557–4562 (2003)

31. Handbook of chemicals and laboratory equipment: Aldrich® (2006)
32. Website: http://www.solvaysolexis.com/static//wma/pdf/5/4/3/4/fom_thin.pdf, last accessed in May (2010)
33. Shen, W.W., Boxer, S.G., Knoll, W., Frank, C.W.: Polymer-supported lipid bilayers on benzophenone-modified substrates. Biomacromolecules **2**(1), 70–79 (2001)
34. Price, G.J., Clifton, A.A., Keen, F.: Ultrasonically enhanced persulfate oxidation of polyethylene surfaces. Polymer **37**(26), 5825–5829 (1996)
35. Sira, M., Trunec, D., Stahel, P., Bursikova, V., Navratil, Z., Bursik, J.: Surface modification of polyethylene and polypropylene in atmospheric pressure glow discharge. J. Phys. D Appl. Phys. **38**(4), 621–627 (2005)
36. Kiguchi, M., Eva, P.D.: Photostabilisation of wood surfaces using a grafted benzophenone UV absorber. Polym. Degrad. Stabil. **61**(1), 33–45 (1998)
37. Freudenberg, U., Zschoche, S., Simon, F., Janke, A., Schmidt, K., Behrens, S.H., Auweter, H., Werner, C.: Covalent immobilization of cellulose layers onto maleic anhydride copolymer thin films. Biomacromolecules **6**(3), 1628–1634 (2005)
38. Satyanarayana, N., Sinha, S.K., Shen, L.: Effect of molecular structure on friction and wear of polymer thin films deposited on Si surface. Tribol. Lett. **28**(1), 71–80 (2007)
39. Miyoshi, K.: Surface design and engineering toward wear-resistant, self-lubricating diamond films and coatings. NASA-TM, 208905 (1999)
40. Bhushan, B.: Self-assembled monolayers for controlling adhesion, friction and wear, in Springer handbook of nanotechnology, 2nd edn, pp. 831–860. Springer-Verlag GmbH, New York (2007)
41. Buchholz, V., Adler, P., Bäcker, M., Hölle, W., Simon, A., Wegner, G.: Regeneration and hydroxyl accessibility of cellulose in ultrathin films. Langmuir **13**, 3206–3209 (1997)

Chapter 11
Localized Lubrication of Micromachines: A Novel Method of Lubrication on Micromechanical Devices

L. Y. Jonathan, V. Harikumar, N. Satyanarayana and Sujeet K. Sinha

Abstract The effective lubrication of Micro-Electro-Mechanical Systems (MEMS) has been a major hurdle for the micro-machine industry, limiting commercially available designs and mechanisms in such machines to mostly non-contacting ones. Silicon (Si) is a common material for fabrication of MEMS devices, which has very poor tribological properties (high wear rates and coefficient of friction). MEMS lubrication techniques usually involve highly expensive and/or complex processes such as vapour deposition and hermetic packaging to extend the wear life. This study presents a novel method of applying lubricant onto a specifically designated location on MEMS, extending its wear life by several orders of magnitudes. In a feasibility study, a fixed amount of PFPE, which is well documented to be a good lubricant in the hard disc industry, providing a very thin monolayer film as protection, was delivered using the proposed method, between two contacting Si pieces which were then subjected to reciprocation sliding at an applied normal load of 0.5 N (approximately 125 kPa of apparent pressure) and a sliding velocity of 5 mm s^{-1}. The tribological properties (coefficient of friction and wear behaviour) were studied to ascertain the effectiveness of this lubrication method, compared both across polished and unpolished Si surfaces, as well as among varying methods of lubrication; namely dip-coating, vapour-deposition and the novel method of localized lubrication. Unpolished surfaces were also used to replicate the practical rough surfaces on MEMS devices, especially sidewalls. The results have revealed that the current localized lubrication method is very effective in reducing the coefficient of friction and increasing wear life in the reciprocating sliding motion between the two surfaces and was then applied to actual devices custom designed and fabricated for tribological testing of sidewalls, which has

L. Y. Jonathan · V. Harikumar · N. Satyanarayana
Department of Mechanical Engineering, National University of Singapore,
9 Engineering Drive 1, (S) 117576, Singapore

S. K. Sinha (✉)
Department of Mechanical Engineering, Indian Institute of Technology,
Kanpur 208016, India
e-mail: sujeet@iitk.ac.in

S. K. Sinha et al. (eds.), *Nano-tribology and Materials in MEMS*,
DOI: 10.1007/978-3-642-36935-3_11, © Springer-Verlag Berlin Heidelberg 2013

shown to reduce both the static and dynamic friction against the sliding sidewalls, as well as the adhesion forces on contact between the sidewalls.

Contents

Introduction

Friction and wear in mechanical devices has always been a concern—as such, lubrication and tribological properties can be considered universal issues that often need to be addressed in any mechanical situation. Particularly in Micro-Electro-Mechanical Systems (MEMS), where stiction and adhesion is predominant in the surface interactions due to the micro-scale of the materials and devices, the tribological properties present a concern and act as the limiting factors in designing or fabricating components for MEMS. The reliability and functionality of such devices depend heavily on these factors [1].

The adhesion between two surfaces in a MEMS device, commonly known as "stiction", has been well documented [2] and is an issue due to the small scale of the components compared to common machines. The micro-scale of MEMS devices lead to a large surface area to volume ratio, and due to the small size, lubrication techniques tend to be complex and costly—requiring specialized packaging and storage of devices particularly when in use [3]. These procedures not only limit the environments in which such devices can be used, but make the application and processing of the devices uneconomical due to increased costs. Commonly used methods for lubrication of Si surfaces are dip-coating, which is not universally viable for MEMS devices as the surface tension forces would bring the components in contact with each other leading to adhesion before the lubricant can be effective, and vapour deposition, which although provides a thin layer of chemically bonded molecules to the surface, also requires careful surface processing and handling conditions with specialized equipment. Often these methods modify the surface conditions of entire exposed plane surfaces on MEMS, possibility reducing the functionality of certain components (e.g. electrical contact pads, micro-gears, sensing mechanisms, etc.) if they are not masked—but masking of the device then introduces another step of careful processing and chemical etching. The need for effective lubrication only at certain locations on the device or component without affecting the rest of the device is therefore evident.

Furthermore, the effect of conventional lubrication methods on MEMS sidewalls is unknown—the inaccessibility of the sidewalls and the size of the gaps ($\approx$10–20 μm for functional parts) limit lubrication methods and the eventual study of the effectiveness on the sidewalls. There are also many differences between the interactions for in-plane surfaces and those between sidewall surfaces in MEMS, primarily due to the processing conditions [4].

Conventional lubrication methods, such as the direct application of oil lubricants, are impractical at the micro scale—one method of improving the friction and wear properties in such cases is to reduce the surface energies of the surfaces involved, thereby reducing the adhesion between them. Hydrophobic surfaces provide low friction and adhesion properties, and the additional application of a lubricant to minimize both contact and sliding drag between the two surfaces will accentuate this effect. In order to achieve this, methods such as Self-Assembled Monolayers (SAMs) are used to lubricate MEMS devices in sliding, and are effective in reducing wear to a small extent, and friction and stiction to a larger extent [5].

Polymer lubricants have gathered extensive interest due to their excellent wear and anti-corrosion properties—many techniques also allow for coating over varying topographies on the surface. These lubricants are cost-effective and have been used widely in industrial applications [6]. Perfluoropolyether (PFPE) lubricants have been used in magnetic hard disks to prevent wear and friction due to the low surface tension, lubricity and hydrophobic properties to the surface, as well as thermal and chemical stability which prevent degrading and decomposition over time [7, 8]. Because of these properties, PFPE is used as a lubricant in this study.

The method of lubrication proposed and to be tested is sought not only to lubricate MEMS sidewalls under sliding contact [9], but also for a localized and predetermined application of lubricant to preserve the overall functionality and surface condition of MEMS devices while providing improved tribological properties where necessary. The proposed method is designed to address the inaccessibility and the difficulty of local application of lubricant onto a MEMS device, particularly a sidewall gap. In order to investigate the effectiveness of both the method and the lubricant used, reciprocating tests were conducted against Si surfaces, comparing polished and unpolished surfaces, and the relative effectiveness of various methods of application of lubricant in improving the tribological properties.

Experimental Materials

Silicon Wafers

N-type silicon (1 0 0) wafers were used in this study, obtained from Engage Electronics (Singapore) Pte Ltd). Both the polished (roughness Ra = 16 nm) and unpolished (roughness Ra = 616 nm) surfaces were used for the investigation on the effects of topography and roughness. The wafers were approximately 525 μm thick, with 12.4 GPa hardness. Base pieces measured approximately 1.5 cm by 1.5 cm, and smaller upper pieces were laser diced into 2 mm by 2 mm (see Fig. 11.1). All Si surfaces were cleaned by thorough washing in ethanol for 1 min, ultrasonic cleaning in ethanol for 1 h, then dried in nitrogen (N_2) gas before cleaning with air plasma using a Harrick Plasma Cleaner/Sterilizer for 5 min, using an RF power of 30 W [10]. Although piranha treatment provides better bonding between polymer lubricant and the silicon substrate [11], this particular treatment was not used in this study due to the complex steps and long duration required, both of which were not feasible in actual MEMS applications and on MEMS devices. Removing this step also leads to a more environmentally friendly process, eliminating the use of hazardous liquids and acids.

PFPE Lubricant

A well-known PFPE lubricant, Z-dol 4000 (monodispersed) was used to lubricate the Si surfaces using the various methods of application viz. dip-coating, vapour deposition, and the proposed "Loc-Lub" method. The PFPE lubricant used has a molecular weight of 4000 g/mol, and has terminal OH groups at its ends, with the chemical formula as follows (p/q ratio is 2/3):

PFPE Z-dol 4000:

$$H-O-CH_2-CF_2-\left(O-CF_2-CF_2\right)_p-\left(O-CF_2\right)_q-O-CF_2-CH_2-O-H$$

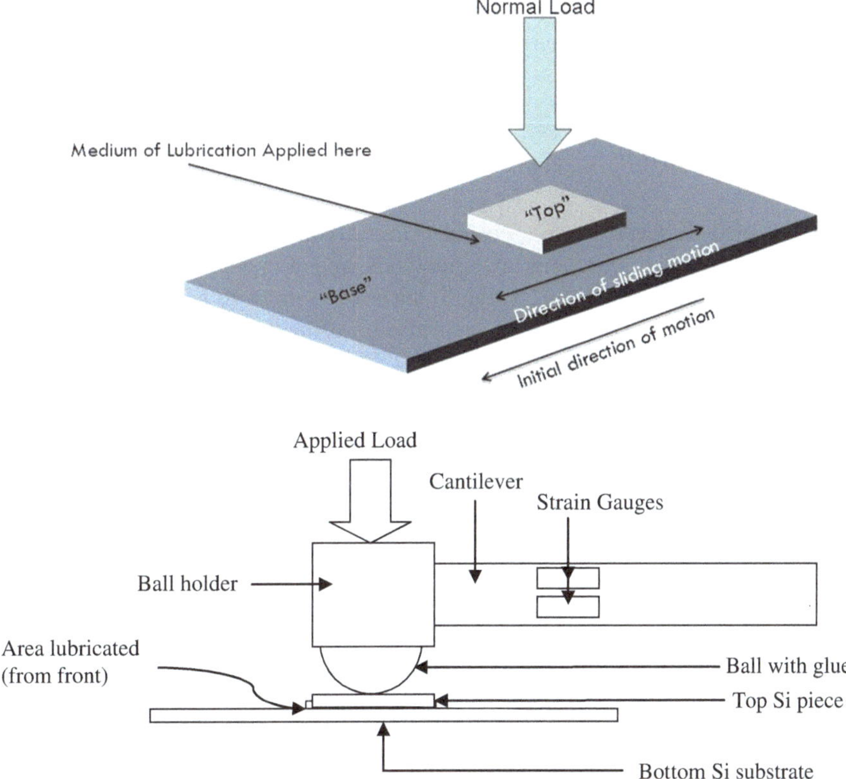

Fig. 11.1 Localized lubrication and reciprocating sliding wear test schematics. Reprinted with permission from Wear 270 (2010) 19–31, ©2012, Elsevier

All concentrations of PFPE solution used in this study were prepared using H-galden solvent, purchased from Ausimont INC. We used 1.0 and 4.0 wt % of PFPE in H-galden solution in this investigation, with further emphasis on 4.0 wt % as it presented more promising results.

Lubrication Methods and Wear Tests

Dip Coating

A simple dip-coating method was implemented with dipping/withdrawal speeds of 2.4 mm $s^{-1}$and submerged duration of 1 min. These parameters were kept constant for all the samples used in this investigation. Dip coating of samples was only done after overnight storage of the cleaned samples, and was further stored in desiccators for 1 day after coating to allow for the surfaces to stabilize prior to surfaces analysis or tribological testing. Only the larger base Si pieces were PFPE coated due to handling and surface area constraints.

Vapour Deposition

Only polished Si surfaces were coated using this method to provide a basis for comparison between the other two methods and a method involving only a chemically bonded monolayer on the Si surface. In order to achieve this, Si surfaces were functionalized with oxygen plasma using a Harrick Plasma Cleaner (PDC-32G) under a vacuum of 18 W. These samples were placed inverted over a dish containing PFPE lubricant solution and left in a vacuum chamber for approximately 5 min without exceeding 200 mTorr, and were then removed from the chamber. Surface analysis using EDS scanning as an additional step was done to ensure that this technique indeed provided a layer of the polymer lubricant on the surface, as the liquid solution at no point had come into contact with the surface and some non-destructive verification was necessary prior to wear tests.

"Loc-Lub": Localized Lubrication

Figure 11.1 illustrates a schematic for the experimental setup for reciprocating sliding as well as for localized lubricant application. In this method, the lubricant solution is delivered directly between the contact planes of the two Si surfaces via a syringe-tube-needle system, in which the needle is placed precisely at the edge of the interface of contact planes. The positioning of the needle is controlled from a metric stage, which provides movement in all three axes of motion. A controlled amount of lubricant was dispensed consistently each time across all experiments involving this method, using a push-dispenser. 2 μl of PPFE solutions was dispensed to the interface on each set, before the application of the load on the upper piece.

Surface Analysis Techniques

Surface Profiling

The topography of the cleaned and various coated Si surfaces were analyzed using a Wkyo NT1100 Optical Profiler (Veeco Instruments Inc.) with optical phase-shifting and white light scanning interferometry with non-contact static measurements, to study the roughness and topography of the various test surfaces prior to wear testing. The vertical measurement range of the profiler is 0.1 nm–1 mm, a resolution of less than 1 Å Ra, and a vertical scan speed of up to 14.4 $\mu m\ s^{-1}$. All profiles were taken in a class-100 clean booth so as to avoid contamination from dust, using an integrated stroboscopic illuminator.

Surface Hydrophobicity

The hydrophobicity influences the tribological properties of surfaces. As part of the study, the static water contact angles of the various surfaces were measured using a VCA Optima Contact Angle System (AST Products, Inc., USA) with deionized water, to verify that the modifications and lubricant coatings were successfully done on the surfaces. The measurements were taken over the same surface at five or six locations, repeated over various batches of specimens to ascertain accuracy and consistency. An average value was taken from all the measured values, with the variation of measurements within $\pm$ 2 °, and error within $\pm$ 1 °.

Film Thickness

Tribological performance and the thickness of the lubricant film are correlated. The thickness of the film was investigated using a variable angle spectroscopic-ellipsometer (VASE, J. A. Woolam. Co., USA) at wavelengths from 400 to 1,000 nm at 10 nm intervals. Incident angles for measurements were 65 °, 70 ° and 75 °. The data was analyzed with WVASE 32 Windows Version 3.352 software, with refractive indices for SiO_2 and PFPE films taken as 1.46 and 1.3 respectively. For each surface modification or lubricant method used, 5 or 6 replicate measurements were carried out, each at different locations or samples, and an average value reported to ensure accuracy and consistency.

Reciprocating Sliding Wear Tests

To investigate the oscillatory sliding wear between the Si wafers shown in Fig. 11.1 and 11.2, a wear tester was built to provide such loading and sliding conditions. The two surfaces were positioned appropriately, with the upper piece glued to a ball held in a holder attached to a cantilever, before a deadweight load was applied directly above the upper piece. The upper piece was glued onto a ball surface to ensure that the loading was distributed across the entire surface of the upper piece. With the exception of "Loc-Lub", which involves the non-loaded contact of the specimens at the time of lubrication, specimens were already modified or coated respectively prior to wear tests.

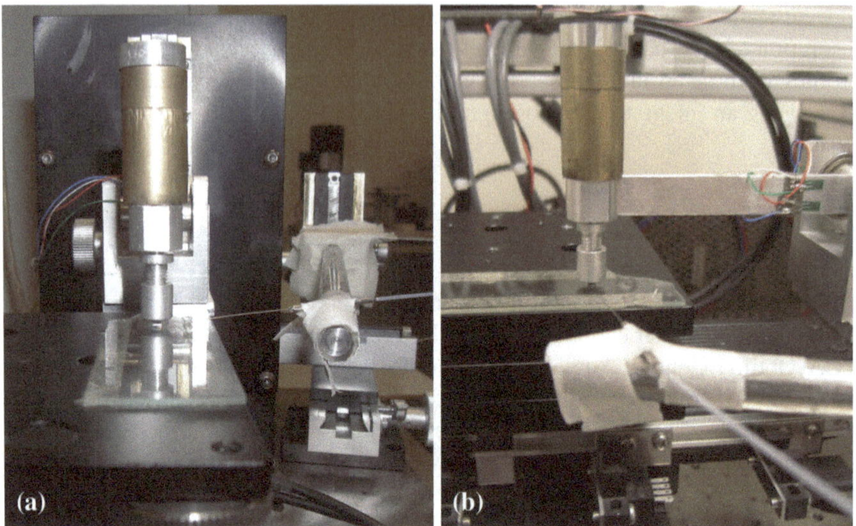

Fig. 11.2 Reciprocating sliding wear test setup as viewed from **a** front and **b** side. Reprinted with permission from Wear 270 (2010) 19–31, ©2012, Elsevier

Table 11.1 Parameters for reciprocating sliding wear tests

Sliding speed	5 mm s^{-1}
Oscillating frequency	2.5 Hz
Sampling rate for data	10 Hz
Amplitude of oscillation	2 mm
Ambient temperature	25 ± 2 °C
Relative humidity	55 ± 5 %

Four strain gauges attached to the cantilever measured the lateral force caused by friction of the two surfaces rubbing against each other; the coefficient of friction (CoF) was calculated as the ratio of the lateral force to the normal dead-weight applied. Parameters for the various factors are shown in Table 11.1.

The initial CoF was defined as the observed CoF in the first four seconds (first 10 cycles) of sliding, and lubrication was considered to have failed when the CoF was noted to exceed 0.3 for a prolonged duration, when massive fluctuations in the recorded CoF was observed, or when wear marks and scratches were noted on the sliding surface of the Si wafer under testing. Wear tests were first carried out for 6 h for feasibility, and specimens that did not fail after 6 h were later subjected to a 60 h extended wear test for wear life determination. Wear life was defined as the point at which the sample failed, and determined by running the wear test on at least ten different samples under the same lubrication method and conditions, with the same experimental parameters, and taking the common average of the data.

Optical and Field Emission Scanning Electron Microscopy and EDS

Optical microscopy was done on the surfaces after wear testing to document the wear track conditions and lubricant movement due to the sliding, as well as debris formed during the wear test. This was followed by electron microscopy performed with a Hitachi S4300 Field Emission Scanning Electron Microscope (FESEM), coupled with an Energy Dispersive Spectrometer (EDS) to characterize the chemical elements on the surface, which serves as an indication of the presence and distribution of lubricant in relative amounts. EDS was also conducted before wear testing to compare the relative concentrations of lubricant due to the different methods, as well as the distribution of the lubricant across the surface. The various methods of lubrication were thought to produce different distributions of the lubricant onto the surface hence greatly influencing tribological performances.

Results and Discussion

Surface Profiling

The 3-D and line profiles of the various surfaces under different methods of lubrication, along with the roughness values (Ra) can be seen in Fig. 11.3. Presence of lubricant lowers the roughness values of the unpolished surface regardless of the method used. However, "Loc-Lub" method reduces the roughness by a larger proportion, as the lubricant has filled up the "valleys" in the asperities which "leveled" out the surface, and this is thought to assist in the self-replenishing action of PFPE which improves the tribological properties [7]. It is thought that the greater roughness reduction in "Loc-Lub" is also due to the greater amount of PFPE present on the lubricated surface compared to those under dip-coating, implying a greater improvement in tribological properties, as will be investigated later.

Surface Hydrophobicity

The various contact angles are shown in Table 11.2, and confirm that the various surface modifications have been successfully made.

A less significant change in the contact angles of the surfaces after vapour deposition was noted, compared to those measured for the other methods. This is possibly due to the least amount of PFPE molecules present on the surfaces, as analyzed later in EDS mapping, and therefore providing a lower degree of hydrophobic modification on the surface. Other methods of lubrication are later shown to have a much higher density of PFPE molecules on the surface, thereby inducing a

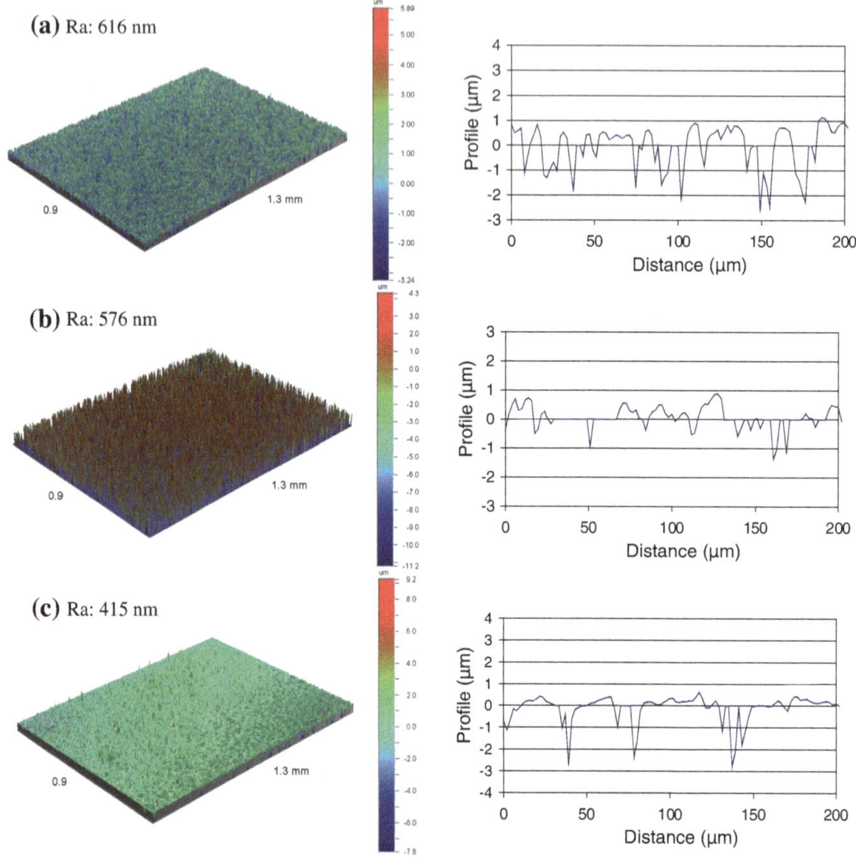

Fig. 11.3 Optical profile images of **a** bare unpolished silicon, **b** dip-coated unpolished silicon and **c** unpolished silicon under "Loc-Lub" with their respective line profiles. Roughness values (Ra) are given for each profile. Reprinted with permission from Wear 270 (2010) 19–31, ©2012, Elsevier

Table 11.2 Water contact angles for various surfaces and modifications. Reprinted with permission from Wear 270 (2010) 19–31, ©2012, Elsevier

Surface modification	Water contact angle, °
Polished Si	5.5
Unpolished Si	5.7
Polished Si-Localized PFPE (4.0 wt %)	55
Unpolished Si-Localized PFPE (4.0 wt %)	38.8
Polished Si-Vapour Deposition (4.0 wt %)	25.2
Polished Si-Dip Coated (4.0 wt %)	38.5
Unpolished Si - Dip coated (4.0 wt %)	30.4

much higher degree of hydrophobicity as PFPE provides a semi-hydrophobic surface. It should also be considered that the texture and roughness of the surfaces have also been known to affect the water contact angles of surfaces [12].

After dip-coating, samples were noted to have dewetting marks and uneven spreading of the lubricant on the surface. This led to a larger variation of water contact angles on the same sample surface at different locations. The same dewetting marks were noted for "Loc-Lub" specimens, without the diverse variation of water contact angles, but were not observed on any of the unpolished surfaces, partly due to the roughness of the surface making such occurrence visibly undetectable. The uneven spreading of the PFPE lubricant could be caused by the increased concentration of PFPE solution used in this study, compared to previous studies on tribology—particularly with magnetic hard discs which use only 0.2 wt %. This uneven spreading has also led to a slightly larger margin of error in the measurement of water contact angles than observed in previous studies. In general, larger increases in water contact angle were noted for polished surfaces than unpolished surfaces, due to the effect of texturing and the presence of asperity valleys for the lubricant solution to fall into, as well as the effect of the rough surfaces on the detection of dewetting marks.

Film Thickness

Specimens under "Loc-Lub" showed the thickest films, measuring 4.07 nm, followed by dip-coated specimens (2.40 nm) and vapour deposition (0.15 nm). As mentioned, the thickness of the PFPE films has a known effect on the coefficient of friction and wear durability of surfaces, which will be further discussed and verified in the section on wear tests. Since the thickness of the lubricant film is also directly correlated to the amount of polymer lubricant on the surface, it is then expected that a thicker film may provide better lubrication properties as long as the thickness is not large enough to interfere with the sliding surfaces producing viscous drag or capillary forces.

Reciprocating Wear Tests

Reciprocating wear tests showed that textured surfaces allowed for better lubrication properties by assisting self-replenishment, as well as the thicker films providing lower coefficients of friction and longer wear lives. The collated results can be seen in Fig. 11.4, with unpolished samples under "Loc-Lub" showing the best tribological properties in this study. A more detailed analysis of the friction coefficient trends for each sample is presented in Fig. 11.5.

Lubrication with PFPE on all surfaces, whether unpolished or polished, showed an improvement in the wear and frictional properties—it can be said therefore that as long as there is some lubrication provided, there will be a drop in the coefficient

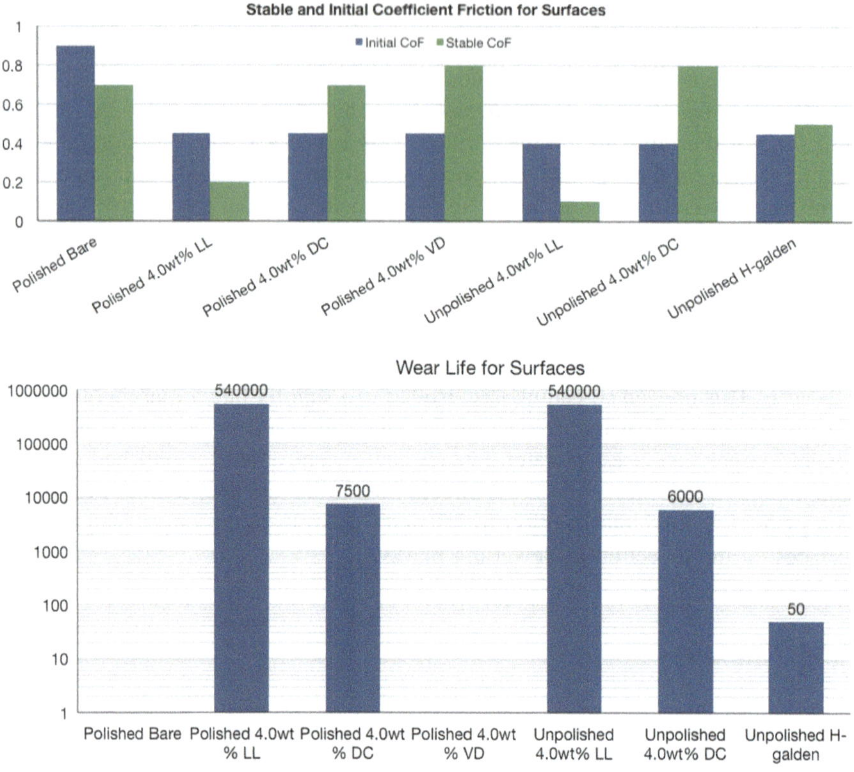

Fig. 11.4 CoF and wear lives for tested surfaces. Reprinted with permission from Wear 270 (2010) 19–31, ©2012, Elsevier

of friction. The solvent (H-galden) was also used on its own as it was necessary to distinguish whether the observed properties were from the application of PFPE or from the solvent itself.

For samples lubricated with PFPE, the initial coefficient of friction reduced quickly enough to prevent any wear until their respective wear lives were reached. Once in the steady state, samples under "Loc-Lub" had enough lubricant at the interface to keep friction at a low level and reduce wear to a minimum during the test, allowing it to last for more than 60 h (540,000 cycles). Dip coated samples had a slightly lower wear life, followed by vapour deposition. The initial coefficients of friction measured were all the same, indicating that the interfacial surface was indeed the same (PFPE) for all surfaces. The different methods of lubrication had vast differences in the eventual coefficient of friction, friction trends and the wear life. Samples under the "Loc-Lub" method of lubrication showed a drop and steadying in the coefficient of friction, compared to other samples which experienced either a gradual or rapid rise, leading to failure, showing that "Loc-Lub" is an effective method of lubrication under these conditions.

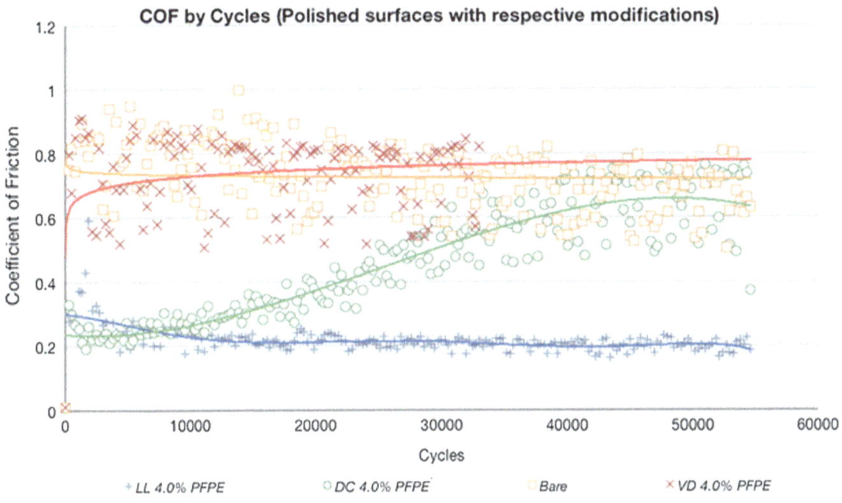

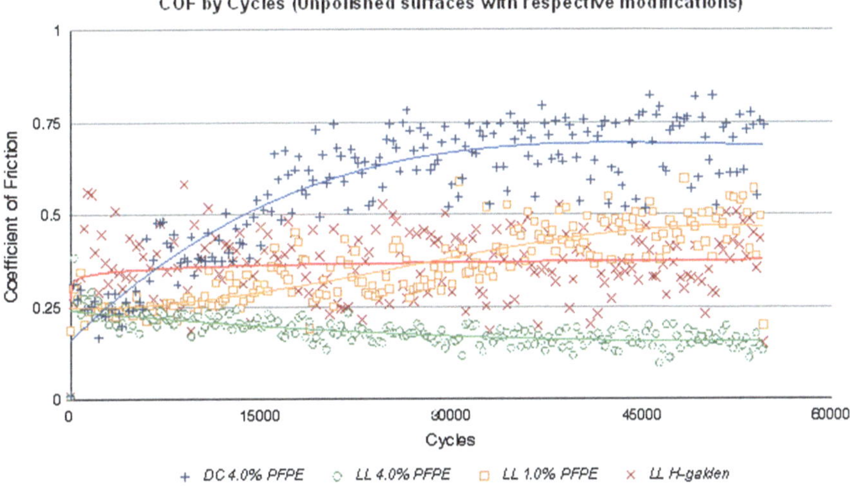

Fig. 11.5 CoF by Cycles for Polished and Unpolished Surfaces

Unpolished surfaces under the same treatment and lubrication method showed better friction and wear properties; a lower coefficient of friction was observed during the tests, particularly for extended wear tests up to 60 h (> 540,000 cycles). Even though both surfaces lasted beyond the time set for the wear tests, unpolished surfaces show lower coefficients of friction in both the initial and stable conditions. The valleys in the rough surface in such a situation have provided reservoirs of mobile PFPE that act as ready sources of replenishment for lubricant that is depleted during sliding. It was also noted that after some tests, some force was required to separate the two polished pieces, implying some form of adhesion between the two surfaces. This adhesion was not observed in the case of unpolished samples—it is suggested that the rough surfaces reduce real contact area and prevent surface tension forces between the surfaces when lubricant is present. However, the actual cause of the adhesion and indeed the role roughness is not well-understood and requires investigation.

The mechanism explaining the difference in wear lives and coefficients of friction is further investigated using EDS to compare the surface conditions of the samples.Table 11.3.

Microscopy

Under simple optical microscopy, vast differences were seen on the various post-testing surfaces. A great deal of scratching and wear and subsequently debris was detected for all bare samples, confirming that Si has very poor tribological properties. There was a visible reduction in wear and scratching with the various methods of lubricant, in tandem to the wear lives and coefficients of friction found in the wear tests, and later found to be correlated to the amount of lubricant detected on the surface later via EDS.

It was noted that the lubricant film could be seen on polished Si surfaces in the form of diffraction marks, similar to a layer of oil on water—for samples under "Loc-Lub", these marks were observed primarily around the perimeter of the wear track and in areas where the lubricant had flowed to but was left undisturbed by the test. This provided visual verification that the lubricant had reached the appropriate areas on the surface for testing, and along with EDS mapping, verifies that the lubricant layer has been in part swept to the side during the wear test.

Scratching was noted to be most severe in unlubricated samples and samples undergone lubrication via vapour deposition. A small amount of scratching was noted for dip coated specimens, but no visible scratches were detected on samples under "Loc-Lub" lubrication. The same trend was noted for tests conducted on unpolished samples—both polished and unpolished "Loc-Lub" samples showed no signs of scratching and a small amount of polymer buildup on the sides in which the lubricant was swept to due to the reciprocating motion. As scratching only occurs when the lubricant film fails to protect the surface exposing it to higher levels of friction and wear, the "Loc-Lub" method has therefore been proven to be the most effective of the selected methods of lubrication using PFPE. It should be noted that unpolished surfaces were also observed to have a smoother surface and fewer asperities on the wear track after testing—possibly due to a polishing effect between the two surfaces. However, this polishing effect seems to have no detrimental effect to the frictional coefficient and wear properties during the wear test. The effect of the wearing of asperities could have also contributed to additional lubricity by releasing of reservoirs of PFPE lubricant stored in asperity valleys, thereby assisting with self-replenishment.

Table 11.3 Optical images for the wear tracks of different surfaces under different lubrication methods, after 6 h of R.S.W test

Method	Polished Si	Unpolished Si
Bare Si	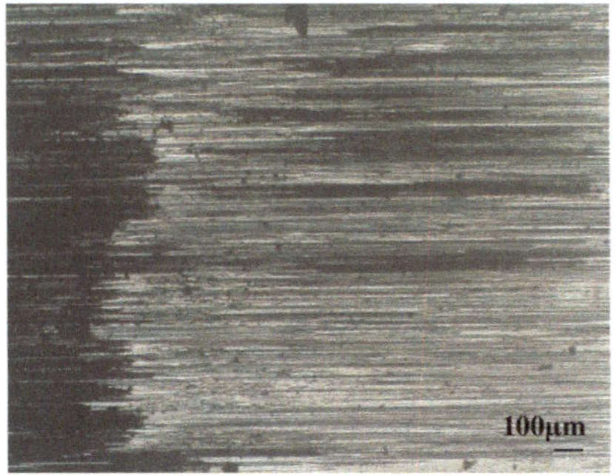	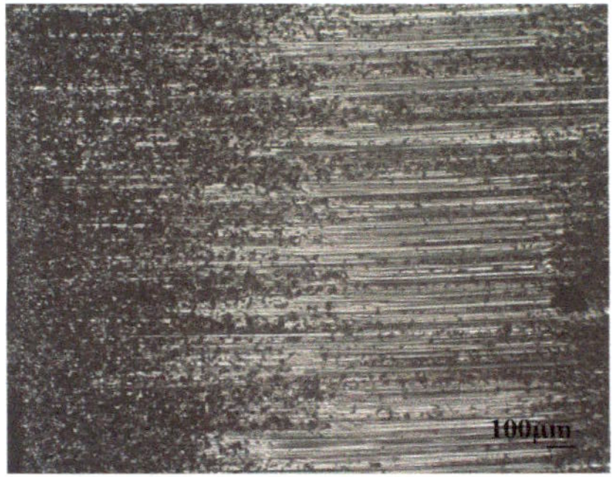
Dip coated	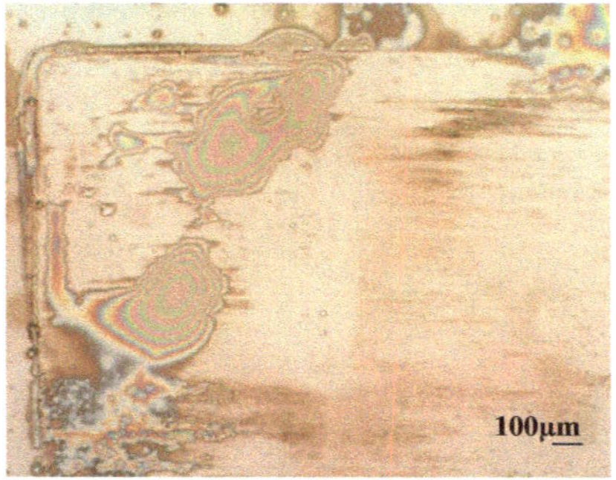	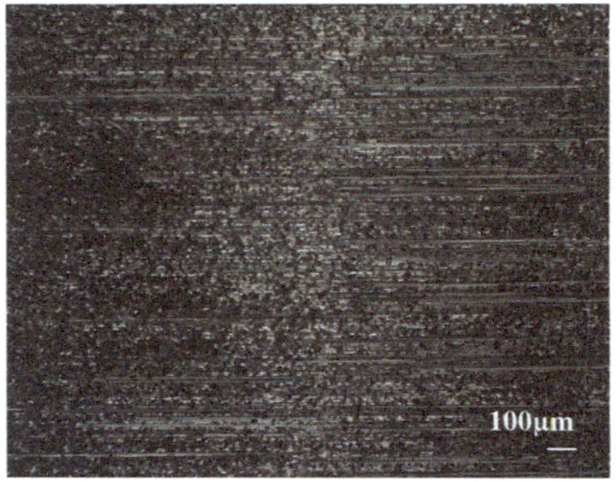

(continued)

Table 11.3 (continued)

Method	Polished Si	Unpolished Si
"Loc-Lub"	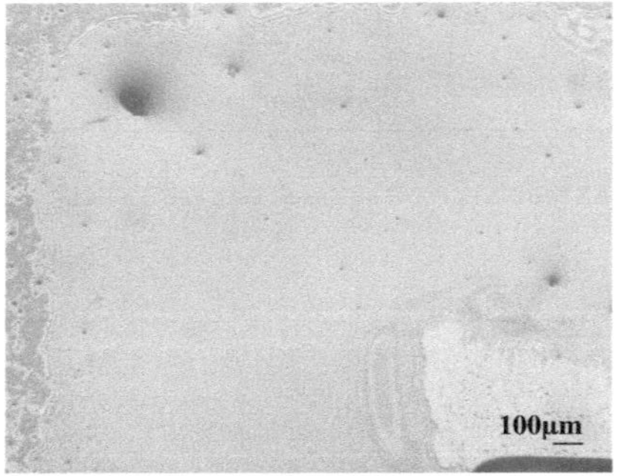 100μm	 100μm
Vapour deposition	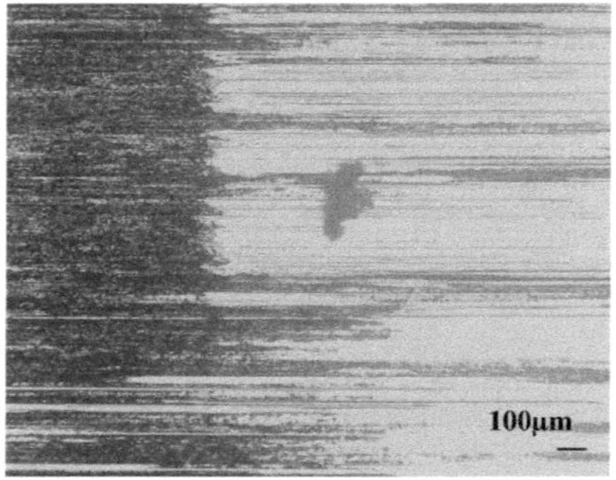 100μm	N.A.

All lubricant concentrations were held at 4.0 wt %. Reprinted with permission from Wear 270 (2010) 19–31, ©2012, Elsevier

FESEM/EDS Mapping

Since fluorine (F) is indicative of the presence of the PFPE molecules (after the solvent has completely evaporated), EDS mapping was done to detect F in order to have a closer look at the distribution of the lubricant under the various methods of lubrication. As shown in Fig. 11.6, the presence of detected lubricant is highest in samples under "Loc-Lub", followed by dip-coating and finally by vapour deposition. No traces of F were found on unlubricated samples.

Tribological properties are expected to be dependent on the density of the lubricant film, as the detected density would also be indicative of the amount of lubricant there. As noted, all surfaces with detectable levels of fluorine (implying significant amounts of PFPE lubricant present) have longer wear lives than bare Si surfaces. The samples under "Loc-Lub" which showed the greatest presence of fluorine also exhibited the longest wear life and the lowest coefficients of friction for wear tests. Other factors in providing the wear durability of the surface include the bonding (both mechanism and strength) of the lubricant to the substrate. For dip-coating, it was noted that there was less PFPE detected on the surface prior to wear tests than those under vapour deposition, even though dip-coated samples showed better properties. This can be explained by the presence of both mobile and bonded phases of PFPE being present on the dip-coated specimens, which provides good frictional properties [7]. Due to the nature of the vapour deposition coating process using oxygen plasma-functionalized surfaces, only a bonded layer exists on the surface, which gives poor friction and wear properties [7, 13].

EDS was done both pre- and post-wear tests, firstly to ascertain that the "Loc-Lub" method has successfully lubricated the contacting interface, and secondly to see the effects of sliding on the distribution and density of the lubricant layer on the surface. In certain instances, a small area in the centre of the contacting interface was not lubricated uniformly (Fig. 11.7). This area was approximately $\leq$ 15 μm in length.

Following this discovery, molecular distribution of three different locations on the sample were compared for analysis: at the middle of the contact area, at the visibly lubricated areas around the contact area, and at a point far away from the

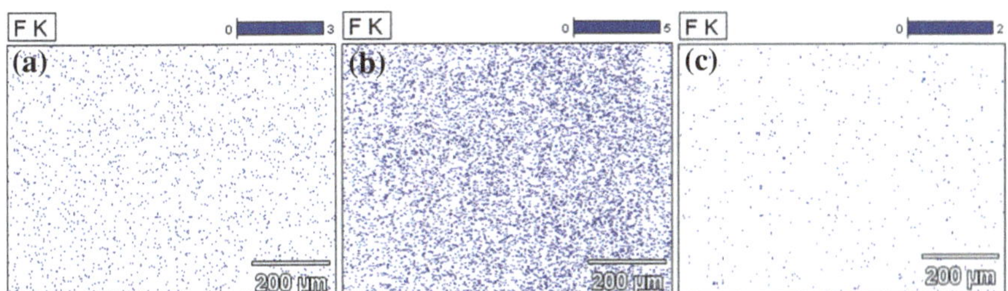

Fig. 11.6 EDS mapping of flourine (F) for unpolished Si samples lubricated with 4.0 wt % PFPE under **a** dip-coating, **b** localized lubrication, and **c** vapour deposition

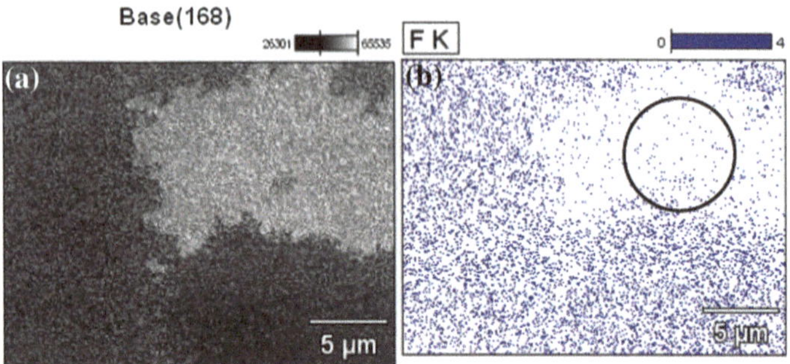

Fig. 11.7 **a** FESEM and **b** EDS mapping images for the presence of fluorine (F), indicating presence of lubricant. The centre of the contact area between the two specimens *circled* is initially not uniformly lubricated

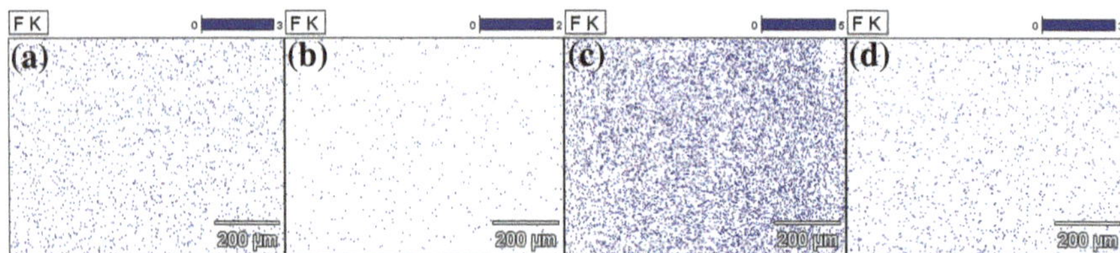

Fig. 11.8 EDS fluorine mapping for wear tracks of unpolished dip-coated Si samples **a** before and **b** after a 6 h wear test; and samples undergone localized lubrication **c** before and **d** after a 6 h wear test. The amount of fluorine is higher for the "Loc-Lub" sample

locally applied lubricant which was assumed to have no lubrication at all. Highest distribution densities were detected in the non-contacting lubricated areas, followed by the contacting but lubricated areas and lastly, no traces were detected at the far away locations.

This pattern of distribution explains the frictional behaviour of the "Loc-Lub" specimens in the reciprocating wear tests—the relatively high initial coefficient of friction is caused from the running-in time in which the lubricant has yet to spread itself to the less coated areas. Once PFPE has rapidly spread to all the portions on the surface, the coefficient of friction lowers as the lubricant layer becomes sufficient to protect the surface. The rapid decrease in friction and replenishment of the PFPE layer is quick enough to prevent excessive wear and surface damage from occurring on the sample. In addition, if the samples were left to stand on their own after "Loc-Lub" application, the lubricant density profile becomes uniform (Fig. 11.6b), thereby reducing the area of any initially unlubricated surfaces to a minimum.

EDS mapping was then carried out on tested surfaces under dip-coating and "Loc-Lub" to investigate the surface conditions and lubrication after reciprocating sliding wear for 6 h (54,000 cycles). Figure 11.8 and Table 11.4 compare the surfaces and amount of PFPE detected both before and after the wear tests were conducted. "Loc-Lub" samples showed a much higher level of PFPE detected after the 6 h tests than dip-coated samples: given that the initial level of PFPE

Table 11.4 Detected percentages of fluorine on surfaces under Dip-coating and "Loc-Lub" methods before and after testing

Surface conditions	Normalized wt % of F	Atom % of F
Dip-Coated (4 wt % PFPE) Untested	9.53	11.36
Dip-Coated (4 wt % PFPE) Tested R.S.W. for 6 h	1.17	1.39
"Loc-Lub" (4 wt % PFPE) Untested	38.37	36.07
"Loc-Lub" (4 wt % PFPE) Tested R.S.W. 6 h	28.90	28.51

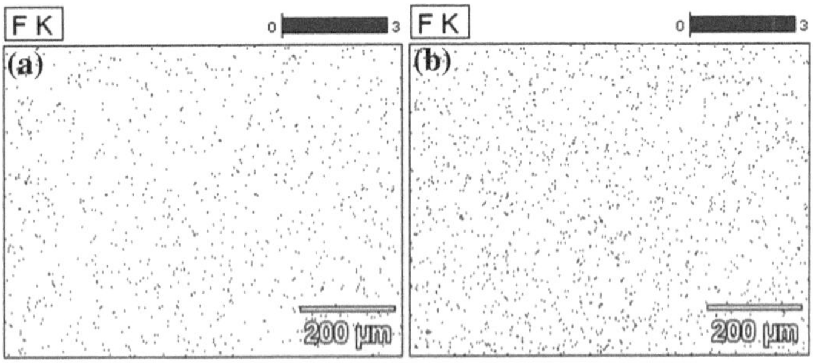

Fig. 11.9 EDS mapping of element fluorine (F) for **a** area near wear track that has an overflow of lubricant, **b** area in the centre of the wear track, both after 540,000 cycles. Reprinted with permission from Wear 270 (2010) 19–31, ©2012, Elsevier

detected was higher for "Loc-Lub" samples, the self-replenishment mechanism attributed to PFPE would be more pronounced in these cases due to the large amount of readily available PFPE mobile phase present. For dip-coated samples, the self-replenishment would not continue for as long a duration, nor would it be as concentrated as in the case of "Loc-Lub" [7, 13]. Even after 60 h (540,000 cycles), "Loc-Lub" samples showed a large amount of PFPE still on the surface (Fig. 11.9), confirming that the lubricant film is still present in significant amount to protect the surfaces during the relative sliding motion, resulting in low coefficient of friction and wear.

Another reason for the high distribution density of lubricant after the wear tests were carried out could be the effect of the textured surfaces—the asperities provided valleys in which small reservoirs of PFPE could reside, assisting the replenishment process and aiding in the easy removal of any debris that may have formed in the polishing process, avoiding third body abrasion. This is supported by the sustained presence of PFPE on the wear track for "Loc-Lub" samples, but a significant drop in the level of PFPE detected on the dip-coated surfaces (Fig. 11.9), even though both surfaces experience a drop in the nominal levels of PFPE detected. Tribological differences were observed between rough and polished Si surfaces used in the study, with rough surfaces showing better friction and wear properties. Adhesion was also observed in separating the polished surfaces under "Loc-Lub", which was not noted in the unpolished surfaces. Effects of texturing for the improvement of tribological and surface properties have been studied and are also thought to reduce the stiction, friction and wear in MEMS devices [12, 14–16].

Discussion

Effects of Use of Higher Concentration

Other than the improvement of the wear lives due to an increased concentration of the lubricant solution, additional effects were also observed. The coating of polished surfaces under dip-coating and "Loc-Lub" showed dewetting effects, causing the PFPE lubricant layer to be unevenly distributed. This effect on the distribution for polished surfaces is a cause for the large statistical variation in the frictional properties exhibited in the wear test, although those lubricated with a higher concentration have a higher wear life and lower measured friction.

Effects of Textured Surfaces

The dewetting effect is not observed in the unpolished surfaces as the lubricant solution has sunk into the valleys of the asperities (from optical profiling images), and the droplets that would otherwise naturally form are spread out by the rough surfaces. This enforced separation of the droplets causes a more uniform distribution of the lubricant film over the entire surface.

Texturing not only provides a lower real contact surface area which has been proven to reduce stiction between surfaces, but in the breaking up of the droplets of lubricant have become reservoirs of temporary lubricant storage, allowing for ease of replenishment due to ready availability of mobile phase PFPE. Studies on PFPE have shown that its lubrication properties are partly due to both a mobile and bonded phase being present; the bonded phase alone provides little lubrication and protection against wear. The uneven surfaces also ensure against complete sweeping of the lubricant to the edge of the wear tracks during the test, as in the case of polished surfaces, by providing enclaves or valleys in which amounts of lubricant can be stored. Polymer buildup around the perimeter of the wear track was found to be much more evident in the case of polished surfaces, indicating that a significant layer of PFPE mobile layer has been removed, reducing the effectiveness of the lubricant.

Load Comparison to Actual MEMS

It was noted that the load used in this study was much higher than that experienced in actual MEMS, which usually ranges from μN to nN. As friction is also dependent on the normal loading, an extreme condition of 0.5 N was used to illustrate the effectiveness of the methods and solution. Under normal conditions in actual MEMS devices, the frictional force experienced would be much lower than those presented in this study, showing that the relative effectiveness of the various methods compared in this study will be representative at the device level.

Effect of Different Lubrication Methods

The effectiveness of PFPE lubricant, as mentioned, depends on the availability of both the mobile and bonded layer and its self-replenishing properties as a result, and the thickness of the film. Vapour deposition involves only the chemically bonded layer of PFPE on the surface, and therefore provides very little prevention against wear and high amounts of friction as the mobile phase is completely absent.

Dip-coating, as a well-known lubricant film application technique, provided better properties—however, the unevenness of the lubricant film with the method applied on polished surfaces leads to further complications and large statistical variations between the tests. The unevenness of the film was avoided on unpolished surfaces, but did not lead to a significant improvement of wear properties as there was only a thin layer on the surface, insufficient to provide the necessary self-replenishment.

"Loc-Lub" seems to provide the best properties among the three, providing both a bound and mobile layer of lubricant, as well as sufficient lubricant in the surroundings to aid self-replenishment. The contacting surfaces were found to be adequately lubricated and better wear properties were observed on unpolished surfaces under this technique.

Practical Application on MEMS Devices

Introduction

Having ascertained that the "Loc-Lub" method is effective in lubricating two sliding surfaces under oscillatory motion, the same method concept was applied to actual MEMS devices. The custom design can be seen in Fig. 11.10. As the radius of the round contact is only 50 μm and the depth of the gap is ~25 μm, the contact conditions can be considered a line contact.

Since the gaps for an actual MEMS device are much smaller and the surrounding areas require functionality to be kept as fabricated, the localized application of PFPE should ideally not have a large spillover area, and focus on lubricating the sidewalls. For this reason, the amount of PFPE dispensed has to be kept under control, as well as carefully positioned. As the device is also structurally fragile, the technique cannot be exactly replicated as the surface tension caused by the simultaneous contact of the droplet, the needle and the device components may exert enough force to break the device by bending it out of plane.

To combat these obstacles, the method was slightly modified—a syringe-tube system was still used, but contact with the surfaces was avoided in the dispensing of PFPE solution. Instead, pressurized air was delivered through a tube to the syringe/needle in a short burst to drop a very small amount of PFPE solution onto a

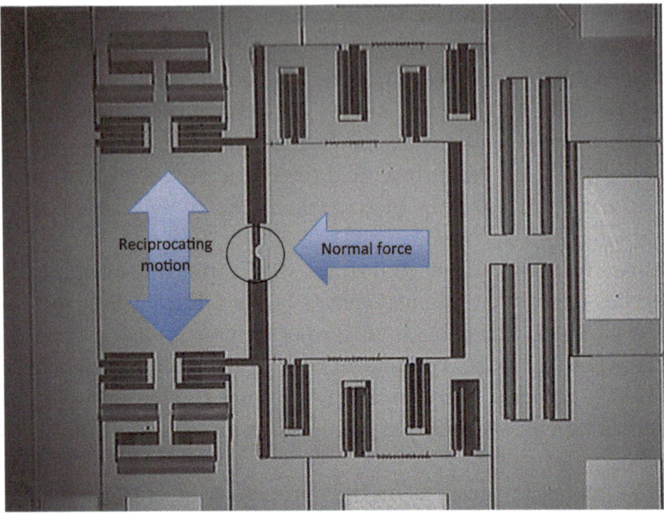

Fig. 11.10 Custom-made MEMS device for testing of tribological properties under reciprocating sliding wear. Round on flat contact is *circled*, and the movement of the components is shown with *arrows*

pre-designated location. The volume of PFPE dispensed could then be controlled by varying the pressure used to eject the solution from the syringe.

For the positioning of the needle/syringe setup and to ascertain that the lubricant was dispensed at the desired location, a video imaging microscope system was used. The system allowed visual magnification and real time video observation of the procedure. An image capture of the lubricated device is shown in Fig. 11.11.

Experiments and Analysis

PFPE lubricant solution was dispensed in extremely small quantities from the needle—in our study 0.119 μl was used in each dispense and lubrication process. As seen in Fig. 11.11, there is some spillover in the areas around the contact point—care needs to be taken in application and design of the MEMS device that a small area around the contact point is not critical and delicate in structural stability.

Friction and adhesion tests were carried out both before and after lubrication to compare the effects of the lubricant on the surface both on contact and withdrawal of the surfaces, as well as upon oscillatory sliding. It was found that PFPE lubrication via "Loc-Lub" not only retains the same level of functionality as an unlubricated device, implying that the general components and electrical contacts

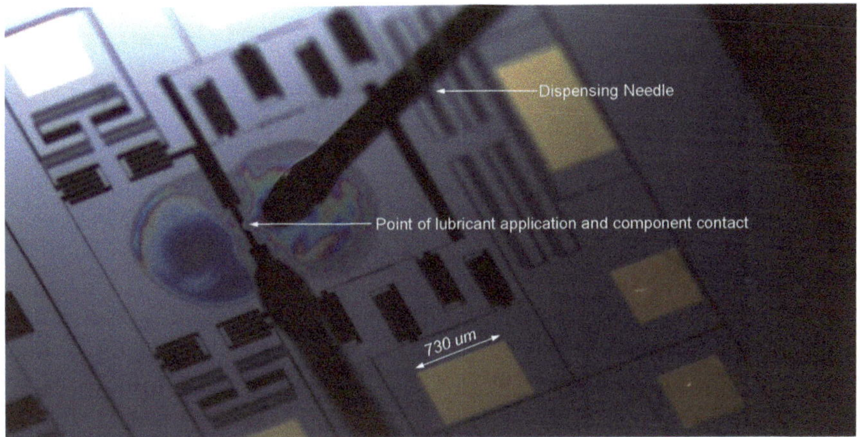

Fig. 11.11 Video still of a lubricating of the device shown in Fig. 11.10. Reprinted with permission from Wear 270 (2010) 19–31, ©2012, Elsevier

remain unmodified, but also shows a reduction of the adhesion and friction properties. Unlubricated surfaces had an adhesive force of approximately 240 μN upon contact, compared to only 32 μN after "Loc-Lub" lubrication. Relative sliding against surfaces was not possible on unlubricated surfaces as the frictional forces were so high that sliding could not be achieved—however, surfaces were able to slide smoothly against each other for a total of 4 μm after lubrication, without experiencing any detectable wear. Surfaces have therefore shown to be successfully modified for improved tribological properties.

The practicality of the lubricant also depends on how long it sustains its lubricity after being exposed to the environment after application—avoiding the need for hermetic packages and costly closed system containments in which the lubricant is recirculated. Tests were re-done on the same device with the same parameters after 2 weeks of storage in a desiccator under a clean room environment, and results showed the same properties and values measured, with no significant changes and within an acceptable margin of experimental error in the measurement of the voltage signal as a relative measurement of the frictional and adhesive force.

Concerns were expressed on the spillover of lubricant and whether they would affect other components such as the comb drives—primarily the concern of surface tension causing the comb drives to stick together upon the evaporation of solvent. To investigate these concerns, PFPE solution in the same concentration used to lubricate the device contacts was dispensed directly onto a device's comb drives. Microscopy investigation showed no signs of contacting between the comb drives at any point, and further testing showed that the sensitivity and functionality was not significantly reduced as the comb drives performed as designed.

Conclusion

Various methods of lubrication were tested in this study, including the novel "Loc-Lub" method, to investigate the effectiveness and applicability of the novel method. From the current friction and wear observations, it can be concluded that the "Loc-Lub" method is the most effective among the three methods investigated. Unpolished surfaces were used to mimic the rough conditions of MEMS device sidewalls.

The local application of lubricant has not only been found to be useful in reducing the friction and wear of reciprocating sliding surfaces, but also in modifying the hydrophobic properties of the surfaces, implying improved tribological properties at the micro-scale as well. The combination of texturing, "Loc-Lub" and usage of PFPE lubricant has shown to provide better tribological properties than the other conditions stated in this study—allowing for lower coefficients of friction and extremely reduced amounts of wear and debris. Under the scientific analysis, the method of lubrication seems viable for application onto actual MEMS devices.

Certain modifications were made with the transition to the micro-scale on actual devices, specifically in the method of dispensing; employing ejection of a droplet to avoid simultaneous contact between the needle and the device. This avoids surface tension forces from causing the device to stick to the needle and therefore damaging the device. The modified "Loc-Lub" method has been found to reduce both the adhesion and friction experienced in contact and relative sliding of the components respectively, verifying that the sidewalls of the MEMS device has indeed been modified and properly lubricated under "Loc-Lub".

Acknowledgments Authors wish to acknowledge the financial support given to this work by the National Research Foundation (NRF), Singapore (Award no. NRF-CRP 2-2007-04).

References

1. Mathew Mate, C.: Tribology on the Small Scale: A Bottom Up Approach to Friction, Lubrication, and Wear. Oxford Scholarship Online. 22 April 2010. Oxford University Press, Oxford (2007)
2. Kim, S.H., Asay, D.B., Dugger, M.T.: Nanotribology and MEMS. Nanotoday **2**, 22–29 (2007)
3. Potter, C.N.: Hermetic MEMS package and method of manufacture. US Patent 7,358,106 B2
4. Ashurst, W.R., de Boer, M.P., Carraro, C., Maboudian, R.: An investigation of sidewall adhesion in MEMS. Appl. Surf. Sci. **212–213**, 735–741 (2003)
5. Satyanarayana, N., Sinha, S.K., Shen, L.: Effect of molecular structure on friction and wear of polymer thin films deposited on Si surface. Tribol. Lett. **28**, 71–80 (2007)
6. Bhushan, B.: Tribology and mechanics of magnetic storage devices. Springer, New York (1990)
7. Tani, H., Matsumoto, H.: Spreading mechanism of PFPE lubricant on the magnetic disks. J. Tribol. **123**, 533–540 (2001)

8. Wang, M., Miyake, S., Matsunuma, S.: Nanowear studies of PFPE lubricant on magnetic perpendicular recording DLC-film-coated disk by lateral oscillation test. Wear **259**, 1332–1342 (2005)
9. Sinha, S.K., Jonathan, L.Y, Satyanarayana, N., Yu, H., Harikumar, V., Zhou, G.: Method of applying a lubricant to a micromechanical device, US Provisional Patent 61/314,627, 17 Mar 2010
10. Samad, M.A., Satyanarayana, N., Sinha, S.K.: Tribology of UHMWPE film on air-plasma treated tool steel and the effect of PFPE overcoat. Surf. Coat. Technol. **204**, 1330–1338 (2010)
11. Liu, H., Bhushan, B.: Nanotribological characterization of molecularly thick lubricant films for applications to MEMS/NEMS by AFM. Ultramicroscopy **97**, 321–340 (2003)
12. Marchetto, D., Rota, A., Calabri, L., Gazzadi, G.C., Menozzi, C., Valeri, S.: Hydrophobic effect of surface patterning on Si surface. Wear **268**, 488–492 (2010)
13. Eapen, K.C., Patton, S.T., Zabinski, J.S.: Lubrication of Microelectromechanical Systems (MEMS) Using Bound and Mobile Phases of Fomblin Zdol®. Tribol. Lett. **12**, 35–41 (2002)
14. Křupka, I., Poliščuk, R., Hartl, M.: Behavior of thin viscous boundary films in lubricated contacts between micro-textured surfaces. Tribol. Int. **42**, 535–541 (2009)
15. Talke, F.E.: An overview of current tribology problems in magnetic disk recording technology. Tribol. Interface Eng. Ser. **38**, 15–24 (2000)
16. Tan, A.H., Cheng, S.W.: A novel textured design for hard disk tribology improvement. Tribol. Int. **39**, 506–511 (2006)

Index

S. K. Sinha et al. (eds.), *Nano-tribology and Materials in MEMS*,
DOI: 10.1007/978-3-642-36935-3, © Springer-Verlag Berlin Heidelberg 2013